MULTIPLE GRAVITY ASSIST INTERPLANETARY TRAJECTORIES

Earth Space Institute Book Series
Editor in Chief: Dr Peter Kleber, Chairman, Earth Space Institute

The Earth Space Institute is a division of the Foundation for International Scientific and Education Co-operation, registered in London, no. 2254798. The Institute is a non-profit organization.

Volume 1
Space Science in China
edited by *Wen-Rui Hu*

Volume 2
Der Mensch im Kosmos
edited by *Peter R. Sahm and Gerhard Thiele*

Volume 3
Multiple Gravity Assist Interplanetary Trajectories
by *Alexei V. Labunsky, Oleg V. Papkov and Konstantin G. Sukhanov*

This book is part of a series. The publisher will accept continuation orders which may be cancelled at any time and which provide for automatic billing and shipping of each title upon publication. Please write for details.

MULTIPLE GRAVITY ASSIST INTERPLANETARY TRAJECTORIES

ALEXEI V. LABUNSKY

Moscow Aviation Institute, Moscow, Russia

OLEG V. PAPKOV and KONSTANTIN G. SUKHANOV

Lavochkin Association, Moscow, Russia

CRC Press
Taylor & Francis Group
Boca Raton London New York

CRC Press is an imprint of the
Taylor & Francis Group, an **informa** business

First published in 1998 by Overseas Publisher Association

Published in 2020 by CRC Press
Taylor & Francis Group
6000 Broken Sound Parkway NW, Suite 300
Boca Raton, FL 3487-2742

First issued in paperback 2020

ISBN-13: 978-0-367-57923-4 (pbk)
ISBN-13: 978-90-5699-090-9 (hbk)

British Library Cataloguing in Publication Data

A catalogue record for this book is available from the British Library.

ISSN: 1026-2660

Cover illustration: Diagram of a spacecraft flight to the orbit of an artificial satellite.
See Figure 5.12, p. 191.

CONTENTS

INTRODUCTION

It is hard to overestimate the scientific and historical significance of the period of practical space exploration since the launch of the first artificial satellites and automated interplanetary probes. It is rightly compared with the period of great geographical discoveries. Now, based on space technology, the discoveries extend far beyond the Earth, almost throughout the solar system.

Systematic space studies are of primary importance for the development of fundamental scientific views of the Universe, Solar System, and our planet. One direction of space studies is associated with the exploration of deep space using automated spacecraft. It embraces a wide range of problems solved by automated space stations on various missions to celestial bodies in the Solar System—planets, their satellites, comets and asteroids—as well as on flights to the near-Solar zone, to the zones beyond the ecliptic plane and to the polar zones of the Solar System.

Flights by automated spacecraft made it possible to directly collect data on the physical nature of bodies in the Solar System. Analyses of the composition of soils, atmospheres, magnetic and radiation fields open a new stage in the exploration of functions of natural mechanisms in planetary complexes. These studies in deep space are of importance not only because they obtain fundamental new knowledge about the Universe, the Solar System, and the planets. They are of no less practical significance, since they make it possible to consider the processes taking place on the Earth and in its vicinity from the general planetary point of view. Progress in astronautics has brought about a series of new scientific directions: comparative planetology, climatology, and meteorology based on comparative studies of the Earth as one of the planets in the Solar System.

Thus, the opportunities opened up by the extending field of space studies account for the considerable interest aroused amongst the international scientific community in new programs of exploration of the Solar System.

The purely practical problem of what direction further studies of the Solar System should follow is closely related to many aspects of the development of science, engineering, and economics. At the same time, it is clear that the large scale of experiments involving space technology requires them to be planned so as to collect maximum data on the Solar System at minimum cost. In other words, improving the efficiency of space studies is one of the most vital problems of their further development.

In terms of selecting the exploration spacecraft routes in the Solar System, this means that the route should minimize the cost of the mission, in addition to widening the scope of studies.

The best solution to this problem appears to be the concept of **multi-purpose space missions** in the Solar System.

By this, we mean spacecraft missions in the Solar System, whose trajectory passes by a series of celestial bodies to be explored, so that the gravity fields of these bodies can be utilized to purposefully change the spacecraft trajectory. Such maneuvers are called gravity assist or perturbation maneuvers.

Thus, multi-purpose missions on the one hand significantly expand the scope of research carried out by the spacecraft in the Solar System by making it possible to explore several celestial bodies in a single mission, and on the other hand, allow the energy costs for the flight to be reduced, relative to direct flights to individual celestial bodies included in the mission. These energy savings are due to the use of the energy of gravity fields of the celestial bodies passed by the spacecraft. These gravity fields serve as "gravity springboards" and allow the spacecraft trajectory to be purposefully changed with no (or minimal) fuel spent. The possible energy savings enhance the usefulness of the mission, because the lower the energy expenditures (and, therefore, the fuel reserves on board), the greater the range of scientific equipment on board the spacecraft, which maximizes the scope of planned research and the efficiency of the mission.

The potential of a planetary flyby and the use of its gravity field energy on the way to the target planet was mentioned by the founders of astronautics, K.E. Tsiolkovskii and F.A. Tsander [34, 35]. F.A. Tsander substantiated the benefits of using the energy of gravity assist maneuvers, but at the same time, pointed to significant difficulties in designing such trajectories. In the 1920s, German researcher W. Hohmann [16] considered these possibilities. Decades later, when the accomplishment of interplanetary missions was considered as a practical problem, researchers once again focused on such maneuvers. In 1956, G.A. Crocco was the first to publish the results of studying spacecraft trajectories to Mars and Venus and back to Earth and methods for their calculation based on simple models [36]. In a series of later publications, other researchers examined such flights based on more rigorous models, and searched for new variants of flight schemes to the planets of the Earth group and distant planets [37–60]. The trajectories suggested in these studies are still of great interest for practical astronautics, e.g. synchronous and symmetrical trajectories (modified Crocco trajectories), combined missions with flybys and delays of the spacecraft near planets, etc.

Later, after the benefits of routes with planetary flybys had become evident, experts in interplanetary missions focused on the examination of new flight schemes to planets and their satellites [111–170], comets, and asteroids [185–231], as well as into less accessible regions of the Solar System [171–184] so as to minimize energy expenditure. To date there are a series of remarkable examples of accomplishing such flights—the US *Voyager* probe mission to Jupiter–Saturn–Uranus–Neptune; the USSR *Vega* probe mission to Halley's comet via Venus; the *Galileo* project, etc. New projects are planned, particularly within the framework of the international New Millennium programme.

The unique scientific results obtained in these flights helped the multi-purpose missions to be recognized as very promising ways of researching the Solar System.

At the same time, the difficulties in calculating and optimizing multi-purpose trajectories, the narrow time intervals for their implementation, their sensitivity to changes in the launch dates and planetary flybys made the researchers regard them as rather peculiar cases of forming spacecraft interplanetary orbits, which require the development of special methods and algorithms.

Indeed, the methods developed to date (mostly numerical) were commonly applicable only to individual schemes or their groups, and successful solution of the problem depends to a great extent on the intuition and experience of the researcher, the successful reduction of the model of the problem in hand, and the application of available numerical methods at different design stages.

A large body of experience has now been accumulated in examining interplanetary missions (including multi-purpose ones), and solutions were obtained for different variants of missions [76–110].

The work of researchers in the field of mission design has determined to a great extent the level of present-day achievements and prospects in the exploration of the Solar System and deep space.

In this monograph, we never intended to embrace all the possible approaches and methods of solving such problems, nor did we intend to analyze the results accumulated in recent years.

The book presents chiefly the methods and algorithms we have developed over a period of more than 20 years of activity in the field of interplanetary missions, and offers the most interesting results of studying and implementation of multi-purpose flight schemes in the Solar System. At the same time, the material included in the book reflects a complete concept of a general methodological approach to the analysis of multi-purpose spacecraft missions and the choice of their optimal characteristics.

We present a series of interrelated methods and algorithms which are applied at successive stages of the initial design studies and which allow the designer to:

- select an efficient multi-purpose spacecraft path;
- form a (design) multi-purpose spacecraft orbit;
- choose the optimal design and ballistic characteristics of the spacecraft to accomplish the selected multi-purpose mission.

Such segmentation of the problem of multi-purpose interplanetary mission design makes it possible to develop mathematical models that match the aims and tasks of the individual stages of the problem under consideration. Successive application of the solutions for each stage will enhance the efficiency of the methods of both the stage and the design and ballistic problem as a whole. This approach is based on the idea of successively narrowing the solution set with a parallel increase in the number of parameters and characteristics determining the optimal design solution. A schematic diagram of this approach, together with the links between the methods and algorithms involved, is given in Fig. I.1.

This view on the problem of multi-purpose flights as a whole is also reflected in the structure of the monograph. The book consists of seven chapters. The first three chapters of the book and the concluding chapter are concerned with the development of models, methods, and algorithms for designing multi-purpose spacecraft missions in the Solar System. The reader will find the description of new approaches and the

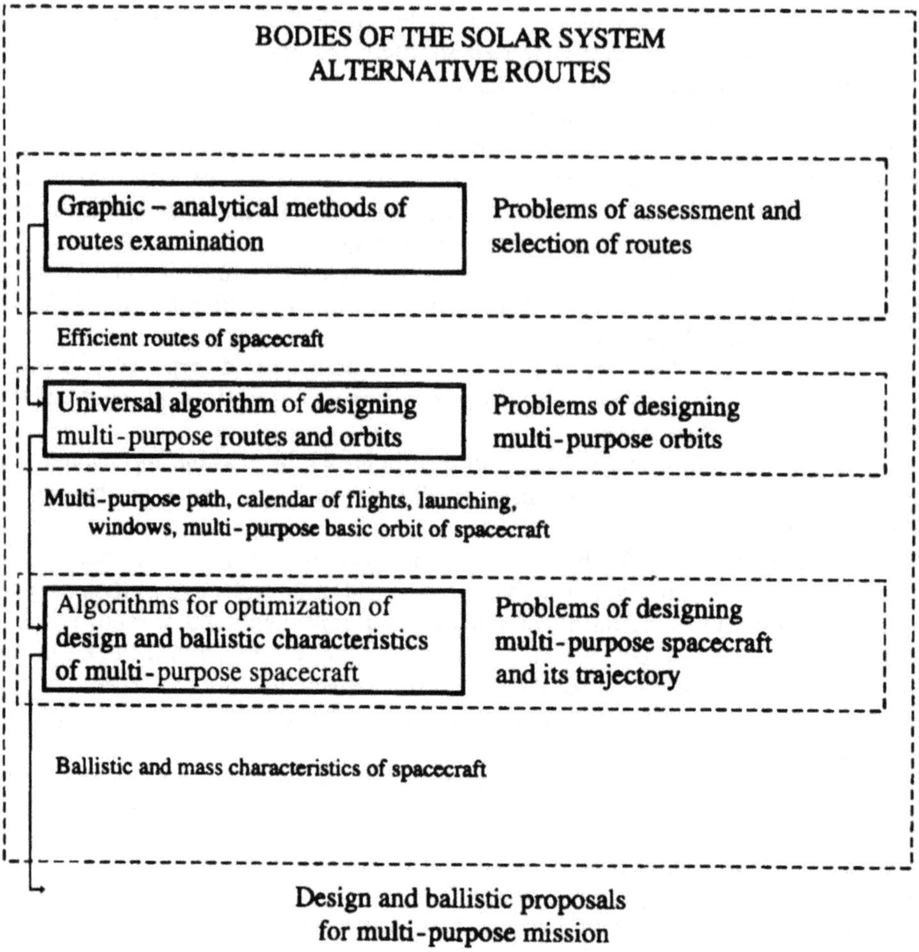

Figure I.1

results of our studies in the engineering methods of examining multi-purpose flights in the Solar System, methods of designing multi-purpose interplanetary spacecraft missions and near-planetary flight segments, as well as examples of their application at various stages of the design procedure. Chapters 4 to 6 of the book are devoted to the practical results of the application of the developed methods: studies of specific interplanetary spacecraft trajectories, the search for new, promising variants of the spacecraft routes in the Solar System, etc.

The first chapter is introductory and deals mainly with the problems of mathematical modeling of the spacecraft flight trajectories in the Solar System using gravity assist maneuvers, describing their dynamic scheme, and assessing the efficiency of such maneuvers.

Chapters 2 and 3 describe the methods of mission analysis and synthesis of multi-purpose trajectories.

The first part of Chapter 2 considers methods for a preliminary analysis of possible multi-purpose paths. The analysis is based on the use of simplified models of the motion of celestial bodies to assess the feasibility of the mission, and the choice of mission schemes and trajectories that are most beneficial in terms of minimum energy expenditures. This stage uses mostly graphic–analytical methods.

A universal algorithm for designing multi-purpose orbits is considered in the second part of Chapter 2. Taking into account that several thousand objects—planets and their natural satellites, asteroids and comets—are known to be registered in the Solar System, a selective approach was used to search for trajectories which are optimal in terms of energy expenditure. To do this, a system of coding all the celestial bodies and maneuvers of the spacecraft was developed. A chain of codes can be constructed to specify virtually any trajectory in the Solar System. The trajectory is calculated within the framework of the central gravitational field, with piecewise approximation of the multi-purpose interplanetary trajectory by segments of Keplerian orbits. An important feature of the algorithm is its multifunctionality. The algorithm makes it possible to search for the optimal trajectory for the given mission (synthesis of the trajectory) and to improve the accuracy of the path in case it is specified incompletely (synthesis of the path) considering the constraints on the mission resources (energy expenditure and mission time, the number of flybys and the type of planets). This stage yields the multi-purpose spacecraft trajectory which is optimal in terms of energy expenditure.

However, experiments show that it is often impossible to use the optimal trajectory for the spacecraft flight. This is due to numerous additional requirements and limitations imposed by both on-board equipment and ground-based tracking and control systems. They affect the selection of ballistic characteristics of trajectories and make the designers deviate from the optimal solutions. Thus, the ballistic design as a component of the whole process of spacecraft design should be carried out at the level of admissible domains of design parameters in the vicinity of the optimal solution in terms of energy expenditure.

These methods are considered in the first part of Chapter 3, which presents a generalized algorithm for constructing isolines. Its computational scheme makes it possible to construct isolines in the space of any dimensionality. With the use of this algorithm, one can construct the admissible trajectory domains to satisfy given design restrictions. This in turn can yield the general pattern of parameter interplay and allows the basic trajectories to be selected as the best compromise solutions.

The second part of Chapter 3 gives the methods and algorithms for a more detailed examination of the basic trajectories, taking into account the length of powered flight segments, radii of planetary spheres of activity, and motion of the planets while the gravity assist maneuvers are being performed. The algorithms allow a combined consideration of heliocentric and planetocentric trajectory segments (the so-called external and internal problems). These studies allow a refinement of the parameters concerning energy, geometry, and kinematics of individual segments of the multi-purpose trajectory that are necessary to assess the functioning of on-board and ground systems involved in the mission, as well as the design and ballistic parameters of the spacecraft.

In essence, the process of designing complex systems has a conceptually contradictory character. The designer will soon find himself working in the Pareto domain, where for a certain parameter to be improved, it is necessary to make other parameters deviate from their optimal values. A moment may come when an improvement is necessary, but the constraints are at their limits. In this case, it is necessary to go back to an earlier stage of the design procedure and to change the generalized criteria to produce a new optimal solution and to examine its neighborhood. The methods developed in Chapters 2 and 3 allow this kind of interactive design of multi-purpose missions.

Chapters 4 to 6 deal with problems of applying the developed methods and algorithms to a wide range of practical problems in multi-purpose mission analysis in the Solar System. Presented here are the results of calculations of multi-purpose trajectories of planets, comets, asteroids, and natural planetary satellites. Ballistic characteristics of typical trajectories are presented together with mission calendars. Direct schemes are correlated with flight schemes involving gravity assist maneuvers. Some methodological peculiarities of computational schemes for individual routes are considered in more detail. The emphasis is on the formation of minimum-energy trajectories using either gravity assist maneuvers or those combined with powered maneuvers. These results are particularly useful for engineers and researchers dealing with spacecraft construction.

Chapter 7 is devoted to the problems of navigation of multi-purpose spacecraft in near-planetary flight segments. The chapter presents methods of designing different types of near-planetary orbits (trajectories for landing on the planet, orbiting as its artificial satellite, gravity assist maneuvers to create interplanetary probes, etc.). All these problems are solved based on a unified approach, which consists in mapping parameters of near-planetary trajectories and interplanetary probes onto the B-plane of the flyby planet. Analytical relationships for such mappings are studied, and examples of their application are given. It is important to note that the methods developed in Chapter 7 are applied not only in the mission design, but also in the ballistic and navigation control of real spacecraft flights.

Thus, the methods suggested in this book embrace all the stages of preliminary design studies. Solutions obtained at this stage can be used as initial estimates of the design characteristics. They may be refined (if necessary) with the use of more accurate methods and models. At the last, working stage of the design, in accordance with the selected path and basic trajectory, a boundary value problem is solved by way of integrating complete differential equations of motion of the spacecraft and celestial bodies taking into account the real characteristics of the space complex (booster, accelerating unit, spacecraft and its components, on-board control system, engines, etc.). However, this is beyond the scope of our study.

By and large, it should be noted that the theory of multi-purpose flights is now rapidly developed.

This monograph presents the main results of more than 20 years of our work in this field, encompassing the design of many accomplished interplanetary missions (e.g. flights to Mars and Venus, the Earth–Venus–Halley's comet international project) as well as future multi-purpose flights, whose design is in progress now (flights to asteroids and comets using gravity assist of Mars or Venus, flight to the Sun via Jupiter, etc.).

This monograph is a result of the joint work of a team of authors. A.V. Labunsky wrote the Introduction, Chapters 1, 2, 3 (except Section 3.1), Chapters 4 (except Section 4.2), and 5. O.V. Papkov wrote the Introduction, Section 4.2, and Chapter 6. K.G. Sukhanov wrote the Introduction, Section 3.1, and Chapter 7.

The authors are grateful to their colleagues and other specialists for the interest they showed in discussing the results of our research. These discussions were of great use in the writing of this monograph.

1. DYNAMIC SCHEME, MODELS, AND METHODS OF ANALYSIS OF SPACECRAFT TRAJECTORIES USING GRAVITY ASSIST MANEUVERS

The motion of the spacecraft in the solar system is controlled by various physical forces. These are the gravitational forces exerted by the sun and natural celestial bodies in the solar system; environmental forces (atmospheric, electromagnetic, forces of solar pressure, etc.); and the jet power of the spacecraft's engines.

Depending on the purpose of the mission, its path, and duration, different physical forces can become decisive in determining the spacecraft's trajectory. In accordance with these circumstances, different models can be used to examine the spacecraft's motion in space [1–4, 7–10, 12, 15, 26].

In the general case, determining the trajectory of spacecraft motion in the solar system under the effect of several attracting bodies is a very complicated computational problem which can be solved only by numerical integration. At the same time, a comprehensive analysis of interplanetary flight trajectories requires examination of a great number of options. This can only be achieved by using approximate models and methods, which result in an effective and clear analysis.

At the initial stage of examining the motion of a spacecraft with high-thrust engines in an interplanetary flight from the departure planet to the target celestial body (planet, comet, asteroid), a model accounting for the thrust and gravitational forces of planets is commonly used. In this model, the spacecraft is regarded as a "zero-mass" body (that is, it experiences the gravitation of other bodies, but does not attract them) [12].

The great distances between the attracting masses in the solar system, and the ratios between the masses of planets and the sun allow the gravity field in the solar system to be represented as the gravity field of the sun (heliosphere) and the gravity fields of the planets, each of which is restricted to the planetary gravisphere. Various methods are known to describe the boundaries of these gravispheres: the planets' spheres of gravitational activity (or Laplace's gravispheres), gravispheres of planet activity, Hill's spheres, spheres of minimum deviations [1, 8, 30, 61].

In such models, the interplanetary trajectory of the spacecraft is represented by a series of segments of undisturbed Keplerian motion in the gravispheres of relevant celestial bodies, while on the boundaries of these segments, the trajectory passes from the gravisphere into the heliosphere and vice versa. In accordance with this division, the so-called "external" and "internal" problems of interplanetary flight are considered (that is, flights in the central fields of the sun and corresponding planets) [3, 12, 17–22, 30].

Studies of interplanetary flight with gravity assist maneuvers are known to deal with cases where the spacecraft on its way from one celestial body to another approaches a third attracting body (e.g. a planet), which brings about a significant change in the spacecraft trajectory. Strictly speaking, in addition to the gravity force of the planet, the spacecraft is also subject to the gravity force of the sun, and the planets of the solar system, as well as other external forces (atmospheric, electromagnetic, etc.). In the case where the spacecraft engine is to be used when passing the planet, powered-gravitational maneuvers are considered.

Different methods and approaches which have been developed to analyze gravity assist maneuvers allow the factors forming the flyby trajectory to be taken into account to a different extent [1–5, 62–70].

Within the framework of the simplified model of piecewise-conical approximation of the spacecraft trajectory, the flyby trajectory segment at a planet can be represented as a hyperbolic orbit in the central field of the planet; the perturbation effect manifests itself in the radical change experienced by the elements of the spacecraft orbit's heliocentric segment after flying by the planet (at the moment of crossing its gravisphere).

1.1 Dynamic scheme

Consider in more detail the simplified dynamic scheme of the spacecraft perturbation maneuver.

Suppose that the location and velocity of the spacecraft in the heliocentric coordinate system $\{\bar{R}_1, \bar{V}_1\}$ (see Figure 1.1) are known at the moment it enters the sphere of activity of a flyby body (e.g. a planet).

In accordance with the conical approximation of the spacecraft trajectory, on the gravisphere, the coordinates are to be changed to planetocentric ones $\{\bar{r}_1, \bar{v}_1\}$, which

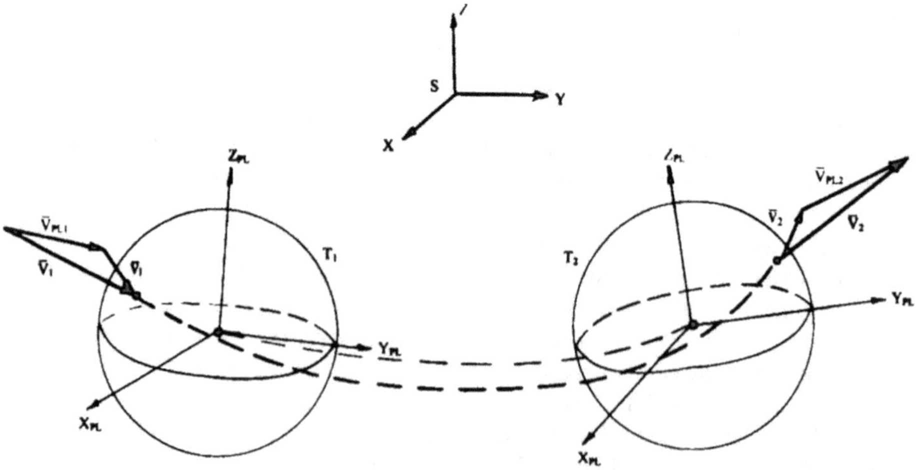

Figure 1.1 Schematic diagram of forming of gravity assist maneuver of a spacecraft.

uniquely determine the hyperbolic trajectory of the spacecraft flight in the central field of the attracting body:

$$\bar{r}_1 = \bar{R}_1 - \bar{R}_{pl}, \quad \bar{v}_1 = \bar{V}_1 - \bar{V}_{pl,1}, \quad |\bar{r}_1| = R_{sph\,pl}.$$

Here $R_{sph\,pl}$ is the radius of the planetary gravisphere.

The location and velocity of the planet $\bar{R}_{1pl}$ and $\bar{V}_{1pl}$ can be found from the well-known relationships for the given date of the gravity assist maneuver T_1

$$\bar{R}_{pl,1} = \bar{R}(E_{pl}, T_1), \quad \bar{V}_{pl,1} = \bar{V}(E_{pl}, T_1),$$

where $E_{pl} = \{\dots\ \dots\}$ are the osculating elements of the flyby body orbit.

The gravity assist maneuver results in a rotation of the spacecraft velocity vector after hyperbolic flyby of the planet. Coordinates of the spacecraft "exit" point from the gravisphere of the flyby planet as well as the coordinates of the spacecraft velocity in the planetocentric coordinate system can be found by transformation:

$$\begin{Bmatrix} x_2 \\ y_2 \\ z_2 \end{Bmatrix} = \Omega(\varphi^*) \begin{Bmatrix} x_1 \\ y_1 \\ z_1 \end{Bmatrix}, \quad \begin{Bmatrix} v_{x_2} \\ v_{y_2} \\ v_{z_2} \end{Bmatrix} = \Omega(\varphi) \begin{Bmatrix} v_{x_1} \\ v_{y_1} \\ v_{z_1} \end{Bmatrix}. \tag{1.1.1}$$

Here $\{v_{x_1}, \dots, z_1\}$ and $\{v_{x_2}, \dots, z_2\}$ are the velocity and location of the spacecraft in the points of entry and exit on the flyby planet gravisphere. The matrix of transformation is determined as

$$\Omega = \begin{bmatrix} \cos\varphi & \frac{c_3}{c}\sin\varphi & -\frac{c_2}{c}\sin\varphi \\ -\frac{c_3}{c}\sin\varphi & \cos\varphi & \frac{c_1}{c}\sin\varphi \\ \frac{c_2}{c}\sin\varphi & -\frac{c_1}{c}\sin\varphi & \cos\varphi \end{bmatrix}. \tag{1.1.2}$$

Here, the components of the vector of areas $\bar{c}\ (c_1, c_2, c_3) = \bar{r}_1 \bar{v}_1$

$$c_1 = y_1 v_{z_1} - z_1 v_{y_1};$$

$$c_2 = z_1 v_{x_1} - x_1 v_{z_1};$$

$$c_3 = x_1 v_{y_1} - y_1 v_{x_1};$$

$$c = \sqrt{c_1^2 + c_2^2 + c_3^2}.$$

The angle of rotation of the spacecraft velocity vector is

$$\varphi = 2\,\mathrm{arctg}\,\frac{K_{pl}}{\beta v_\infty^2}. \tag{1.1.3}$$

To determine the coordinates of the spacecraft exit point on the planetary gravisphere we use the relationships:

$$\varphi^* = \pi + \varphi - 2\gamma; \quad \varphi = \arcsin \frac{\beta}{R_{sph.pl.}}, \tag{1.1.4}$$

see Figure 1.2.

Here, K_{pl} is the gravitational parameter of the flyby planet; $\beta = \bar{r}_1 \bar{v}_1 / v_1$ is the aiming point distance of the spacecraft flyby from the planetary center; $v_\infty = (v_1^2 - 2K_{pl}/r_1)^{1/2}$ is the hyperbolic excess of the spacecraft velocity at the planet.

Note that the following restriction should be imposed on the aiming point distance β in the flyby of a planet:

$$\beta \leqslant \beta_{min} = r_{0min} \sqrt{1 + \frac{2K_{pl}}{r_{0min} v_\infty^2}}, \tag{1.1.5}$$

where r_{0min} is the minimum admissible distance in the pericenter of the flyby hyperbola; it is determined taking into account the radius of the planet and the height of the atmosphere (if one is present),

$$r_{0min} = r_{pl} + h_{atm}.$$

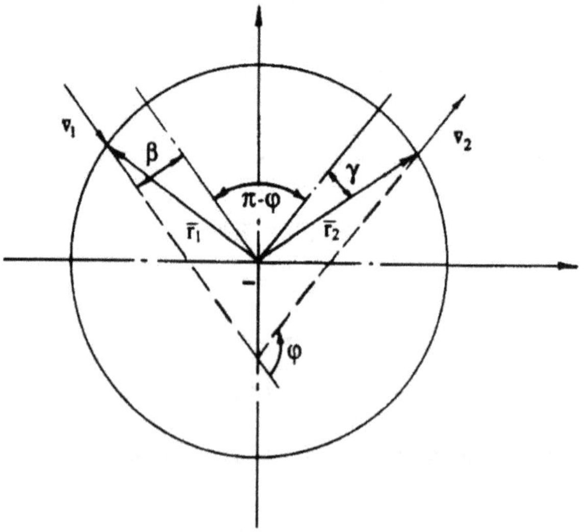

Figure 1.2 Gravity assist rotation of the spacecraft velocity in the gravisphere of a planet.

The duration of the spacecraft motion in the flyby planet gravisphere can be determined from

$$\Delta t = 2\sqrt{\frac{a^3}{K_{pl}}}\left(\operatorname{cosec}\frac{\varphi}{2}\operatorname{sh}H - H\right),$$
$$\operatorname{ch}H = \left(1 + \frac{R_{sph\,pl}}{a}\right)\sin\frac{\varphi}{2}, \tag{1.1.6}$$
$$a = \left(\frac{v_1^2}{K_{pl}} - \frac{2}{R_{sph\,pl}}\right)^{-1}.$$

Thus, flying by an attracting body, the location and velocity of the spacecraft at the moment it leaves the gravisphere $T_2 = T_1 + \Delta t$ in the heliocentric ecliptic coordinate system can be found as follows:

$$\bar{R}_2 = \bar{R}_{2pl} + \bar{r}_1\Omega(\varphi^*), \quad \bar{V}_2 = \bar{V}_{2pl.} + v_1\Omega(\varphi) \tag{1.1.7}$$

or

$$\bar{R}_2 = \bar{R}(E_{pl}, T_1 + \Delta t) + \bar{r}_1\Omega(\varphi^*),$$
$$\bar{V}_2 = \bar{V}(E_{pl}, T_1 + \Delta t) + v_1\Omega(\varphi). \tag{1.1.8}$$

The above relationships represent the dynamic model of the spacecraft's perturbation maneuver considering the size of the flyby body's gravisphere and the motion of the attracting body itself during the perturbation maneuver.

In parallel with such a model for examining trajectories with gravity assist maneuvers, a model with a far simpler representation of this maneuver is also used. The thing is that the comparison of the sizes of attracting bodies and radii of their orbits allows the following significant assumption: the sphere of influence of the attracting body can be considered infinitesimal as compared to the radius of its orbit and infinitely large relative to the size of the attracting body itself (this assumption is to a certain extent valid for both the planets in the solar system and their satellites).

In accordance with the assumption that the flyby body's sphere of influence has zero radius, the gravity assist maneuver can be approximated by an instantaneous rotation of the velocity vector at infinity $\bar{v}_\infty$ with respect to the center of the flyby body, and can be represented in accordance with the vector diagram of the velocities of the spacecraft and the attracting body (Figure 1.3). In this case, the duration of the gravity assist maneuver $\Delta t = T_2 - T_1 = 0$.

In the diagram, $\bar{V}_1$ is the vector of the spacecraft's incoming velocity towards the flyby body; $\bar{v}_{\infty 1}$, $\bar{v}_{\infty 2}$ are respectively the initial and final hyperbolic velocity excesses of the spacecraft (they are directed along the incoming and outgoing asymptotes of the flyby hyperbola), and

$$|\bar{v}_{\infty 1}| = |\bar{v}_{\infty 2}|.$$

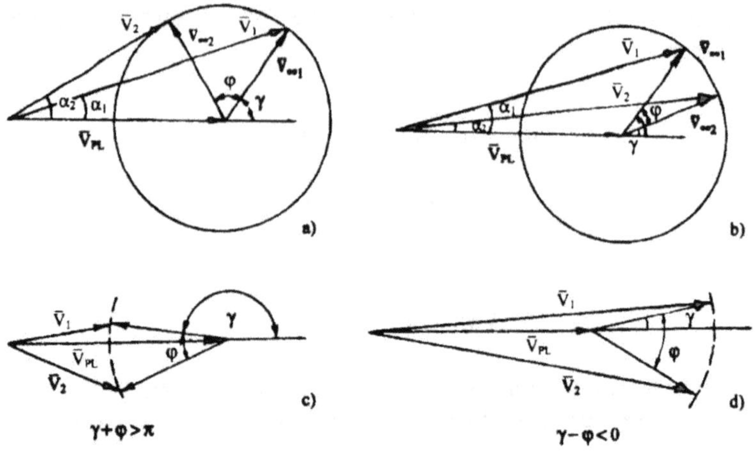

Figure 1.3 Vector diagram of the gravity assist maneuver.

$\bar{V}_2$ is the vector of the spacecraft's outgoing velocity after the gravity assist maneuver.

The efficiency of the gravity assist maneuver depends on the rotation of the velocity vector $\bar{v}_{\infty 1}$ by the angle φ that is determined by the gravitational constant of the flyby body and the flyby altitude relative to its surface. In accordance with Equations (1.1.3) and (1.1.5) we can write:

$$\varphi = 2\arcsin\left(\frac{K_{pl}}{K_{pl} + r_{0min} v_{\infty 1}^2}\right). \tag{1.1.9}$$

It is evident that the value of angle φ is limited by the size of the flyby body—the angle is maximum when the spacecraft trajectory is nearest to the surface of the attracting body.

As can be seen from the diagram, the rotation of $\bar{v}_\infty$ may be accompanied both by an increase $|\bar{V}_2|>|\bar{V}_1|$ (acceleration perturbation maneuver) and by a decrease $|\bar{V}_2|<|\bar{V}_1|$ (deceleration perturbation maneuver) in the spacecraft's outgoing velocity. In the case where the spacecraft passes before the flyby body (i.e. the spacecraft passes through the point of intersection of the orbits before the flyby body), angles φ and γ add up ($\bar{v}_{\infty 1}$ rotates counter-clockwise), and the spacecraft velocity decreases (case *a*). If the spacecraft flies behind the flyby body, the angles φ and γ are subtracted ($\bar{v}_{\infty 1}$ rotates clockwise), and the spacecraft velocity increases (case *b*).

From Figure 1.3 it can be seen that the minimum value of the post-perturbation velocity V_2 can be attained when $\varphi+\gamma=\pi$, and the maximum value can be reached when the angles are subtracted $\varphi-\gamma=0$. In both cases, the spacecraft's post-perturbation velocity will be collinear to the planet's velocity vector.

$$V_{2min} = V_{pl} - v_{\infty 1}, \qquad V_{2max} = V_{pl} + v_{\infty 1}. \tag{1.1.10}$$

It should be noted that special cases may exist when the change in the spacecraft post-perturbation velocity (increase or decrease) will be the same whatever the direction of rotation of $\bar{v}_{\infty 1}$. This takes place when the spacecraft's incoming velocity vector is collinear (or almost collinear) to the velocity vector of the flyby body. If $V_1 < V_{pl}$, then only acceleration perturbation maneuvers can occur; and if $V_1 > V_{pl}$, a deceleration maneuver takes place. This is illustrated in Figure 1.3 (*c* and *d*, respectively).

When analyzing such cases, one should take into account that, depending on altitude of the spacecraft flyby relative to the planetary surface, the angle of vector rotation $\bar{v}_{\infty 1}$ can vary from 0 to φ_{max} (according to Equation 1.1.9).

Thus, the effect of a perturbation maneuver during the flyby of an attracting body manifests itself in changes both to the value and direction of the spacecraft velocity. One of the criteria in assessing such maneuvers is the value of the maximum change in the spacecraft velocity vector that can be obtained in a flyby of the given body in the solar system [26].

As can be seen from the diagram, the spacecraft velocity change is equal to the vector

$$\Delta\bar{V} = \bar{V}_2 - \bar{V}_1.$$

In accordance with Figure 1.3 and using Equation 1.1.9 we can write

$$\Delta V = 2v_{\infty 1}\sin\frac{\varphi}{2} = \frac{2v_{\infty 1}K}{K + r_\pi v_{\infty 1}^2}, \tag{1.1.11}$$

where K is the gravitational constant of the flyby body and r_π is the pericentral radius of the flyby hyperbola.

The maximum change in the velocity ΔV for a given value of v_∞ in the flyby of a given body (e.g. a planet with a radius of r_{pl}) is obviously attained when $r_\pi = r_{pl}$. It is evident that this maneuver is impossible to implement, but it is of certain interest in assessing the potentialities of the flyby body.

There is no difficulty in finding the values of hyperbolic velocity excess v_∞ that provides maximum possible change in the module of the spacecraft velocity vector for any flyby body. From the condition $\partial\Delta V/\partial v_{\infty 1} = 0|_{r_\pi r_{pl}}$ it can be established that ΔV_{max} is attained when $v_\infty = (K/r_{pl})^{1/2}$, i.e. the hyperbolic velocity excess should be equal to the local circular velocity at the surface of the flyby body $(\Delta V_{max} = (K/r_{pl})^{1/2} = 0|_{r_\pi r_{pl}})$. Table 1.1 gives the values of ΔV_{max} for all nine planets of the solar system to show the perturbation potentialities of the planets.

Table 1.1 Maximum possible values of the spacecraft velocity increase due to planetary flybys, ΔV_{max} km/s.

Mercury	Venus	Earth	Mars	Jupiter	Saturn	Uranus	Neptune	Pluto
3.01	7.33	7.91	3.55	42.73	25.62	15.18	16.75	1.10

The suggested criterion for assessing perturbation maneuvers gives only a partial estimate of the maneuver, because it primarily allows an estimate of the effect of vectorial rotation of the spacecraft heliocentric velocity as a result of flying by an attracting body. At the same time, when analyzing interplanetary missions with planetary flybys, it is of greater interest to estimate the effect of perturbation maneuvers on the flight trajectory parameters (orbital constants or elements of the post-perturbation orbit), which determine the further motion of the spacecraft to the target planet.

1.2 Effect of perturbation maneuvers on the spacecraft orbital characteristics

A gravity assist maneuver at an attracting body can be used to implement a controlled change of the spacecraft orbital characteristics.

Generally speaking, the elements of post-perturbation spacecraft orbit can be estimated directly based on the model of gravity assist maneuvering in space taking into account the motion of the attracting body itself and size of its sphere of activity (see Figure 1.1) and using relationships (1.1.8). In this case, the effect of this kind of maneuvering manifests itself both in the relative motion of the spacecraft in the field of the flyby planet, and in the spacecraft motion together with the planet in the central field of the sun. It is obvious that such an estimate of the effect of perturbation maneuvering on the spacecraft trajectory can be made only based on numerical calculations taking into account all the factors of such interrelated motion of the spacecraft and the flyby planet.

The accepted model of perturbation maneuvering, in which the gravisphere of the flyby body is assumed to have zero radius, makes it easier to assess the perturbation effect of a planet on the characteristics of spacecraft post-perturbation orbit. In this case, the perturbation effect on the trajectory will demonstrate itself only in a rotation of the vector of the spacecraft's planetocentric velocity by an angle φ within the plane of the spacecraft flyby hyperbola, which is uniquely determined by the incoming and outgoing spacecraft velocity vectors at the flyby body. This makes it possible to conduct analysis on a vector diagram of velocities in the center of the celestial body (see Figure 1.4). Such an approach is quite common and is widely used by researchers.

Thus, the parameters of the post-flyby orbit (both angular parameters and those determining the size and form of the spacecraft's new orbit) will depend only on the parameters of the spacecraft hyperbolic trajectory being formed in the gravitational field of the flyby body. By varying the arrival angles and velocities, and the altitude at the flyby body, the parameters of the post-flyby orbits can cover a relatively wide range. Consider the functional representation of variation in the spacecraft orbital constants resulting from a perturbation maneuver.

The scheme of perturbation maneuver will be considered within the framework of the model described above. Suppose that a spacecraft and an attracting body with a gravitational constant K_1 move along coplanar orbits in the central gravitational field of a body (with a gravitational constant K). Denote their locations and

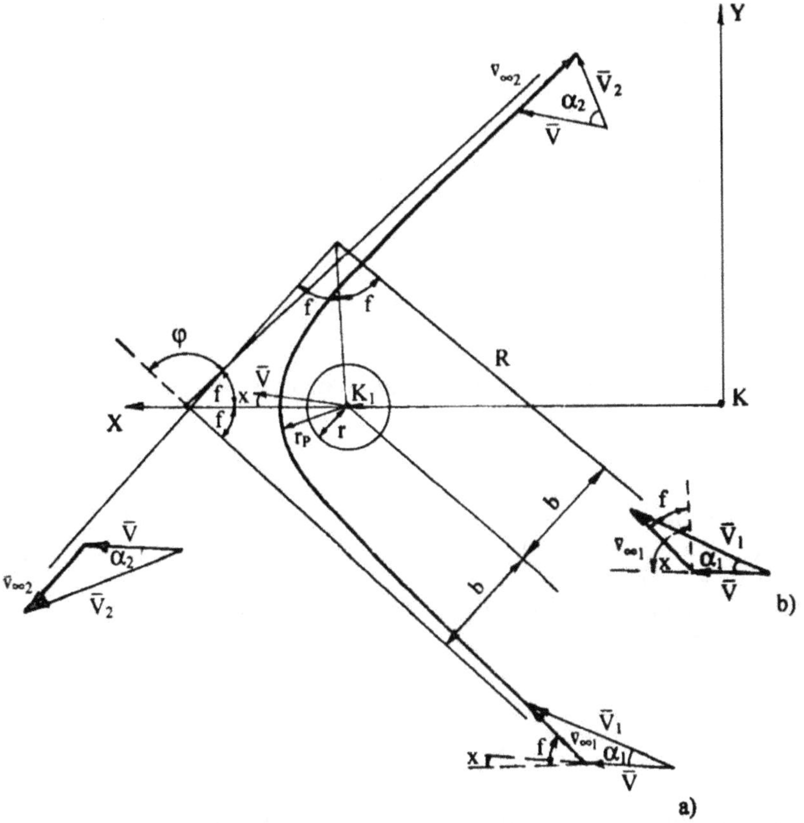

Figure 1.4 Schematic diagram of gravity assist maneuver of the spacecraft.

velocities at the moment of perturbation maneuver by $\bar{R}$ and $\bar{V}$ for the celestial body and $\bar{R}$ and $\bar{V}_1$ for the spacecraft. After the gravity assist maneuver, $\bar{V}_1$ transforms into $\bar{V}_2$—outgoing spacecraft velocity—as a result of an instantaneous rotation of the hyperbolic velocity excess $\bar{v}_{\infty 1}$ by an angle φ. In this case, $|\bar{v}_{\infty 1}|=|\bar{v}_{\infty 2}|=v_{\infty}$. Projections of the spacecraft incoming $\bar{V}_1$ and outgoing $\bar{V}_2$ velocities onto axes X and Y (Figure 1.4) can be written as

$$V_{1x}=v_{\infty}\cos f+V\cos\chi;$$

$$V_{1y}=v_{\infty}\sin f+V\sin\chi;$$

$$V_{2x}=V\cos\chi+v_{\infty}\cos(\pi-f);$$

$$V_{2y}=V\sin\chi+v_{\infty}\sin(\pi-f).$$

The change in the integral of the spacecraft orbital energy resulting from the flyby maneuver is determined as

$$\Delta h = V_2^2 - V_1^2 = -4v_\infty V \cos f \cos \chi.$$

As noted above, two cases can be considered:

(a) a flyby before the attracting body $\chi = f - \gamma$ (the perturbation maneuver reduces the spacecraft energy);
(b) a flyby behind the attracting body $\chi = f + \gamma$ (the perturbation maneuver increases the spacecraft energy).

We write the change in the integral of the spacecraft orbital energy as

$$\Delta h = -4v_\infty V \cos f \cdot \cos(f \mp \gamma). \tag{1.2.1}$$

In this case, the angle f between the asymptote and axis of the flyby hyperbola is correlated with φ—the angle of rotation of $\bar{v}_\infty$, which can be found from Equation 1.1.9: $f = (\pi - \varphi)/2$.

From triangles of velocities we also determine the angle γ between the velocity vectors of the flyby body $\bar{V}$ and spacecraft $\bar{v}_\infty$:

$$\sin \gamma = \frac{V_1 \sin \alpha_1}{v_\infty}, \quad \cos \gamma = \frac{V_1 \cos \alpha_1 - V}{v_\infty}.$$

Here α_1 is the angle between the vectors of velocity $\bar{V}$ and incoming velocity $\bar{V}_1$ of the spacecraft.

Next, using the formula for rotation angle φ, we write

$$\cos f = \sin \frac{\varphi}{2} = \frac{1}{1 + r_p v_\infty^2 / K_1} = (1 + mn^2)^{-1},$$

$$\sin f = n(m^2 n^2 + 2m)^{1/2} \cos f,$$

$$v_\infty = (V_1^2 + V^2 - 2V_1 V \cos \alpha_1)^{1/2}.$$

Here, $m = r_p / r$ is the relative pericentral distance of the hyperbola expressed in terms of radii of the flyby body, $n = v_\infty / v_{cr}$ is the relative hyperbolic excess of the spacecraft velocity (in terms of local circular velocity at the surface of flyby body $v_{cr} = (K_1 / r)^{1/2}$).

Returning to Equation 1.2.1 and using the relationships obtained, we can write the expression for the spacecraft post-flyby energy for any values of the initial energy h_1 and the spacecraft incoming angles α_1 as follows:

$$h_2 = h_1 - 4V(1 + mn^2)^{-2}[V_1 \cos \alpha_1 - V \pm V_1 \sin \alpha_1 (2mn^2 + m^2 n^4)^{1/2}]. \tag{1.2.2}$$

Here, the upper sign refers to the decrease, and the lower sign to the increase, in the spacecraft orbital energy.

The change in the integral of areas of the orbit as a result of the perturbation maneuver in the coplanar model considered here, is determined as

$$\Delta C = R(V_1 \cos \alpha_1 - V_2 \cos \alpha_2) \tag{1.2.3}$$

see Figure 1.4.

Determining cosines of the spacecraft's incoming and outgoing angles from velocity triangles

$$\cos \alpha_1 = \frac{V_1^2 + V^2 - V_{\infty 1}^2}{2V_1 V}; \quad \cos \alpha_2 = \frac{V_2 + V^2 - v_{\infty 2}^2}{2V_2 V}$$

we get

$$\Delta C = R\frac{V_1^2 - V_2^2}{2V} = -\frac{R\Delta h}{2V} = 2Rv_\infty \cos f \cos \chi. \tag{1.2.4}$$

The sum of collinear vectors $\bar{C}_1$ and $\Delta\bar{C}$ yields the post-flyby vector constant of areas

$$\bar{C}_2 = \bar{C}_1 + \Delta\bar{C}.$$

Using Equations 1.2.2 and 1.2.4 we write the final expression for the post-flyby constant of areas of the spacecraft orbit for any values of the spacecraft initial velocities V_1 and ingoing angles α_1 as follows:

$$C_2 = C_1 + 2R(1 + mn^2)^{-2}[V_1 \cos \alpha_1 - V \pm V_1 \sin \alpha (2mn^2 + m^2 n^4)^{1/2}] \tag{1.2.5}$$

Finally, we determine the perturbation change in the vectorial Laplace integral, controlling the location of the apsides of the spacecraft orbit in the plane of its orbital movement:

$$\Delta\bar{\lambda} = \bar{\lambda}_1 - \bar{\lambda}_2 = -(\bar{C}_2 \times \bar{V}_2) - \frac{K\bar{R}_2}{R_2} + (\bar{C}_1 \times \bar{V}_1) + \frac{K\bar{R}_1}{R_1}. \tag{1.2.6}$$

Since the rotation of the velocity vector in the flyby point is assumed instantaneous, then $\bar{R}_1 = \bar{R}_2$, and

$$\Delta\lambda = \bar{C}_1 \times \bar{V}_1 - \bar{C}_2 \times \bar{V}_2.$$

Therefore, the angle of rotation of apsides line δ of the orbit resulting from the perturbation maneuver is equal to the angle of rotation of the spacecraft velocity vector, and can be found as follows:

$$\cos\delta=\frac{V_1^2+V_2^2-\Delta V^2}{2V_1V_2},\quad \text{where } \Delta V=2v_\infty\cos f=\frac{2v_\infty}{1+mn^2};$$

$$\cos\delta=\frac{(V_1^2+V_2^2)(1+mn^2)-2v_\infty}{2V_1V_2}. \tag{1.2.7}$$

It is evident from (1.2.2) and (1.2.5) that with the use of the well-known relationship we can obtain the value of the post-flyby Laplace integral for any values of the spacecraft's initial energy and incoming angles

$$\lambda_2=(K^2+h_2C_2^2)^{1/2}. \tag{1.2.8}$$

Using the obtained values of constants h_2, C_2, λ_2 it is easy to obtain the Keplerian elements of the post-flyby orbit from the well-known relationships of celestial mechanics [1, 2].

As can be seen from relationships (1.2.1) and (1.2.4), the conditions of maximum change in the constants of integrals of energy and areas are satisfied at the same values $v_\infty=\sqrt{K_1/r_p m}$ and $f=60°$. In this case, the values of the angle χ are

$\chi=0$ in the case of decrease in Δh and increase in ΔC,
$\chi=\pi$ in the case of increase in Δh and decrease in ΔC.

The optimal angles $\alpha_{1\text{opt}}$ for approaching the attracting mass at a given value of the spacecraft orbital energy h_1 can be selected based on the condition of maximizing the increment of the energy constant:

$$\Delta h_{\max}=\max_{\alpha_1}\Delta h(K_1,r,R,V_1,\alpha_1,m)\quad V_1=\text{const.}\ \ m=\text{const.} \tag{1.2.9}$$

A computational algorithm was developed to select optimal angles for the spacecraft to approach the flyby body and to determine the maximum spacecraft energy change for incoming orbits. The algorithm allows these characteristics to be determined in a wide range of initial energies.

Figures 1.5 and 1.6 present curves $\Delta h_{\max}=f(h_1)$, which characterize the maximum spacecraft energy increase attainable from perturbation maneuvers at each of the solar system planets. The corresponding optimal angles $\alpha_{\text{opt}}=f(h_1)$ of the spacecraft approaching the planet are shown in Figures 1.7 and 1.8 for $m=1$.

The given results show that Jupiter has the highest potentialities of all the planets ($\Delta h_{\max}\approx 1100\,\text{km}^2/\text{s}^2$). Venus, Saturn, and Earth display lesser potentialities ($\Delta h_{\max}\approx$ 400 to 500 km^2/s^2), and even lower potentialities are typical of Mercury, Mars, Uranus, and Neptune ($\Delta h_{\max}\approx$ 150 to 250 km^2/s^2). It is worth mentioning that for the planets of the Jovian group, the values of $\Delta h_{\max}$ lie in the zone of high-energy incoming orbits (hyperbolic or parabolic), and their peaks are not distinct. For the

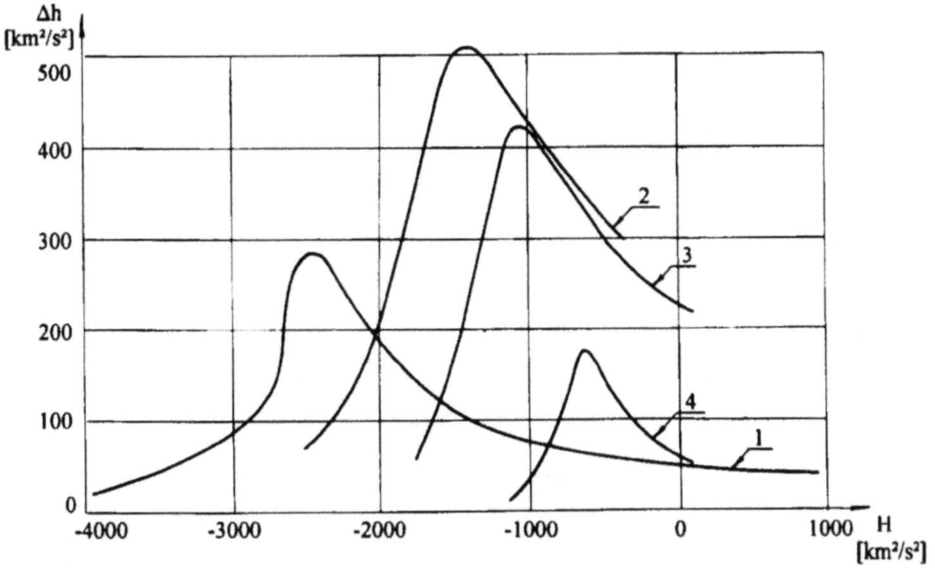

Figure 1.5 Maximum change in the spacecraft orbital energy constant at planets. 1—Mercury, 2—Venus, 3—Earth, 4—Mars.

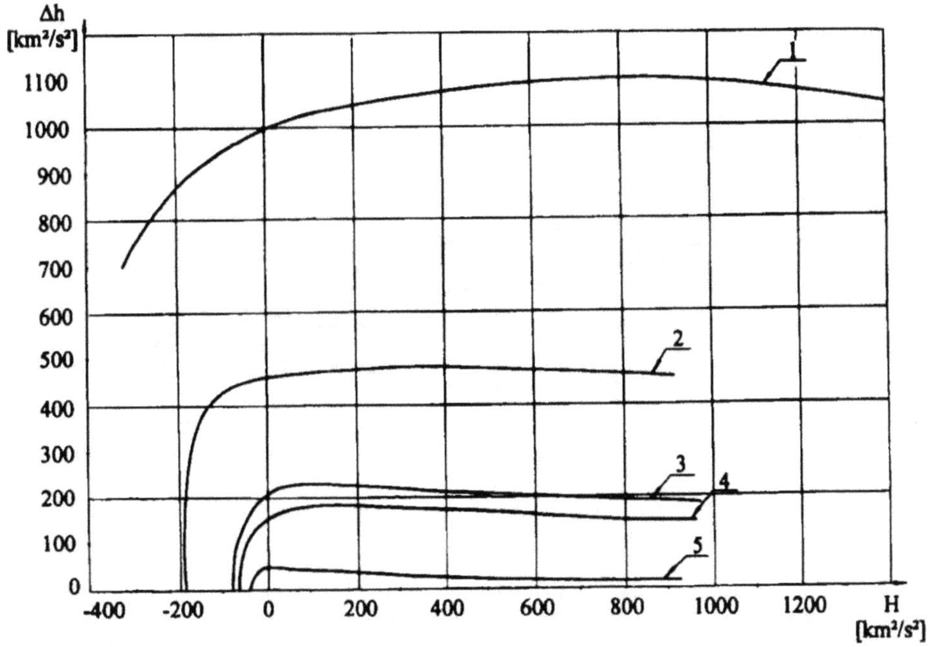

Figure 1.6 Maximum change in the spacecraft orbital energy constant at planets. 1—Jupiter, 2—Saturn, 3—Uranus, 4—Neptune, 5—Pluto.

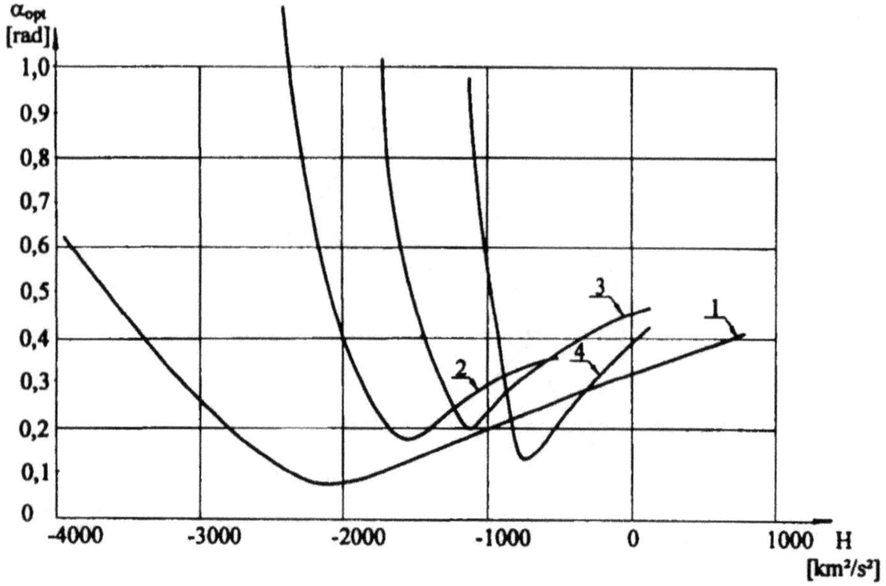

Figure 1.7 Optimal angles of approaching planets to perform a gravity assist maneuver: 1—Mercury, 2—Venus, 3—Earth, 4—Mars.

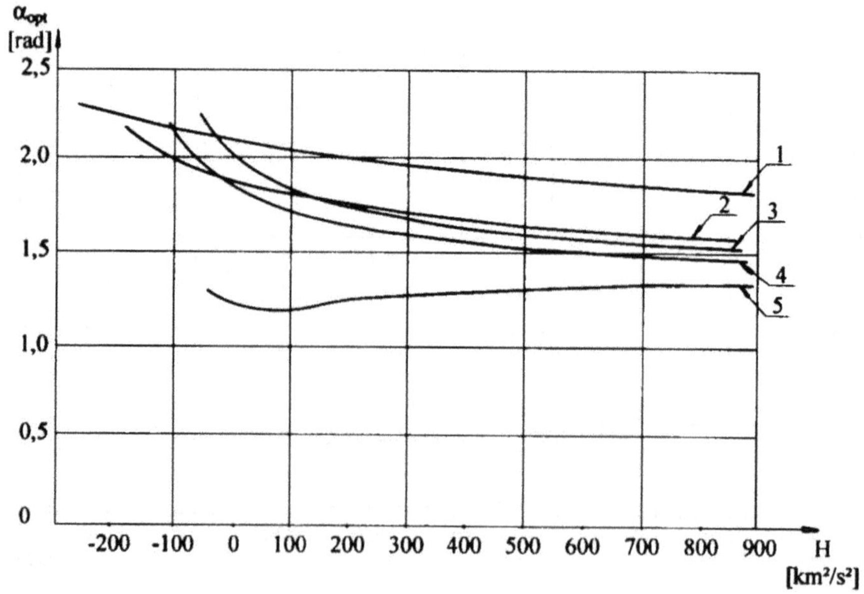

Figure 1.8 Optimal angles of approaching planets to perform a gravity assist maneuver: 1—Jupiter, 2—Saturn, 3—Neptune, 4—Uranus, 5—Pluto.

planets of the Earth's group, these maxima are more clearly defined, and are located in the zone of elliptic incoming orbits.

The optimal angles of the spacecraft approaching the flyby planet are rather high for the planets of the Jovian group ($\alpha_{opt} \approx 1.5$ to 2.0 rad) and show small variations within a wide range of incoming energies. The planets of Earth's group have clearly defined minima (0.1–0.2 rad) in the zone of extreme change in spacecraft energy, and increase drastically (up to 1 rad), especially in the case of a shift into the zone of orbits with low incoming energies.

The above results were obtained by numerical solution of Equation 1.2.9 and refer to the flyby where the pericentral altitude of the flyby hyperbola coincides with the radius of the flyby body. This obviously limiting case cannot be accomplished in practice (that is, the condition $m>1$ must hold). An increase in the relative pericentral altitude m obviously results in a reduction in the efficiency of the gravity assist maneuver (in accordance with Equation 1.2.2). An examination of the effect of changes in m on the maximum values of the perturbation increment in energy showed that throughout the range of incoming energies, there is virtually no change in the pattern of curves (the growth in m results in a monotonic decrease in max Δh), while the peaks of the curves move slightly toward lower incoming energies. The behavior of curves $\alpha_{opt}=f(h_1, m)$ is similar. The optimal angles of the spacecraft approaching the flyby body show insignificant variations (not more than 1°–2°) as the relative flyby altitude m increases from 1 to 2. Thus, the values of α_{opt} show almost no dependence on deviations of m from its minimum value.

It is very interesting to use gravity assist maneuvers of the spacecraft to change the spatial characteristics of its orbit—inclination of the orbital plane, longitude of the ascending node, and longitude of the pericenter. As was mentioned earlier, to be performed in full measure, these estimates require a more rigorous model which allows for the mutual effects of different factors on the formation of the spacecraft's new orbit (Equation 1.1.7).

However, parameters as important as the inclination of the orbital plane can be estimated based on the simplified model considered here.

The essence of the gravity assist rotation of the spacecraft orbital plane lies in the fact that the maneuver allows the spacecraft velocity vector $\bar{V}_2$ to rotate relative to the plane of the initial incoming orbit. The angle of this rotation depends on both the gravitational capacity of the flyby body and on the location of the point at which the spacecraft enters the body's sphere of activity.

Suppose that the spacecraft approaches the flyby body, whose velocity is $\bar{V}_p$, within the body's orbital plane. Let the spacecraft velocity be $\bar{V}_1$. The gravity assist maneuvers transform the spacecraft hyperbolic velocity excess $\bar{v}_{\infty 1}$ into the outgoing hyperbolic velocity $\bar{v}_{\infty 2}$. For clarity, Figure 1.9 shows the sphere of possible locations of the end of vector $\bar{v}_{\infty 2}$ after the gravity assist maneuver. It is evident that the domain of possible maneuvers is restricted by a spherical and angular sector, which is determined by the angle of rotation of vector $\bar{v}_{\infty 2}$ relative to the attracting center.

Figure 1.9 allows us to write the expression for the maximum change in the angle of inclination of the spacecraft orbital plane in the case of a single flyby of the

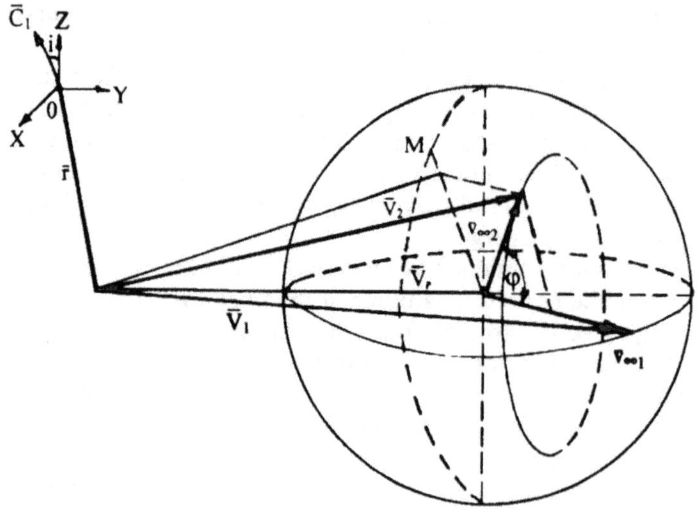

Figure 1.9 Changes in the inclination angle of the spacecraft orbit as a result of gravity assist maneuver.

attracting body

$$\sin \Delta i = \frac{v_\infty \sin \varphi}{V_p}. \tag{1.2.10}$$

Here Δi is the change in the inclination of the spacecraft orbital plane; v_∞ is the spacecraft hyperbolic velocity excess at the flyby body; φ is the angle of spacecraft rotation as a result of a gravity assist maneuver at an attracting body with the gravitational parameter K.

The angle of rotation φ can be found from the formula

$$\sin \varphi/2 = \frac{1}{1+(r_m v_\infty^2/K)},$$

where φ_{max} will be attained at $r_m = r_p$.

In the case where $\varphi_{max} > \pi/2$, the maximum increment in the angle of inclination of the spacecraft orbital plane should be calculated by the formula

$$\sin \Delta i = \frac{v_\infty}{V_p}. \tag{1.2.11}$$

Figures 1.10 and 1.11 give the curves of maximum change in the spacecraft's orbital plane inclination resulting from a single flyby of different planets in the solar system. The data show that the highest potentialities for changing the spacecraft orbital plane in a single flyby is typical of those large planets of the Jovian group—Jupiter,

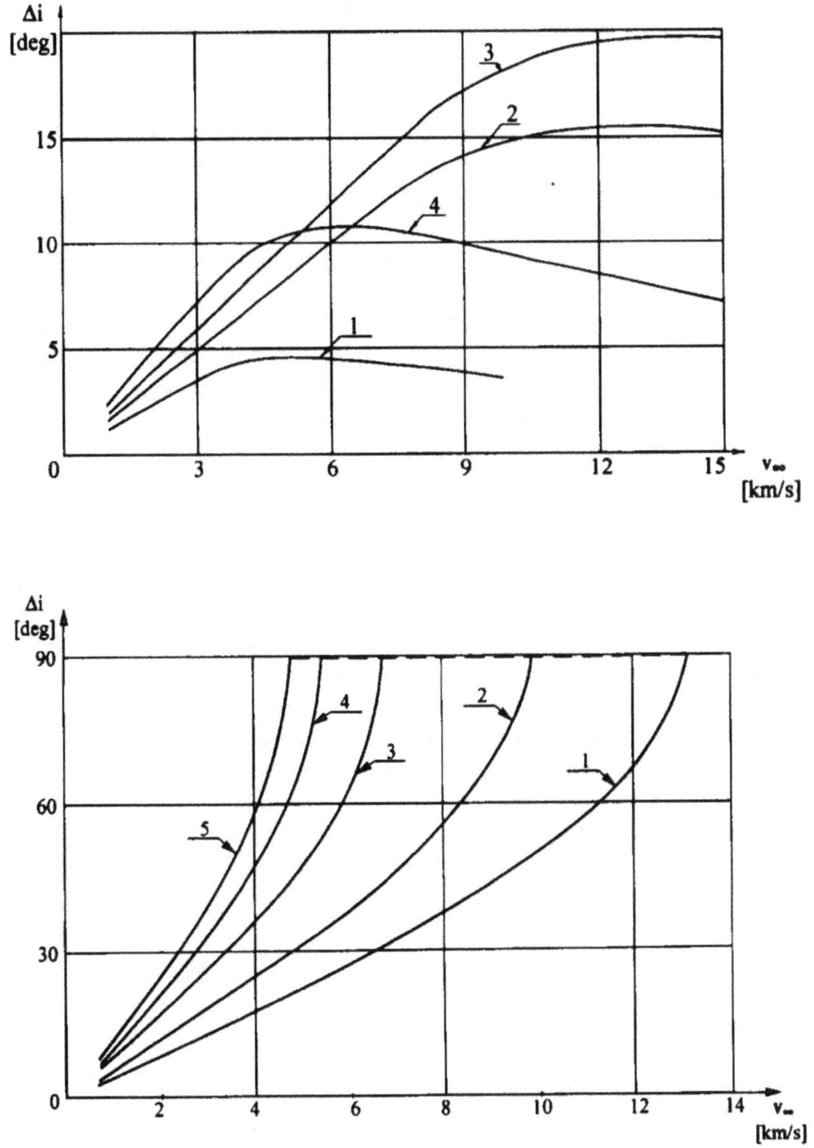

Figure 1.10–1.11 Maximum changes in the inclination angle of the spacecraft orbit as a result of gravity assist maneuver at planets.

Saturn, Uranus, Neptune, Pluto (Figure 1.11). The potentialities of the Earth group planets—Earth, Venus, Mars, Mercury—are somewhat lower (Figure 1.10).

Multiple flybys of the attracting body allow an even greater increase in the inclination angle of the spacecraft orbital plane by way of additional rotation of $v_{\infty 2}$ during each flyby. The maximum value of the inclination angle of the spacecraft orbit i_{lim} (relative to the initial orbit) can be determined from Equation 1.2.11. The position of the point M on the sphere of possible positions of $\bar{v}_{\infty 2}$ (Figure 1.9) corresponds to the maximum possible change in the spacecraft's orbital plane due to a gravity assist maneuver at the given $\bar{v}_{\infty}$.

It is evident that orbits with large inclination angles can be obtained. To do this, it is necessary to perform flybys with large $\bar{v}_{\infty}$, though the efficiency of the gravity assist maneuver in this case will decrease (in accordance with Equation 1.1.9, growth in v_{∞} results in a reduction in the rotation angle φ). This, in turn, will require the number N of flybys of the attracting body to be increased so as to attain the necessary i_{lim}.

The above results demonstrate that the gravity assist maneuvers of spacecrafts in the solar system allow significant changes in the spacecraft's orbital characteristics to be made at the expense of gravity forces. A relevant flyby of an attracting body results in changes in the orbital constants of the spacecraft trajectory: its size and the shape of its orbit, and its spatial orientation.

Gravity assist maneuvers can be either accelerating or decelerating. They can be used to reach distant zones and objects in the solar system (e.g. planets, comets, asteroids) or near-solar zones.

Gravity assist maneuvers in space make it possible to significantly change the inclination of the spacecraft's orbital plane. This can be particularly important in exploration beyond the ecliptic plane (e.g. in flights over the sun), or in flights to celestial bodies whose orbits are notably inclined to the ecliptic plane.

Similar problems can be considered for the spacecraft maneuvers in the vicinity of planets that have attracting natural satellites. Gravity assist maneuvers of the spacecraft in such planetary-satellite systems notably extend the scope of scientific problems which can be resolved by missions to these planets.

Thus, the gravity assist maneuver is a significant factor in the design of multi-purpose interplanetary missions in the solar system. Special methods have to be developed for designing and monitoring such missions, and for assessing of their feasibility and results.

1.3 Flight paths; methods of study

A wide variety of space missions can use gravity assist maneuvers. The maneuvers are determined by the purpose of the mission, the celestial bodies at which the maneuvers are to be performed, the number of celestial bodies, and the opportunity to use powered maneuvers involving the spacecraft engine.

An interesting classification of the interplanetary mission trajectories in terms of the purpose of the mission is suggested in [32].

Some missions using gravity assist maneuvers have been intensively studied and even fully accomplished, whereas others are still being studied and discussed or have yet to be discovered.

The simplest mission involving a gravity assist maneuver is a flyby of an attracting body after which the spacecraft orbits as an artificial planet with known orbital characteristics. This kind of flight was first accomplished by the Soviet automatic interplanetary probe "Luna-1", which entered a heliocentric orbit of an artificial planet after flying by the Moon.

Different conditions can be imposed on the characteristics of the post-flyby trajectories (depending on the purpose of the mission and the perturbation potentialities of the celestial body bypassed). There can be missions where the spacecraft leaves the ecliptic plane, where it flys over the polar near-solar regions, and also flights directly to the Sun.

Gravity assist maneuvers can be used in interplanetary missions to reach distant planets (e.g. a planet of the Jovian group or Mercury), in flights between planets of the Earth's group, and in missions with the spacecraft returning to Earth.

Also of interest is the problem of using gravity assist maneuvers in missions to small celestial bodies (comets, asteroids). The orbits of small celestial bodies (especially comets) are in most cases notably elliptic, and the plane of their motion does not generally coincide with the ecliptic plane (the inclination angles can be quite significant). The use of gravity assist maneuvers in such missions helps to significantly improve the energy characteristics of such flights.

An important group of multi-purpose missions are flights in the spheres of activity of planets with large natural satellites (e.g. Jupiter or Saturn). Gravity assist maneuvers at natural satellites of planets extend the potentialities of the exploration probe.

Missions with gravity assist maneuvers can include two or more flybys of one or more attracting bodies. Such maneuvers are obviously more difficult to study and accomplish. Examples of these are exploration missions, for instance, multiple-flyby recurrent missions within the planets of the Earth's group, or the "Grand Tour" mission, where successive flybys of several planets make it possible to reach the most distant planets of the solar system.

Depending on the mission, various requirements can be imposed concerning energy, navigation, exploration, etc. The combined effect of these factors determines the final pattern of the spacecraft to implement the mission.

However, the most important condition allowing the route with a flyby to be accepted is that it should reduce energy consumption relative to an analogous direct flight. Therefore, from the first stages of analysis of missions with gravity assist flybys, the criterion of minimum total energy expenditure is commonly used. The criterion can be based on different approximate models of the flight depending on the requirements of accuracy and detail of the analysis.

The simplest model that can be found in works by the founders of astronautics—K.E. Tsiolkovskii and F.A. Zander—represents interplanetary trajectory as an arc of Keplerian trajectory in the central field of the Sun, which begins and terminates on the orbits of the departure and destination planets (the orbits are considered circular and coplanar). The energy expenditure for the interplanetary flight is estimated based on the criterion of the total characteristic velocity needed for the spacecraft to fly from the orbit of departure to that of the destination planet.

Later the model was extended to include more complicated schemes of flights with gravity assist maneuvers [2, 12, 15, 17, 30]. In this model, the orbits of departure, flyby, and destination planets are considered circular and coplanar. Phasing of planets on their orbits is neglected, and the gravity assist maneuver is approximated by an instantaneous rotation of the spacecraft velocity at the moment a planet with a gravitational constant K is passed.

In this case, the optimization problem is reduced to the computational procedure of searching for the minimum of the multivariable function

$$\begin{aligned} &\operatorname{opt} V_\Sigma = \min V_\Sigma(T_1, r_{\pi 1}, m_\pi, K); \\ &V_\Sigma = |\vec{V}_1 - \vec{V}_{\text{pl}1}| + |\vec{V}_{\text{pl}.2} - \vec{V}_K|; \\ &\vec{V}_1 = V(T_1, r_{\pi 1}, R_1), \quad \vec{V}_2 = V(T_K, r_{\pi K}, R_K). \end{aligned} \tag{1.3.1}$$

Here T_1 and $r_{\pi 1}$ are the orbital parameters (rotation period and pericentral altitude) of the starting leg of the flight; m_π is the relative altitude of the spacecraft flight (in the pericenter of the flyby hyperbola) over the planetary surface $m_\pi = R_\pi / R_{\text{pl}} > 1$; $\vec{V}_{\text{pl}1}$, $\vec{V}_{\text{pl}2}$ are the velocities of the departure and destination planets; $\vec{V}_1$, $\vec{V}_K$ are the spacecraft velocities at the start and arrival points; R_1, R_K are the orbital radii of the departure and destination planets; V_Σ is the total characteristic velocity of the flight.

The orbital characteristics of the post-perturbation leg of the flight $(T_K, r_{\pi K})$ are thus determined only by the altitude of the spacecraft flight m_π relative to the surface of the flyby planet.

Despite its simplicity this model helps us to obtain some important approximate estimates for interplanetary flights with gravity assist maneuvers. In particular, it allows an estimate of the lower bound of energy requirements for such flights (similar to Hohmann's scheme). Thus, investigations of the Earth–Mars–Jupiter path conducted with the use of this model [137] showed that the optimal trajectory in terms of fuel expenditure is unique, and should meet the condition of tangency of the orbits of departure and destination in the apsides $(r_{\pi 1} = R_1, r_{\pi K} = R_{2K})$. In addition to that, the angular distances of both the legs of the flight were found to be much less than π, and the optimal asymptotic velocity $v_{\infty \text{flb}}$ of a flyby of Mars to be rather high. It is evident that similar studies can be also conducted for other options of interplanetary missions with planetary flybys.

The simple model above is rather convenient and useful for approximate estimation of different paths for interplanetary missions. However, it should be mentioned that the search for optimal solutions, even based on a model as simple as that, require numerical calculations with the use of methods and algorithms of nonlinear programming.

A more rigorous model of interplanetary flights is far more widely used. It approximates the spacecraft trajectory by segments of nonperturbed Keplerian movement, as does the previous model but represents the planetary movement in more detail, making allowance for the planetary orbits being elliptic and non-coplanar, and taking into account the phasing of the planetary movement along the orbits. In this case the analysis is made for specific dates of interplanetary flights (or their intervals) with estimates of minimum energy expenditure for the mission [1–4, 10, 12, 17–20, 26, 30].

This approach to the analysis of interplanetary flight schemes utilizing gravity assist maneuvers is used by many researchers. Its essence is as follows: within the framework of the technique of the spheres of activity, segments of heliocentric motion from Earth to the flyby planet and from the flyby planet to the destination planet are constructed. These segments of the interplanetary trajectory are joined, based on the vectors of asymptotic incoming and outgoing velocities relative to the flyby planet ($\bar{v}_{\infty i}^-$ and $\bar{v}_{\infty i}^+$), with possible application of a powered maneuver. In the case of multiple flybys, a similar construction is made for subsequent segments of the trajectory. The optimal trajectories are sought based on the criterion of minimum total characteristic velocity, which can be represented in the general form as

$$\operatorname{opt} V_\Sigma = \min[V_0 + \sum_i V_i + V_K], \quad i=1,\ldots,N, \tag{1.3.2}$$

where V_0 is the starting impulse from the near-Earth orbit with radius R_0; V_i is the velocity impulse applied during the flyby of the ith planet; V_K is the characteristic velocity of deceleration at the destination planet to enter a near-planetary orbit; N is the number of the flybys of planets.

Here

$$V_0 = V_0(T_1, T_0, \bar{R}_0),$$
$$V_i = V_i(T_{i-1}, T_i, T_{i+1}, \bar{\rho}_i),$$
$$V_K = V_K(T_N, T_K, \bar{R}_K),$$

where T_0 is the date of the spacecraft launch; T_i are the dates of intermediate planetary flybys; $T_\Sigma = T_K - T_0$ is the total flight time; T_K is the date of arrival at the destination planet; $\bar{R}_0$ and $\bar{R}_K$ are the vectors determining the departure and destination orbits; $\bar{\rho}_i$ is the radius vector of the point of passing from the incoming to outgoing hyperbolas in the flyby of the ith planet.

Other variants of formulation of the criteria function V_Σ are evidently possible depending on the peculiarities of the problem in question.

Thus, the problem of finding the optimal interplanetary trajectory with gravity assist maneuvers is reduced to the minimization of the mission characteristic velocity as a multivariate function

$$\operatorname{opt} V_\Sigma = \min V_\Sigma(T_0, T_i, T_K, \bar{R}_0, \bar{\rho}_i, \bar{R}_K), \; i=1,\ldots,N. \tag{1.3.3}$$

In the minimization process, the problems of finding heliocentric and planetocentric legs of the flight path are considered separately within the framework of the method of planetary spheres of activity.

The problem of optimization of a planetocentric flyby segment implies minimization of energy expenditures for the interhyperbolic transfer [30, 63, 64], and is regarded as an internal problem with respect to the external heliocentric pre- and post-perturbation segments, which determine the values of hyperbolic velocity excess in the initial and end points of the flyby trajectory segment

$$\bar{v}_{\infty i}^{-}=\bar{v}(T_{i-1},T_i),\quad \bar{v}_{\infty i}^{+}=\bar{v}(T_i,T_{i+1}),\ i=1,\ldots,N.$$

They uniquely determine the plane of the spacecraft planetocentric maneuver

$$[\bar{v}_{\infty i}^{-}\times\bar{v}_{\infty i}^{+}]\cdot\bar{\rho}_i=0. \qquad (1.3.4)$$

The calculation of the velocity impulse to be applied during flyby $\min V_i$ $(0\leqslant\bar{\rho}_i\leqslant\infty)$ in the case of a single-impulse maneuver (if $v_{\infty i}^{-}\neq v_{\infty i}^{+}$) reduces to finding of the impulse and the point of transfer $\bar{\rho}_i$ between the incoming and outgoing hyperbolas by solving a 4th-order nonlinear algebraic equation [30, 63]. If the maneuver is physically impracticable ($\rho<\rho_{pl}$) ρ_{pl} is the minimum admissible radius of the planetary flyby), the optimal impulse value is determined by a one-dimensional search for the minimum V_i on the outgoing segment of the flyby trajectory specified by the values $v_{\infty i}^{-}$ and ρ_{pl} [141].

Other approximate methods can be used to determine the value of impulse V_i close to the optimal one (for example, in the case of pericentral interhyperbolic transfer $\rho_\pi\geqslant\rho_{pl}$) [5].

In the case where $v_{\infty i}^{+}=v_{\infty i}^{-}$ and $\rho_\pi\geqslant\rho_{pl}$, a purely ballistic maneuver of the spacecraft is performed during the planetary flyby

$$V_i=0,\ i=1,\ldots,N.$$

In terms of computations, the problem of search for the optimal multiple-flyby interplanetary trajectory can be reduced to the successive minimization of a multivariate function

$$\mathrm{opt}V_\Sigma=\min_{T_1}\left(\min_{T_2}\cdots\cdots\left(\min_{T_{K-1},T_K}V_\Sigma\right)\right)$$

with internal optimization of the components V_i at each stage. This problem can be solved by one of the nonlinear programming methods on the manifold of variables $T_1,\ldots,T_{K-1},T_\Sigma,\bar{\rho}_i,\bar{R}_0,\bar{R}_K$ [11].

Depending on the specificity of the problem in question, various techniques can be used for its reduction, e.g. decomposition of the problem, reduction of its dimensions, etc.

The approach suggested is successfully used during the analysis in terms of energy expenditures of the optimal interplanetary trajectories of this kind for flights to planets, comets, and asteroids. The total characteristic velocity is taken as a quantitative estimate of the entire interplanetary mission. This criterion is implied when speaking about optimal interplanetary trajectories and minimum-energy trajectories; the notion of minimum characteristic velocity corresponds to minimum launching weight of the spacecraft [10, 28, 30]. It is worth mentioning, however, that sometimes the solutions obtained from this kind of model fail to satisfy the designer, because they provide no data on the design parameters of the spacecraft itself or its movement on the powered segments of its trajectory, because the segments of motion with the engine turned on are replaced by instantaneous changes of velocity.

On the other hand, using the notion of minimum characteristic velocity as a criterion for optimization of the interplanetary mission as a whole is not quite sufficient either at the design stage or when ballistic parameters are being set, because the interplanetary flight with given V_Σ, aimed at delivering a specified payload in the vicinity of the target planet can be implemented using different control regimes and different design parameters, and therefore, with a different spacecraft launching weight.

Therefore, there arises the problem of selecting optimal design and ballistic characteristics to minimize the spacecraft's launching weight [14].

There is a need to improve the estimate of launch weight and main design parameters based on the solution of several separate problems of optimization of spacecraft design and ballistic parameters at each of the flight segments (external and internal problems, respectively). This need is due to the fact that the spacecraft's heliocentric velocities at the beginning and end points of the trajectory segments, which are obtained from the solutions of external problems and serve as initial data for internal problems, refer to the mass centers of the corresponding planets and not to the entry and exit points on the gravispheres of the planets, whose position cannot be found from the solution of external problems. Moreover, the calculation procedure ignores the movement of the planet itself during the gravity assist maneuver.

In other words, the initial data for internal problems will be used with a certain error that will result in a corresponding error in selecting optimal control regimes, principal design parameters, and in the estimate of the launching weight of the entire spacecraft. Again, the entire interplanetary trajectory of the spacecraft formed from the separate segments obtained in this way will not be optimal [31].

Therefore, when optimizing the design and ballistic parameters of a spacecraft for a mission including gravity assist maneuvers, it is reasonable to consider internal and external problems of interplanetary flight in combination [33]. Given a multipurpose mission M, we can search for optimal trajectories, control regimes, and principal design parameters based on the criterion of minimum weight of the spacecraft on the near-Earth parking orbit

$$\min G_0 \Rightarrow G_0(M, T_0, T_t, \bar{\varphi}_i, \bar{f}_0, \bar{f}_t, \bar{a}_j), \tag{1.3.5}$$

taking into account all constraints and conditions $\bar{f}_m = 0$ for the mission.

Here T_0 and T_t are the dates of the spacecraft's departure and arrival; $\bar{\varphi}_i$ is the vector function of control on the powered flight segments; $\bar{f}_0$ and $\bar{f}_t$ are boundary conditions, determining the orbits of the spacecraft at the start and the finish; $\bar{a}_j$ is the generalized vector of the principal spacecraft design parameters to be selected.

Mathematically, this problem is formulated as a multipoint boundary-value problem, which, with some simplifying assumptions, can be reduced to the problem of searching for the conditional extremum of a multivariate function with equality constraints [83, 88, 172]. The estimates and solutions obtained can be used as starting guesses at the subsequent stages of the project, when the trajectory and design characteristics of the spacecraft are calculated more accurately by solving complete differential equations of motion of the spacecraft and celestial bodies based on real parameters of the booster and the spacecraft itself.

The models and methods considered above, as well as other methods which make allowance for the perturbing impacts on the spacecraft trajectory, currently form the multi-purpose mission designer's set of tools [76–110].

A large number of schemes of interplanetary flights with gravity assist maneuvers have been examined so far. Some of them have been implemented already, whereas others are still being discussed and are included in long-term plans for the study of the solar system.

At the same time, schemes including gravity assist maneuvers still attract the attention of researchers. There are many interesting objects and zones in the solar system (distant planets with their natural satellites, asteroids, the sun, distant zones of the solar system) that can be reached only in missions using gravity assist maneuvers combined with powered maneuvers when necessary (e.g. low cost missions).

Examination of this variety of options for the space flight require both prompt estimates (aimed at revealing the most beneficial variants) and more detailed analysis of design and ballistic characteristics of interplanetary missions. Therefore, it appears reasonable to combine different models and methods as early as at the first stage of the study, starting from approximate estimates of energy potentialities of multi-purpose missions based on simplified calculation models and moving on to more rigorous numerical examination of the spacecraft optimal design and ballistic characteristics for specific mission schemes (Figure I.1 of Introduction).

This has determined the methodical and scientific directions of this monograph, which contains both methods and algorithms developed by the authors for the calculation and analysis of multi-purpose missions and trajectories of spacecraft at the initial stage of design works. The results of studies of the most interesting and promising multi-purpose missions in the solar system are also included.

2. METHODS AND ALGORITHMS FOR THE DEVELOPMENT OF OPTIMAL MULTI-PURPOSE MISSIONS

One of the main stages in multi-purpose mission analysis is the selection of the optimal trajectory and the formation of multi-purpose mission schemes. The role of this stage in the procedure of designing a spacecraft mission is very important. It is at this stage that the feasibility of the selected trajectory is assessed and its principal parameters are determined considering the requirements of the scheme and the conditions of its optimization.

In the first part of this section, a method for approximate analysis of multi-purpose spacecraft trajectories is discussed, taking into account the effect of perturbation maneuvers on the parameters of the trajectories being generated. The method is based on a graphical interpretation of the multi-purpose trajectory parameters and allows one to quickly analyze alternative trajectories, assess their energy and kinematic characteristics and select the most promising of them.

The other part of the section presents a universal algorithm for generating complex multi-purpose trajectories; the algorithm is intended to enable the study of promising trajectories of spacecraft in the solar system in more detail. It allows one to generate the optimal multi-purpose trajectory for the selected spacecraft mission, to create a calendar of flights using the selected trajectory, and to determine the most expedient windows for the spacecraft launch. A model is also provided for the algorithm to form optimal paths given the number and classes of celestial bodies affecting the spacecraft trajectory.

2.1 Graphic-analytical method

Graphical methods are known to be most effective at the early stages of mission analysis, provided that graphical interpretation is possible for variations in the spacecraft trajectory parameters [71–75].

Graphical analysis allows some characteristics of the space flight to be assessed quickly, though approximately, and variants for more detailed studies to be selected. It is also important that many complex mathematical relationships, when visualized in geometric form, allow the researcher to perceive the situation as a whole pattern and to reveal specific relationships that otherwise could be neglected.

Considered below is the method for graphical examination of spacecraft trajectory with perturbation maneuvers which makes it possible to approximately assess the trajectory in a planetary or planetary-satellite system in terms of energy; the method requires no calculations. A simplified model of the study was used in the development of the method. It was assumed that the orbits of all attracting bodies in the system under consideration are circular and coplanar, the gravispheres of the bodies are reduced to points, and the spacecraft trajectory is approximated by successive segments of Keplerian motion in the central body field. At the flyby point, the spacecraft trajectory was assumed to have instantaneous velocity changes (breaks) in accordance with the rotation of the velocity vector in the center of the body the spacecraft is passing by. The rotation is exclusively due to the gravity forces of this body; there are no powered-gravitational maneuvers. No allowance is made in the model for the phasing of celestial bodies on their orbits.

The simplified model is quite acceptable as long as it is not necessary to accurately design a specific interplanetary mission. The model allows the energy and kinematic characteristics of the trajectory, including perturbation maneuvers, to be determined as early as the initial design stages.

The graphic-analytical method of analyzing trajectories with gravity assist maneuvers is based on the invariance of a characteristic of spacecraft trajectory: the magnitude of the spacecraft asymptotic planetocentric velocity does not change during passive flyby of any jth gravitating body $|\bar{v}^-_{\infty j}| = |\bar{v}^+_{\infty j}|$.

The essence of the method consists in using the property of invariance of gravity assist maneuvers to combine the segments of pre- and post-flyby flight on the plane of spacecraft trajectory parameters with the use of the isoline $v_{\infty j} = const$. Combining the fields of isolines for several attracting bodies of a planetary or planetary-satellite system will allow us to analyze different variants of multi-purpose missions that use maneuvers. Graphical means have been developed for the analysis of such trajectories. Fields of isolines of relative velocities were constructed for the orbits of planets and their natural satellites along with the fields of isolines of changes in the spacecraft trajectory parameters due to gravity assist maneuvers. Combining different trajectories of the spacecraft, we can use the combinations of the relevant fields to examine pre- and post-flyby segments of the orbit.

In this section, the suggested approach is illustrated by a nomogram for a planetary-satellite model of Jupiter, which has a system of attracting natural satellites. The approach can be applied to any planetary or planetary-satellite system for studying flights along trajectories containing gravity assist maneuvers.

2.1.1 Isolines of relative velocities and their application to the problems of space flights

Let us consider a central gravitational field and the entire set of periodical coplanar orbits in it, comprising two subsets of circular and elliptic orbits

$$M = M_c \cup M_e.$$

Each element of the set M (orbit $N_j \in M$) can be mapped onto the plane of orbital parameters, which determine the size and form of this orbit. Any combination of

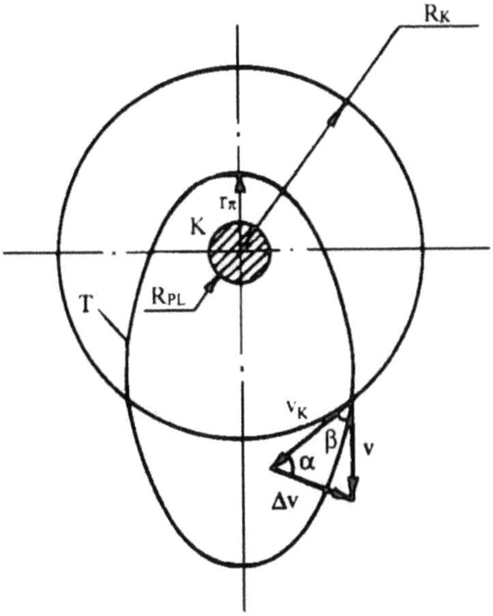

Figure 2.1 Schematic diagram of single-impulse interorbital transfer.

these parameters can be considered, for example, the period of the orbit T and pericentral distance r_π

$$N_j(T, r_\pi) \in M.$$

In the case of circular orbits we have

$$N_j(R) \in M_c \subset M.$$

Let us determine a circular orbit with a given radius R_k

$$\{N(R_k)\} = M'_c \subset M$$

and select the set of elliptic orbits sharing at least one point with the given circular orbit

$$M'_e \subset M_e \subset M.$$

Let us consider the problem of constructing the isoline field of relative velocities in the point of intersection with the circular orbit $N(R_k)$ for all orbits in the set M'. This problem can have very interesting applications, because many of the orbits being studied in astronautics are circular or approximately circular (e.g. the orbits of planets, natural and man-made satellites, and asteroids). Indeed, the velocity Δv (see Figure 2.1) is the relative velocity of transfer from a certain elliptic orbit to the given circular one (or vice versa); in accordance with the theory of interplanetary

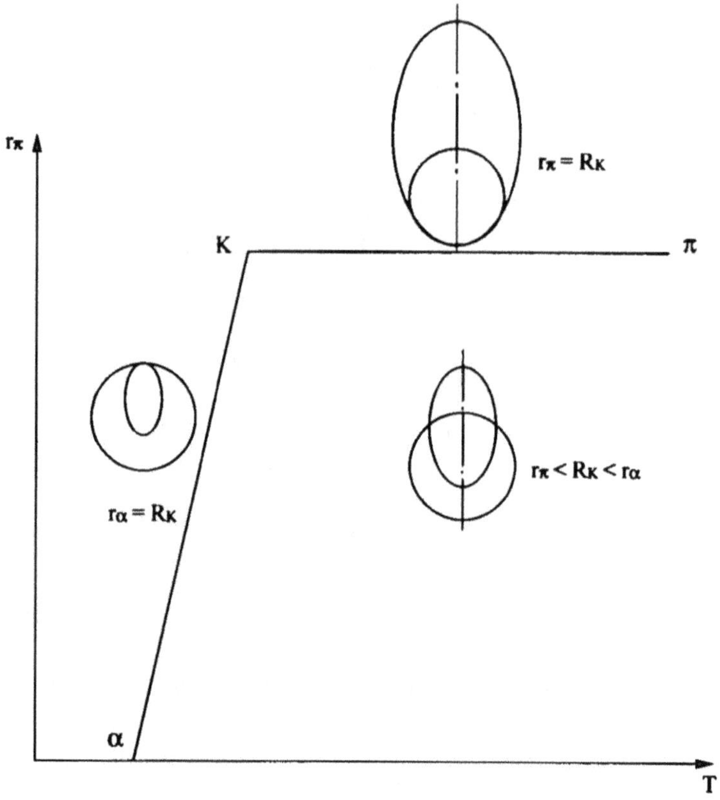

Figure 2.2 The domain of isolines $\Delta v = const$ of interorbital transfer.

flights, it can also be interpreted as the hyperbolic excess of the spacecraft velocity v_∞, if the point of intersection of the orbits coincides with the attracting body.

The problem of graphical representation of the isolines of relative velocities for the selected circular orbit $N(R_k)$ consists in finding the domain in the plane r_π, T, into which set M'_e is mapped, and constructing the isoline field

$$\Delta v = |\bar{v} - \bar{v}(R_k)| = f(r_\pi, T) = const$$

within this domain. Here, $\bar{v}$ and $\bar{v}(R_k)$ are the orbital velocities in the point of intersection of the elliptical and circular orbits.

Thus, the boundaries of the entire domain under consideration (Figure 2.2) are as follows: on the left is the line αK of the orbits tangent to the given circular orbit in their apocenters (i.e. $r_\alpha = R_k$, the entire orbit lies within the given orbit, and the only shared point is the point of tangency of the orbits); the upper boundary is the line $K\pi$ of orbits tangent to the given circular orbit in the pericenter (the only shared point is the point of tangency $r_\pi = R_k$); from below the domain is bounded by the axis $r_\pi = 0$, that is the line of rectilinear orbits (degenerate ellipses); the right boundary is the line of parabolic orbits moved to infinity. The first two lines have a characteristic shared point "k", corresponding to a circular orbit that coincides with the given orbit (all points are shared, $\Delta v=0$).

Let us consider in more detail the limiting lines and the position in them of the boundary points of the isolines $\Delta v = const$. The line αK can be determined analytically from the condition

$$T=2\pi\left(\frac{a^3}{K}\right)^{1/2}, \quad a=0.5\,(r_\pi+r_\alpha).$$

Inasmuch as $r_\alpha = R_k = const$ (from the condition of tangency), we can write

$$T=\pi\sqrt{\frac{(r_\pi+R_k)^3}{2K}};$$

solving for r_π we obtain

$$r_\pi=\left(\frac{T}{\pi}\right)^{2/3}(2K)^{1/3}-R_k,$$

i.e. the line αK is a weakly convex line (close to a straight line); its segment in the graph (Figure 2.2) is limited by points $r_\pi=0$ and $r_\pi=R_k$. The values of Δv are determined from $\Delta v=v_\pi-v_\alpha$, therefore, the position of the point Δv (left ends of isolines $\Delta v=const$) within the segment αK can be determined as

$$r_\pi=2a-R_k=\frac{2K}{2V_k^2-\Delta v^2}-R_k;\quad V_k=\sqrt{\frac{K}{R_k}}.$$

The line $K\pi$ is parallel to the abscissa axis and is determined from the condition $r_\pi=R_k$; the location of the right ends of isolines $\Delta v=const$ on this line can be determined from the conditions

$$\Delta v=v_\pi-v_k$$

$$T=2\pi K[2V_k^2-(\Delta v+v_k)^2]^{-3/2}.$$

Thus, the domain under consideration is determined (Figure 2.2).

It should be noted that each of the points lying within the domain $\alpha K\pi$ corresponds to the orbit $N(r_\pi, T)$ that has two shared points (intersections) with the circular orbit $N(R_k)$ under consideration. Isolines $\Delta v=\Delta v(r_\pi, T)=const$ can be constructed using the relationships of the triangle of velocities and formulas describing undisturbed Keplerian motion of the spacecraft

$$\Delta v^2-2v_k\left(1-v^2+\frac{C^2}{R_k^2}\right)+v_k^2-v^2=0,$$

where

$$v^2 = \frac{2K}{R_k} - \left(\frac{2\pi K}{T}\right)^{2/3},$$

$$C^2 = 2r_\pi K - r_\pi^2\left(\frac{2\pi K}{T}\right)^{2/3}.$$

Figure 2.3 presents the isoline field of relative velocities in the field of the central body with a gravity constant $K=1$ for a circular orbit with a radius of $R_k=1$. Each isoline of this field $\Delta v = const$ represents a set of periodic orbits, for which the condition still holds that the relative velocity of transfer to the orbit with unit radius $R_k=1$ is constant. As can be seen from the graph, the parameters of these orbits (r_π and T) change monotonically from the values corresponding to the orbits with apocentral tangency to the values of pericentral tangency to the orbit $R_k=1$. In this case all points of the curve $\Delta v = const$, except for the boundary points, correspond to the orbits sharing two points with the circular orbit under consideration.

Let us consider the several assembled circular coplanar orbits R_{kj}, $j=1,\ldots,n$ and use the isoline field $\Delta v = f(r_\pi, T) = const$ to analyze the conditions of interorbital transfer in the system of these orbits of different altitude.

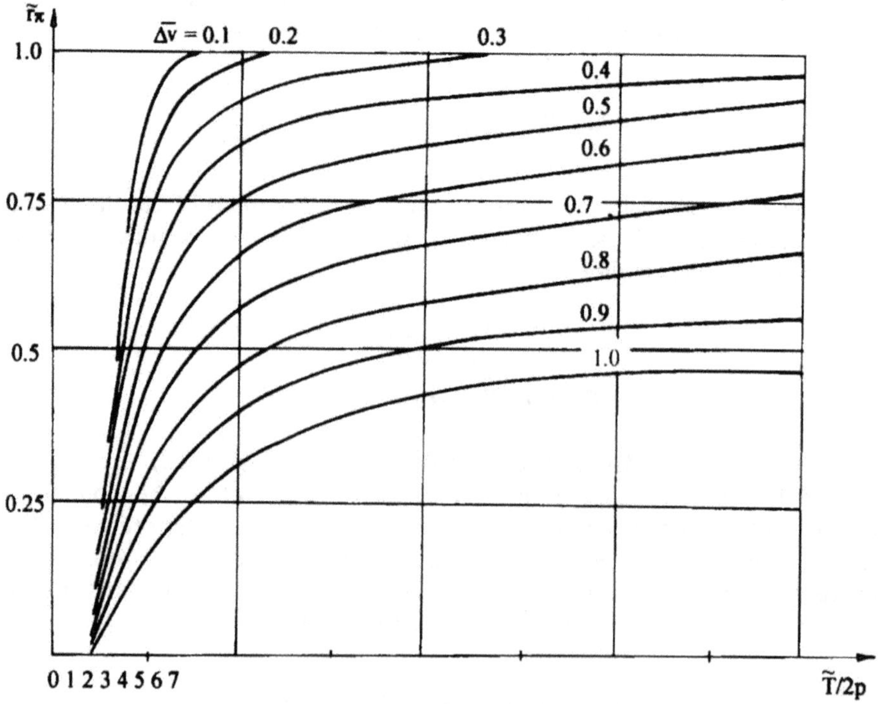

Figure 2.3 Isolines of relative velocities of single-impulse transfer to the circular orbit $R_k=1$.

Each of these orbits will be represented by a point $N(R_j)$ on the plane of orbital parameters r_π and T. This kind of model can be applied to the orbits of planets and natural or artificial satellites. Obviously any number of orbits can be selected. Assume for example that four circular coplanar orbits $N(R_j)$, $j=1,\ldots,4$ are considered. Similarly to the above line of reasoning, the appropriate subsets of orbits can be selected for each $N(R_j)$

$$M_j \subset M, \; j=1,\ldots,4$$

and the isolines of equal relative velocities can be constructed. For all orbits of this subset we have

$$\forall N_j(r_\pi, T) \in M_j: \; N(R_j) \cap N_j(r_\pi, T) \neq 0, \quad j=1,\ldots,4.$$

Figure 2.4 shows the location of the domains of isolines $\Delta v_j = f(r_\pi, T) = const$ in coordinates r_π, T for all the four orbits.

The position of the image of point $N \in M$ in the plane r_π, T of this nomogram allows one not only to judge the feasibility of a certain interorbital transfer in the system of orbits under consideration, but also to make an assessment in terms of energy. Isolines $\Delta v_j = const$ passing through a certain point N^* will determine the

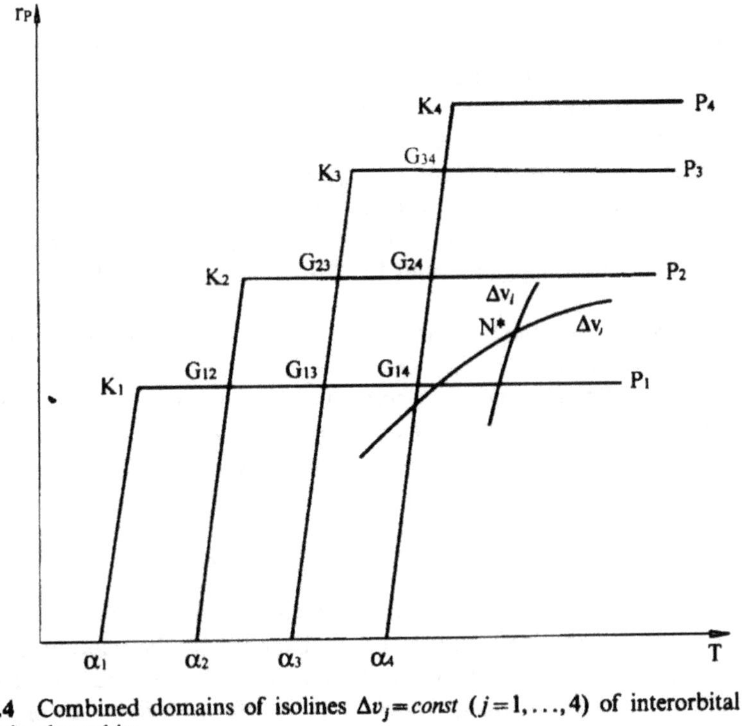

Figure 2.4 Combined domains of isolines $\Delta v_j = const$ $(j=1,\ldots,4)$ of interorbital transfers between circular orbits.

velocity required for this transfer, and the coordinates of the point $N^*(r_\pi, T)$ determine the transfer orbit.

Let us consider possible positions of the image point $N(r_\pi, T)$ on the nomogram:

1. $N \in \bigcap_{j=1}^{4} M_j$ implies that the selected orbit shares points with any given orbit. The interorbital transfer along the orbit N is possible in all cases;
2. $N \in M_i \cap M_j \cap M_k$ $\forall$ $i,j,k = 1,\ldots,4$; $i \neq j \neq k$ implies that the transfer along the orbit N is possible between the three given orbits;
3. $N \in M_i \cap M_j$ $\forall$ $i,j = 1,\ldots,4$; $i \neq j$ implies that the transfer along the orbit is possible between the two given orbits;
4. $N \in M_j$ or $N \notin M_j$ $j = 1,\ldots,4$ implies that the interorbital transfer along the orbit N is impossible.

In addition, specific positions of the image point are possible:

1. $N \in \alpha(M_j)$ $j = 1,\ldots,4$ implies that the orbit has an apocentral tangency with the jth orbit;
2. $N \in \pi(M_j)$ $j = 1,\ldots,4$ implies that the orbit has a pericentral tangency with the jth orbit;
3. $N = K_j$ $j = 1,\ldots,4$ implies that the orbit coincides with the jth orbit;
4. $N = \Gamma_{ij} \in \alpha(M_i) \cap \pi(M_j)$ $i = 2,3,4$; $j = 1,2,3$ $i \neq j$ implies that the transfer orbit is tangent to the orbits of start and finish.

Points Γ_{ij} on the nomogram determine Hohmann transfer orbits between the circular orbits R_j $j = 1,\ldots,4$ being considered.

This kind of nomogram can be used for approximate analysis in terms of energy of transfers between the orbits under consideration and also to select optimal transfer options. Indeed, for any image point $N^*(r_\pi^*, T^*)$ located within the domains under consideration (see Figure 2.4), one can find an isoline $\Delta v_j = const$, $j = 1,\ldots,4$ that passes through this point and determines the velocity of a single impulse transfer between the selected orbit (r_π^*, T^*) and the jth circular orbit (in the point of their intersection).

In the case where the point N^* is located in the zone of overlap between two domains, one more isoline $N^* \in \Delta v_i \cap \Delta v_j \forall i,j = 1,\ldots,4$, $i \neq j$, can be found passing through this point, determining the velocity of a single-impulse transfer between the selected orbit (r_π^*, T^*) and another circular orbit. Thus, for any orbit selected on the nomogram, the characteristic velocity of a two-impulse transfer between two circular orbits can be determined:

$$\Delta v_1 = \Delta v_j;\ \ \Delta v_2 = \Delta v_i,\ \ \Delta v_\Sigma = \Delta v_1 + \Delta v_2,\ \ i \neq j.$$

In a similar way, more complicated trajectories in the system of orbits under consideration can be examined, and the optimal variant can be selected from them.

Incidentally, this kind of nomogram clearly demonstrates that the Hohmann transfers provide minimum energy expenditure for the transfer from one orbit to another. The extreme left and top position of the points Γ_{ij} in the appropriate domains provides the minimum possible values both for Δv_1 and Δv_2 (and, hence, for the total characteristic transfer velocity Δv_Σ).

The above procedure can evidently be applied to the assessment of different space flights in terms of energy expenses (between the orbits of planets and their satellites, between orbits of planetary artificial satellites with differing altitudes, interplanetary flights, etc.). To do this it is sufficient to construct combined isoline fields $\Delta v_j = const$ for relevant orbits.

This kind of nomogram can be also applied at the design stage to assess the flights with gravitational maneuvers, but the nomograms used in this case must represent isolines of the orbital parameters changing as a result of gravitational maneuvers.

2.1.2 Isolines for the analysis of the spacecraft orbital parameters after gravity assist maneuvers

Gravity assist maneuvers in the vicinity of an attracting body make it possible to change the elements of the spacecraft orbit and in particular, to increase or decrease the orbital period T and the pericenter distance r_π.

Let us consider the peculiarities of formation of the new orbit after a gravity assist maneuver near some attracting celestial body. Consider, for example, a natural satellite of a planet with gravitation constant K_s and radius r_s rotating along a circular orbit with a radius of R in the central field of a planet with gravitational constant K. We assume that the orbital planes of the natural satellite and spacecraft are coplanar, and the spacecraft orbit is determined by its period T, pericentral distance r_π, pericentral longitude ω.

The efficiency of a gravity assist maneuver is known to be determined by the conditions of the spacecraft approaching the celestial body: the spacecraft velocity $\vec{V}$ on approaching the celestial body, the velocity of the celestial body $\vec{V}_s$, and the minimum distance of the spacecraft from the body r_p. Figure 2.5 represents a vector diagram of this gravity assist maneuver, reflecting the formation of the spacecraft post-maneuver velocity $\vec{V}^*$ as a result of rotation of the asymptotic hyperbolic velocity v_∞ by the angle 2φ between the asymptote of the descending and ascending branches of the flyby hyperbola.

Thus, the flyby near an attracting body within the plane of its motion will cause a change in the parameters determining the size and shape of the spacecraft trajectory as well as the position of the orbit in the flight plane. That is, as the result of the perturbation maneuver with given r_p, the set of incoming trajectories $M(r_\pi, T, \omega)$ is mapped into the set of post-perturbation orbits $M_1(r_{\pi 1}, T_1, \omega_1)$. This mapping associates each incoming orbit $O(r_\pi, T, \omega)$ with a new (transformed) orbit $O_1(r_\pi^*, T^*, \omega^*)$. The subset of orbits $O_1^*(O_1^* \in M_1)$ with equal values of T^*, selected in the set M_1, allows an isoline $T^* = const$ to be determined in coordinates r_π and T. The isoline will represent the totality of all incoming spacecraft trajectories that will be transformed by the gravity assist maneuver near a celestial body into isoperiodical orbits (i.e. orbits with equal given period $T^* = const$). Constructing the family of these isolines for different values of $T_j^* = const, j = 1, \dots, n$ will yield a general pattern

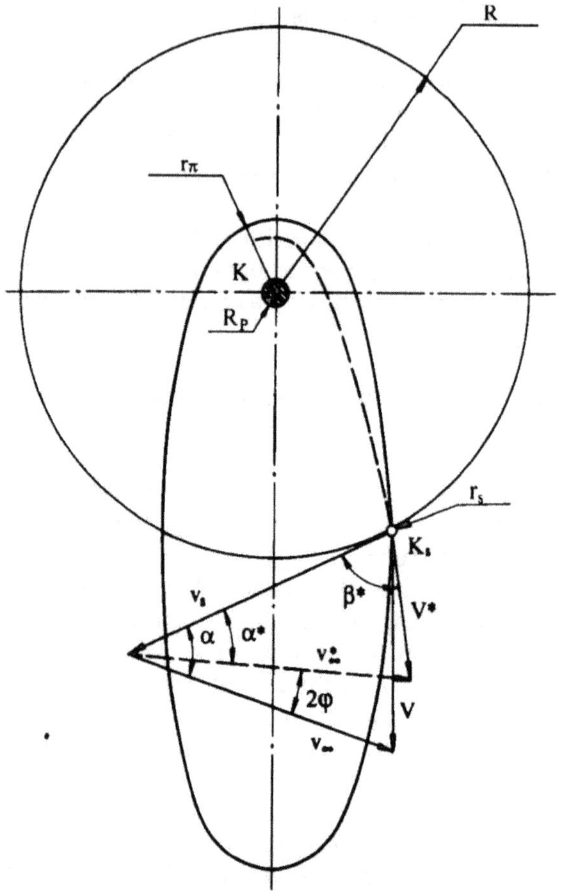

Figure 2.5 Schematic diagram of formation of post-perturbation orbit of the spacecraft.

of the correspondence between the incoming and outgoing trajectories in gravity assist maneuvers near the given celestial body.

Let us consider an algorithm for the construction of isoperiodic curves $T^* = const$ in the coordinates of parameters of initial orbits r_π and T.

Let a certain value be specified for the period of all the post-perturbation orbits T^*. Then, in the model in question, the set of all the post-perturbation orbits with given T^* will be determined (Figure 2.5) by the value of outgoing velocity

$$V^* = \left[\left(\frac{2\pi K}{T^*} \right)^{2/3} - \frac{2K}{R} \right]^{1/2}$$

and the angle β^* between the vectors $\bar{V}_s$ and $\bar{V}^*$ ranging within $0 \leqslant \beta^* \leqslant \pi$.

Given one of the values of β^*, we can determine the asymptotic velocity of the spacecraft at the infinity for departure from and approaching the natural satellite of the planet

$$v_\infty^* = (V_s^2 + V^{*2} - 2V_s V^* \cos\beta^*)^{1/2},$$

$$v_\infty = v_\infty^*.$$

The corresponding angles of the directions of asymptotes of incoming and outgoing directions of flyby with respect to $\bar{V}_s$ can be determined from the formulas

$$\alpha^* = \arccos\left(\frac{V_s^2 + v_\infty^2 - V^{*2}}{2V_s v_\infty}\right);$$

$$\alpha = \alpha^* \pm 2\varphi;$$

where $\varphi = \arcsin[K_s/(K_s + r_p v_\infty^2)]$; 2φ is the angle of rotation of $\bar{v}_\infty$ through the perturbation maneuver.

Here, the plus sign corresponds to a decrease in the spacecraft energy (passing before the attracting body), and the minus sign corresponds to an increase in the spacecraft orbital energy (passing behind the attracting body).

Now let us define the parameters of the incoming orbit of the spacecraft:

1. the velocity of approach to the natural satellite of the planet

$$V = (V_s^2 + v_\infty^2 - 2V_s v_\infty \cos\alpha)^{1/2},$$

2. semi-major axis a and eccentricity e

$$a = \left|-\frac{K}{h}\right|, \quad e = \left(1 + \frac{hC^2}{K^2}\right)^{1/2},$$

where

$$h = V^2 - \frac{2K}{R}, \quad C = VR\cos\beta, \quad \beta = \arcsin\left(\frac{v_\infty}{V}\sin\alpha\right),$$

3. pericentral distance r_π and period T of the incoming orbit of the spacecraft

$$r_\pi = a(1-e), \quad T = 2\pi\left(\frac{a^3}{K}\right)^{1/2}.$$

Thus, the incoming orbit of the spacecraft for the given T^* and β^* has been determined. Varying the values of β^* within $[0, \pi]$, we can construct the isoline $r_\pi = f(T)$ for $T^* = const$, $0 \leqslant \beta^* \leqslant \pi$ in coordinates r_π and T, and a field of isolines $T^* = T^*(r_\pi, T)$ can be constructed for different values $T_j^* = const$, $j = 1, \ldots, n$.

Figure 2.6 presents an example of graphical dependence $T^* = T^*(r_\pi, T)$, calculated using the above algorithm with β ranging within $[0, \pi]$ for the case of the spacecraft attaining isoperiodic orbits ($T^* = 30$ days) after a gravity assist flyby of Jupiter's natural satellite Ganymede. As can be seen from Figure 2.6, the pattern of isoline obtained is rather complex. It consists of two parts, left and right, which correspond to the approaching spacecraft orbits requiring acceleration and deceleration maneuvers respectively for transfer to an isoperiodic orbit $T^* = 30$ days.

The isolines share a single point ($\beta = 0$) that corresponds to the spacecraft orbit with a period of $T = 30$ days and pericentral distance r_π equal to the radius of the bypassed body's orbit. Both branches have points of tangency to the abscissa axis, which determine the degenerate rectilinear elliptic orbits of the spacecraft ($\beta = \pi/2$). As can be seen from the graph, the manifestation of the perturbation effect is very strongly controlled by the selection of the incoming trajectory. Thus, the maximum perturbation effect (both decelerating and accelerating) is observed on the incoming orbits with $r_\pi \approx 0.95\,R$. One can see this from analyzing the behavior of the isolines' branches. It should be mentioned that the isolines are constructed for $m = 1$ (where $m = r_p/r_s$), and in the above example, they reflect the maximum possible perturbation maneuver near the given celestial body. A decrease in the spacecraft flyby altitude relative to the surface of the body will result in a reduction in the perturbation effect with the corresponding "narrowing" of the domain of possible incoming trajectories. Figure 2.7 presents a family of isolines $T^* = 30$d for a relative flyby distance varying within $1 \leqslant m \leqslant 2$.

Thus, the isoline constructed for $m = 1$ can be regarded as a boundary of the domain of incoming orbits that can be transformed by the perturbation maneuver at the given body into isoperiodical orbits with a given period.

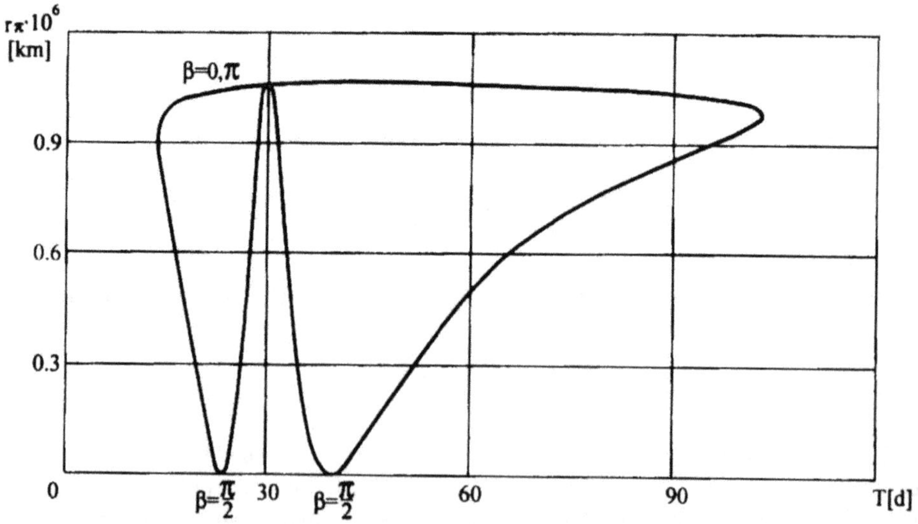

Figure 2.6 Isoline $T^* = 30$days for Ganymede ($m = 1$).

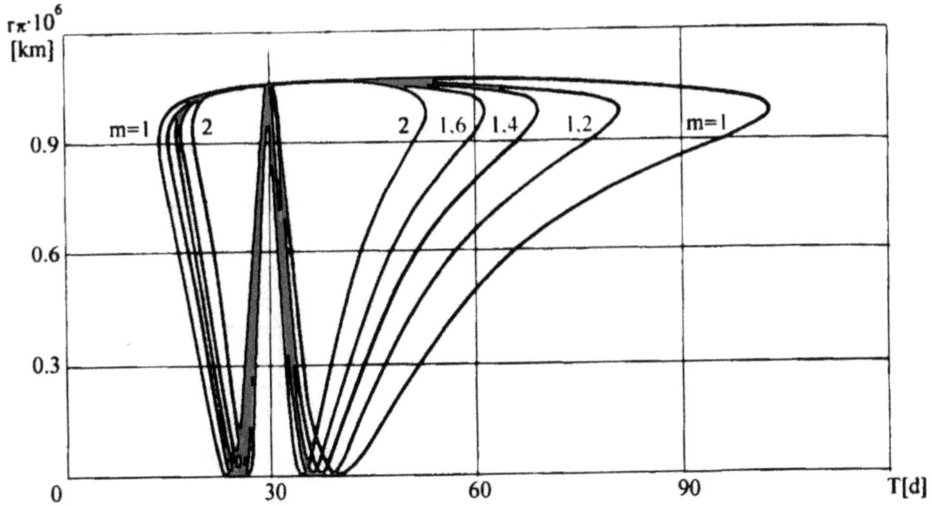

Figure 2.7 The family of isolines $T^* = 30$ days for Ganymede ($1 \leqslant m \leqslant 2$).

This circumstance holds for most planets and planetary satellites in the solar system. However, in the case of bypassing a planet with a strong gravitational field (e.g. Jupiter, where the rotation angle can be $\varphi_{max} > \pi/2$), the perturbation effect can be maximum for the values $m > 1$ determined from Equation 1.2.2.

A comprehensive idea of the efficiency of perturbation maneuvering can be derived from the isoline field for the parameter under study. By varying the values of $T^* = const$ we can construct the field of isolines (isochrones) of the spacecraft post-perturbation trajectories for the flyby of any attracting body. Figure 2.8 shows a field of isoperiodical orbits constructed using the above algorithm and representing the perturbation effect due to Ganymede (for $m = 1$) on the spacecraft orbital periods.

In a similar way, the relationships from Section 1.2 can be used to construct isolines reflecting the post-perturbation orientation of the orbit (isolines of the rotation angles of apsides and the angles of maximum rotation of the orbital plane with respect to its initial position as a result of the perturbation maneuver. Figures 2.9–2.10 present isolines $\Delta\omega = f(r_\pi, T) = const$ and $\Delta i_{max} = f(v_\infty)$ for Ganymede. It is evident that the perturbation maneuver can result in considerable changes in the orientation of the orbit in its plane as well as in the orbital plane rotation.

Constructing fields of isolines of orbital parameters for solar system planets and their largest satellites will allow us to apply graphical analysis of the perturbation maneuvering to the study of different paths of space flights.

2.1.3 Graphical analytical methods for studying the trajectories including gravity assist maneuvering

Trajectories including gravity assist maneuvers possess an important property of invariance—the energy of the flyby trajectory of the spacecraft in the coordinate

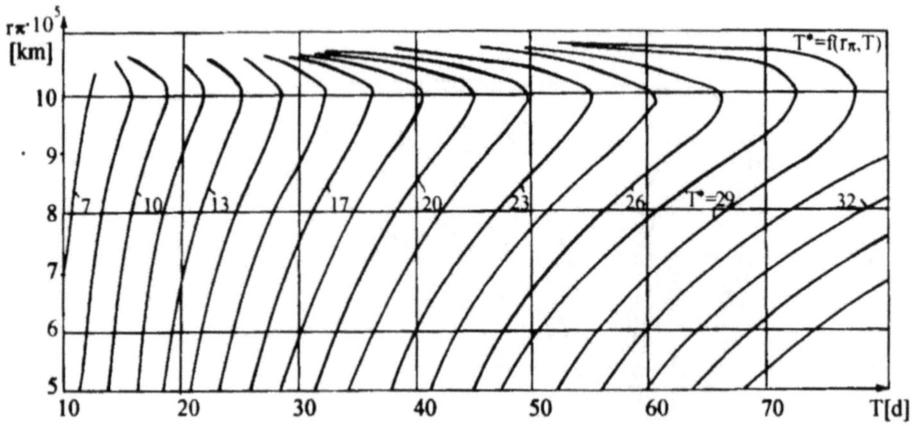

Figure 2.8 Field of isoperiodic curves for Ganymede ($m=1$).

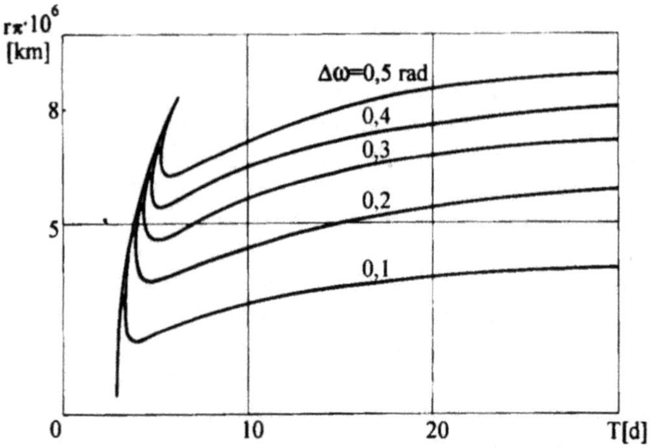

Figure 2.9 Isolines of angles of rotation of the orbital apsides as a result of gravity assist maneuver at Ganymede ($m=1$).

system associated with the planet is constant, and therefore, the values of hyperbolic excess of velocity near the celestial body before and after the perturbation maneuver are the same. This property can be used for graphical studies of this type of mission.

The above-mentioned isolines of equal relative velocities and periods are rather useful in such studies. In this case, any periodical orbit in the given central field will be represented in the nomogram in coordinates r_π, T by a single point N, whereas in the case of a trajectory with a perturbation maneuver, the initial point in the nomogram will change its position, and the trajectory will be represented by two points, corresponding to the pre- and post-perturbation segments of the spacecraft orbit.

The property concerning the invariance of the trajectories with perturbation maneuvering implies that any incoming trajectory uniquely determined by its

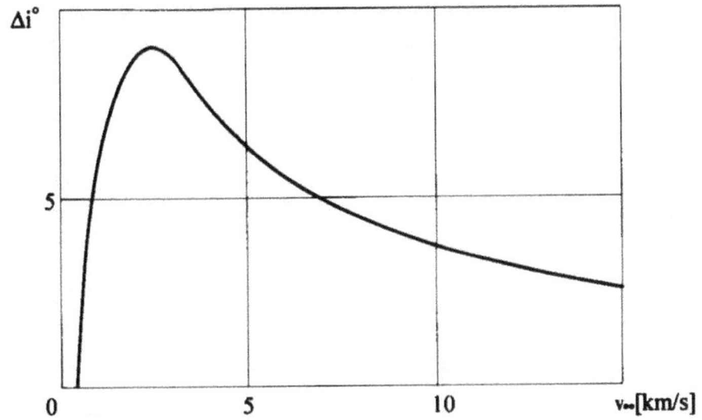

Figure 2.10 Maximum angles of rotation of the orbital plane as a result of gravity assist maneuver at Ganymede ($m=1$).

parameters $r_{\pi 1}, T_1$ (which corresponds to a point on an isoline $v_{\infty K}=f(r_\pi, T)=const$ $N(r_{\pi 1}, T_1)\in V_{\infty K}$ will transform as a result of a coplanar flyby of an attracting body into a new orbit $r_{\pi 2}, T_2$ with $M(r_{\pi 2}, T_2)\in v_{\infty K}$.

The position of the point $M(r_{\pi 2}, T_2)$ in the nomogram can be determined either analytically or using the field of isolines $T^*=T^*(r_\pi, T)$ for the given attracting body:

$$T_2=T^*(r_{\pi 1}, T_1), \quad r_{\pi 2}=f(v_{\infty K}, T_2).$$

We will denote such transformation of the image point in the nomogram by $g:N\to M|_{v_{\infty K}=const}$ for relative flyby altitude $m=1$ (or $g_m:N\to M|_{v_{\infty}=const}$ for values $m>1$).

For any repetition of a gravity assist maneuver at the same body (in the case where the condition of divisibility holds for the orbital periods of the spacecraft and celestial body), the representing point in the nomogram will transform in accordance with the invariance condition: $g:M\to M'\to M''\to\ldots\to M^n$ (n is the number of flybys of the attracting body).

Thus, based on the known regularities of perturbation-induced changes in the orbital characteristics $T^*=T^*(r_\pi, T)$ as well as on isolines $v_\infty=v_\infty(r_\pi, T)=const$, we can determine the parameters of the new, outgoing spacecraft trajectory for any selected incoming trajectory to a given attracting body.

The invariant characteristic of such a trajectory is determined by the isoline $v_{\infty K}=const$ passing through the points representing the transfer $N\to M\to M'\to M''\to\ldots$.

Consider the potential of the suggested graphical analysis of the trajectories with a perturbation maneuver, using as an example a coplanar flyby of Ganymede, one of the natural satellites of Jupiter.

Suppose that the spacecraft occupies an initial pre-flyby orbit with a period of $T=20$ days and a pericentral distance of $r_\pi=6.9\times10^5$ km (it is represented by the point N in the field of isolines $v_\infty=const$, Figure 2.11). The effect of Ganymede will make the spacecraft pass to the flyby hyperbola with $v_\infty=7$ km/s (which can be seen from the position of the initial representing point N in the isoline field, Figure 2.11). Performing an acceleration gravity assist maneuver (a flyby behind the satellite), the spacecraft passes to a new, post-perturbation orbit (with parameters $r_{\pi 2}, T_2$) whose values are to be found.

Now we use the isolines $T^* = T^*(r_\pi, T) = const$ for Ganymede (Figure 2.12) to evaluate the period of the new orbit of the spacecraft. The position of the point with coordinates $r_\pi = 6.9 \times 10^5$ km, $T = 20$ days in the field of isolines determines the period of the spacecraft orbit after the deceleration gravity assist maneuver (the point L lies

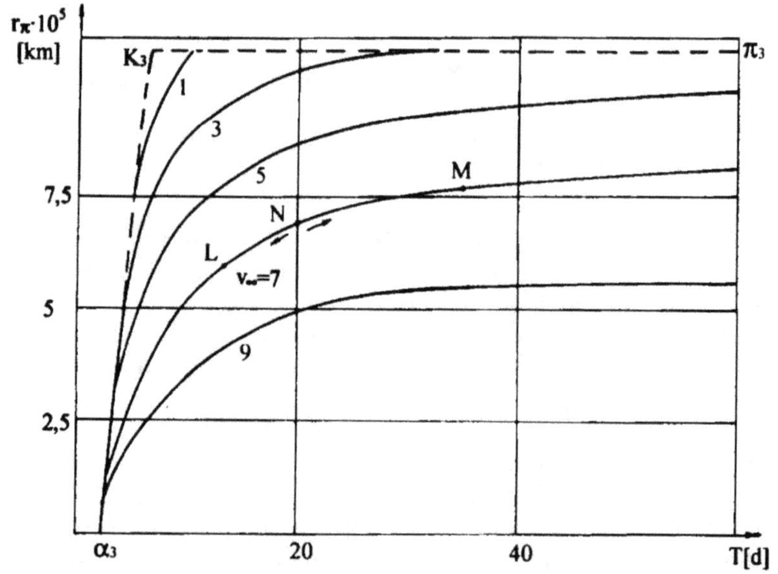

Figure 2.11 Determination of the parameters of the spacecraft post-perturbation orbit.

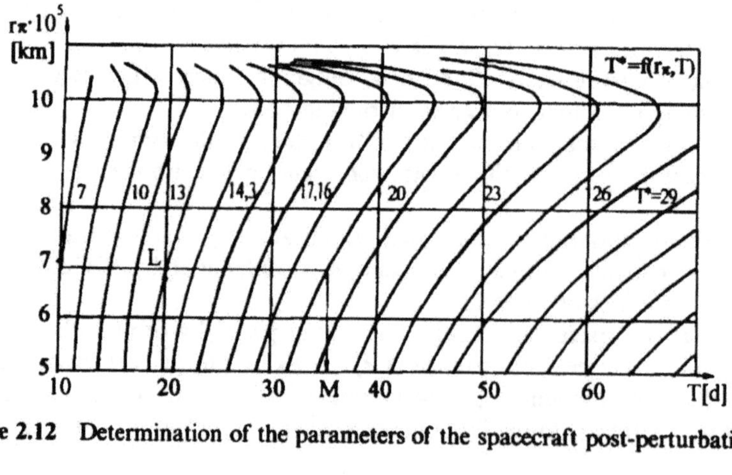

Figure 2.12 Determination of the parameters of the spacecraft post-perturbation orbit.

on the isoline $T_1^* = 13$ days). Following the hyperbolic velocity isoline $v_\infty = 7$ km/s (Figure 2.11) up to the abscissa $T = 13$ days (position of the mapping point L) we find the pericentral distance $r_{\pi 1} = 5.9 \times 10^5$ km. Thus, the parameters of the new orbit, to which the spacecraft passes by means of the perturbation maneuver, are determined.

In a similar way, we can determine the parameters of the spacecraft orbit after an acceleration gravity assist maneuver at Ganymede (the point M is on the isoline $v_\infty = 7$ km/s).

It should be noted that the position of mapping points M or L in the isoline $v_\infty = const$ shown in Figure 2.11 corresponds to the flyby of the attracting body with the relative distance of $m = (r_p/r_s) = 1$. In the example under consideration, any points belonging to the segments "N–M" or "N–L" correspond to the spacecraft orbits resulting from the maneuver for $v_\infty = 7$ km/s, but with other values of m from the interval $1 < m < \infty$.

Variations in other orbital characteristics can be assessed in a similar way. The gravity assist maneuver performed at Ganymede from the selected incoming orbit brings about a deflection of the apsides $\Delta w = 17.66°$ (see Figure 2.9). Possible variations in the inclination of the spacecraft incoming orbital plane are illustrated by the plot $\Delta i = f(v_\infty)$ (Figure 2.10). As can be seen from the figure, a deflection of the spacecraft orbital plane resulting from the gravity assist maneuver at Ganymede can be as large as $\Delta i = 5°$.

The above analysis can also be made for different orbits in space by using similar isoline fields for attracting bodies of the Solar System (planets and their satellites). This approach has great potential in studying paths including gravity assist maneuvers, and can be used for preliminary analysis of different variants of developed schemes, and for selecting the best among them.

Practical applications of the graphic-analytical method are considered in Chapters 4 and 5. The possibility of selecting domains of optimizing solutions is discussed with a view to using them for approximate design and ballistic estimates at the preliminary stage of mission design studies.

2.2 Application of the graphic-analytical method to multi-purpose mission analysis

The graphic-analytical method suggested here for examining orbits with a gravity assist maneuver can be applied to approximately analyze the different problems facing the space flight, and in particular to analyze the interorbital transfers in the central gravitation field.

Variants of the analysis of spacecraft direct interorbital transfers (including optimal) using the nomogram were considered earlier in Section 2.1. The potential for using the proposed method in studies of spacecraft interorbital transfers with the use of gravity assist maneuvers is discussed below.

Consider several circular coplanar orbits with respect to a central body (for the sake of argument, let it be the orbits of Galilean satellites of Jupiter: I—Io, II—Europa, III—Ganymede, IV—Callisto).

To analyze the transfers from the orbit of one of these satellites to another using a gravity assist maneuver let us consider the nomogram in Figure 2.4. For this case, we combine on a single plane in the coordinates of the transfer orbit parameters r_π, T the isoline fields of relative velocities $\Delta v = const$ for each of the four orbits under consideration (Figure 2.13).

Suppose that we are to study the transfer from the orbit of Europa to that of Callisto using Ganymede's gravitational field (let us denote the transfer as II–III–IV). In this case, the set of all initial transfer orbits with points in common with the Europa orbit (hereafter denoted as M_{II}), can be divided into domains $D_1(\alpha_4\Gamma_{24}\pi_2)$, $D_2(\alpha_3\Gamma_{23}\Gamma_{24}\alpha_4)$, $D_3(\alpha_2K_2\Gamma_{23}\alpha_3)$.

Consider the options for the interorbital transfer II–III–IV depending on the location of the mapping point N_1 in nomogram 2.4, which determines the characteristics of the initial segment of the transfer:

1. $N_1 \in D_1, D_1(\alpha_4\Gamma_{24}\pi_2) \subset M_{II} \cap M_{III} \cap M_{IV}$ (the point N_1 represents an orbit that has points in common with orbits II, III, IV). In this case, the spacecraft can perform either a direct II–IV transfer (the point in the domain corner corresponds to the Hohmann solution) or a transfer with either an acceleration or a deceleration maneuver at Ganymede.
2. $N_1 \in D_3(\alpha_2K_2\Gamma_{23}\alpha_3) \subset M_{II}$. The II to IV transfer is possible only through an acceleration perturbation maneuver at satellite III.
3. $N_1 \notin D_3(\alpha_2K_2\Gamma_{23}\alpha_3) \subset M_{II}$. The transfer through II–III–IV is impossible.

The formation of different transfer options following the studied path and their energy estimate can be made using the nomogram 2.13. Let a certain point N_1 be specified in the nomogram, which determines the parameters of the initial transfer segment: N_1 ($r_\pi = 5.8 \times 10^5$ km, $T = 12.25$ days). As can be seen from the nomogram, there are three isolines of different domains passing through the point. $N_1 \in \Delta v^{II}(6) \cap \Delta v^{III}(7) \cap \Delta v^{IV}(4,84)$. This disposition implies the following three transfer options (see Figure 2.14):

(a) without flying by Ganymede; the position of N_1 completely determines the transfer orbit from II to IV. From the nomogram the starting and final impulses can be found to equal $\Delta V_{II} = 6$ km/s, $\Delta V_{III} = 4.84$ km/s.
(b) with an acceleration gravity assist maneuver at Ganymede (which corresponds to the transformation $g: N_1 \rightarrow M_1[\Delta v_0^{III}(7)]$, $M_1 \in \Delta v^{III}(7) \cap \Delta v^{IV}(5,8)$. Points N_1 and M_1 determine the pre- and post-flyby segments of the transfer, respectively;
$M_1(r_\pi = 6.7 \times 10^5$ km, $T = 17.6$ days). The transfer energy parameters are $\Delta v_{II} = 6$ km/s, $\Delta v_{IV} = 5.8$ km/s.
(c) with a deceleration gravity assist maneuver at Ganymede $g: N_1 \rightarrow M_2\ [\Delta v^{III}(7)]$, $M_2 \in \Delta v^{III} \cap \Delta v^{IV}(3,1)$. Points N_1 and M_2 determine pre- and post-flyby segments of the transfer, respectively, $\Delta v_{II} = 6$ km/s, $\Delta v_{IV} = 3.1$ km/s.

(In variants (b) and (c), the relative velocity of approach to Ganymede is interpreted as the velocity at infinity.) Thus, we obtained solutions for the II–III–IV interorbital transfer.

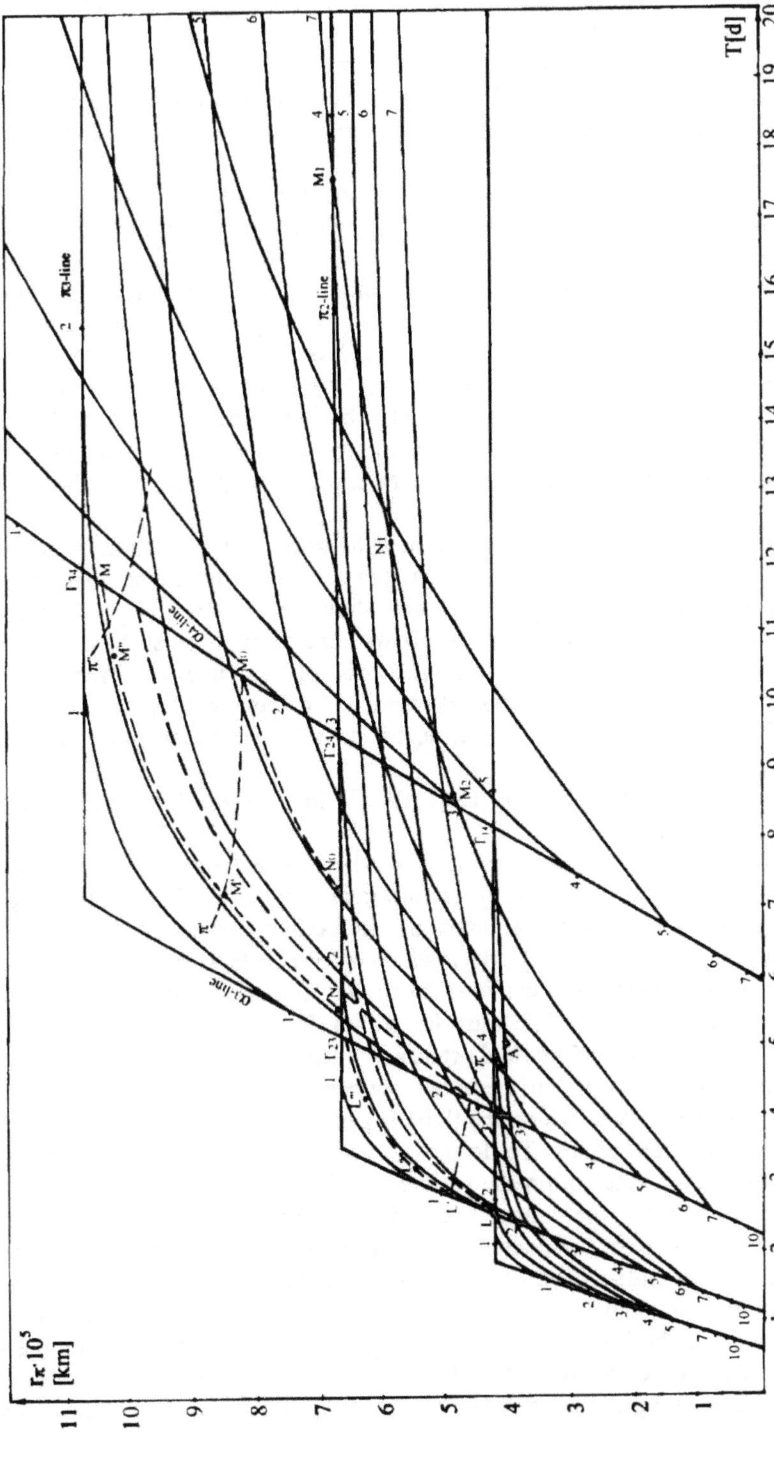

Figure 2.13 Nomogram of combined isolines of transfers between the orbits of Galilean satellites.

The nomogram under consideration allows us to perform direct numerical analysis of possible solutions with a selection of domains containing the best of them. As can be seen from the nomogram, the solutions obtained are not optimal in terms of the required energy expenditure. Thus, the direct Hohmann transfer represented by the point Γ_{24} provides better transfer characteristics $\Delta V_{\Sigma}=5.24$ km/s ($\Delta V_{\text{II}}=2.98$ km/s, $\Delta V_{\text{IV}}=2.26$ km/s).

However, as the nomogram shows, it is possible to improve the characteristics of the transfer in question with the use of Ganymede's gravitational field. This can be achieved by using flyby orbits with lesser values of v_{∞}^{III} which corresponds to a leftward and upward shift of the initial mapping point N_1 within the nomogram. This will result in a reduction in the transfer energy expenses, because the isolines passing through points N and M determine lesser impulse values ΔV_{II} and ΔV_{IV}.

The point Γ_{24} (lying on the line $\Delta v^{\text{IV}}=5.4$ km/s) allows us to select the domain of initial transfer orbits from which the perturbation maneuver is inexpedient in terms of energy. As can be seen from the nomogram, the flyby following any of the hyperbolas $v_{\infty}^{\text{III}} \geqslant v_{\infty}^{\text{III}}(\Gamma_{24})$ results in energy expenditure greater than those in a direct II–IV Hohmann transfer.

Thus, turning to the problem of how to select the best solution for the path under consideration, we can establish that the initial mapping point N_1 must lie within the domain $\alpha_3\Gamma_{23}\Gamma_{24}\alpha_4$ not lower than the isoline $v_{\infty}^{\text{III}}(\Gamma_{24})$. A gravitational acceleration maneuver at Ganymede allows a transformation of the initial orbits that have no points in common with the orbit of satellite IV (from the domain $D_2(\alpha_3\Gamma_{23}\alpha_4)D_2 \not\subset M_{\text{IV}}$) into the domain of orbits with such points (domain $\varepsilon_2(\pi_2\Gamma_{24}\Gamma_{34}\pi_3)\varepsilon_2 \subset M_{\text{IV}}$. In this case, the transfer is implemented on the isolines $v_{\infty}^{\text{III}} < v_{\infty}^{\text{III}}(\Gamma_{24})$.

Optimization of solutions for the II–III–IV flight in this domain can also be made using the graphical method.

Considering the reciprocal position of the isoline fields in the nomogram presented, one can easily see that the condition of minimizing flight energy expenditure brings about the following requirement: the point representing the initial orbit N_0 must lie within the segment $\Gamma_{23}\Gamma_{24}$ (which corresponds to the condition of tangency of the starting orbit) and must lie as far to the left as possible, whereas the point M_0 representing the final orbit must lie within $\Gamma_{23}\Gamma_{34}$ (which corresponds to the condition of tangency of the final orbit) and must lie as high as possible. The position of the points N_0 and M_0 in the above segments is determined by the perturbation capacity of the body at which the flyby is performed.

Let us find these points. To do this we construct the line of post-flyby orbits for the initial orbits lying within the segment $\Gamma_{23}\Gamma_{24}$.

We will apply the above procedure $g:\pi(D_2)\to\pi'$ to use the isolines $T^*=T^*(r_\pi,T)=const$, to construct a line π', each point of which represents a mapping of the initial orbit of pericentral tangency resulting from the acceleration gravity assist maneuver at Ganymede ($m=1$). As can be seen from the plot, the intersection of the line π' with segment $\Gamma_{24}\Gamma_{34}$ in the point M_0 yields the solution for transfer along the path II–III–IV. In this case, the flyby of satellite III is implemented along the hyperbola $v_{\infty}^{\text{III}}=4.1$ km/s, and the flight energy expenditures

$\Delta v_{II} = 2.4$ km/s and $\Delta v_{IV} = 1.81$ km/s are determined by the values of the corresponding isolines, passing through the obtained boundary points. Energy expenditure along this path $\Delta V_{\Sigma} = 4.21$ km/s is more than 1 km/s less than that for the transfer along Hohmann's orbit. (Note that in the domain $4.1 \leqslant v_{\infty}^{III} \leqslant v_{\infty}^{IV}(\Gamma_{24})$, solutions for the II–III–IV transfer are possible with greater energy expenditure for relative distance $m > 1$.)

The solutions being discussed here, which have boundary points on the segments $\Gamma_{23}\Gamma_{24}$ and $\Gamma_{24}\Gamma_{34}$, are of interest due to the fact that they correspond to flights with tangential application of impulses in their starting and end points. Such flights can be called quasi-Hohmann, i.e. paths which still have the features of Hohmann transfer (tangency in the starting and end points), but fail to match the angular distance of the flight $\phi = \pi$ (Figure 2.14).

Similarly, one can find the solution for any other variant of interorbital transfer using a gravity assist maneuver. In doing so, one can use the nomogram to approximately analyze the solutions obtained, to select the best among them, and to assess the optimal solutions. It should be noted that the variant considered belongs to the case of transfer for $R_0 < R_k$. For variants of transfer from higher to lower orbits (that is, for paths like IV–III–II), all the solutions obtained are applicable, but with allowance made for the fact that the impulses ΔV_{II} and ΔV_{IV} to be determined are retroimpulses, and the flyby is performed before the attracting body III (i.e. the perturbation maneuver is decelerating).

Thus, the nomogram makes it possible to select the domain of interorbital transfer solutions where the application of gravitational maneuver is expedient ($v_{\infty}^{III}(\text{opt}) \leqslant v_{\infty}^{III} \leqslant v_{\infty}^{III}(\Gamma_{24})$).

The introduction of such flyby maneuvers for the value $v_{\infty}^{III}(\text{opt})$ (in the case under consideration, $v_{\infty}^{III}(\text{opt}) = 4.1$ km/s, $\Delta V_{\Sigma} = 4.21$ km/s) allows a significant improvement of the characteristics of Hohmann interorbital transfer (by more than 20%).

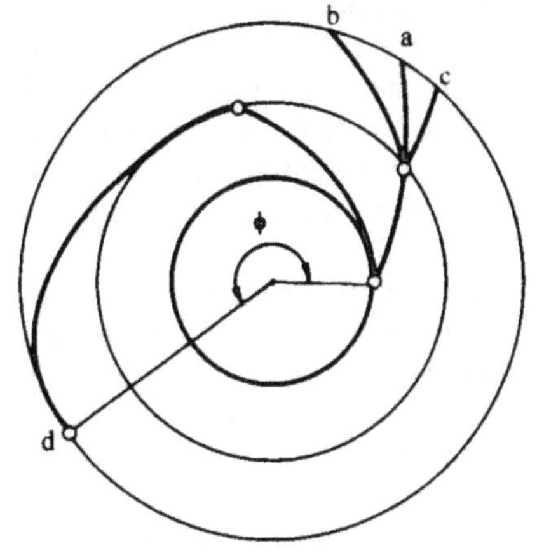

Figure 2.14 Scheme of variants of forming orbits with a flyby of Ganymede along path II–III–IV: (a) without gravity assist maneuver; (b) acceleration gravity assist maneuver; (c) deceleration gravity assist maneuver; (d) "quasi-Hohmann" transfer.

In the solution domain obtained here, the II–III–IV transfer can be implemented with a single flyby of the attracting satellite III. As for the domain $v_\infty^{\rm III} < v_\infty^{\rm III}(\text{opt})$, $N \in \Gamma_{23} N_0$, $M \in \Gamma_{34} M_0$, a single maneuver is insufficient here, and the II–III–IV transfer will require two or more maneuvers with respect to III. As can be seen from the nomogram (Figure 2.13), the first flyby results in the point N moving to the line π', and it is only after the second flyby of the satellite III that the mapping point can be placed on the line α_4, which makes the trajectory tangential to (or crossing) the orbit IV. Such transfers also result in a reduction of energy characteristics of the flight and can be used to analyze complex multi-purpose missions in the solar system.

By and large, the examples considered above do not exhaust all the potential of the graphic-analytical method. Based on combined isoline fields, different variants of multi-purpose missions in the solar system can be studied, including various combinations of celestial bodies, one- and multi-revolution orbits with flybys of one or more attracting bodies.

Studies of specific paths in the solar system based on the developed approach will be considered in the following chapters.

2.3 A universal algorithm for the synthesis of spacecraft multi-purpose missions

The graphic-analytical method developed for studying spacecraft trajectories with gravity assist maneuvering is based on a simple model and allows a quick and clear assessment of different schemes of interplanetary flights. At the same time, the estimates obtained are quite approximate: they determine the lower limit of the flight energy expenditure, thus allowing one to find out whether the variant in question is worthy of further consideration, but they fail to provide a comprehensive idea of the characteristics of the scheme under study.

A more profound design and advanced ballistic studies of the selected variant will require more rigorous models and methods as well as numerical algorithms to seek optimal solutions. The development of these methods and algorithms should both make the mathematical models for studying spacecraft multi-purpose interplanetary trajectories more accurate and comprehensive and universalize the computational algorithm in order to cover a wider set of possible schemes and paths of the spacecraft.

The model of point-impulse approximation is widely used in different problems dealing with flights in a central gravitational field. It is also applied to the problem of interplanetary flight with gravity assist maneuvers. These algorithms are very involved and are developed as a rule for individual schemes (the complexity of such algorithms grows with the number of the planetary flybys). At the same time it can be noted that these models have a common unified base of computational elements, which makes the problem of developing a universal and flexible algorithm quite realistic. This will allow one to apply the algorithm to a far wider set of problems with a minor adjustment, and to pass from one type of problem to another by specifying a certain index.

Indeed, many interplanetary flight problems are based on the model of movement in the field of a single attracting center (two-body problem), and the solution in such problems is frequently sought from two positions (Lambert's problem in classical celestial mechanics).

On the other hand, the range of objects being studied in the Solar System is rather wide; it includes planets, comets, asteroids, planetary natural satellites, and, finally, artificial celestial bodies (if flights to these bodies are also to be considered).

We can define the set of maneuvers performed by the interplanetary spacecraft (within the framework of the model under consideration). This set can include spacecraft acceleration from the given low-altitude orbit of the planetary artificial satellite, powered-gravity assist flyby of the planet, positioning at the orbit of a planetary artificial satellite, and sounding flight near the target celestial body. Clearly, the list can be extended according to the investigator's requirements.

Thus, the appropriate selection of the element base of the algorithm allows a formalization of both the object in the Solar System and the spacecraft maneuvers performed near it. The multi-purpose spacecraft path being studied can be represented as a series of codes determining the sequence of celestial bodies and spacecraft maneuvers. This approach allows one to study the specified trajectory of the spacecraft, to form the optimal path in terms of energy expenditure given only the number of flybys, or to form a multi-purpose spacecraft path given the energy resource (or flight-time resource). The range of problems that can be solved in this formulation is clearly very wide.

2.3.1 Classification of the Solar System bodies

When developing the universal algorithm for the synthesis of complex multi-purpose paths of a spacecraft, the algorithm was assumed to be applicable, with a minimum necessary adjustment, to any problems of space flight between any celestial bodies in the central gravitational field of any body in the Solar System.

For this purpose, the Solar System objects were classified with respect to the algorithm being developed. Five object classes were defined: (1) planets; (2) planetoids (asteroids); (3) comets; (4) natural satellites of the planets (NSP); (5) artificial satellites of the planets (artificial planets—AP; artificial satellites of planets—ASP; artificial satellites of planetary natural satellites—ASPNS).

Lists of numbered specific objects are compiled within each class:

- for planets N_p; $p=1,\ldots,9$ (1—Mercury, 2—Venus; 3—Earth; 4—Mars; 5—Jupiter; 6—Saturn; 7—Uranus; 8—Neptune; 9—Pluto);
- for asteroids, $N_a=1,\ldots,K_a$ (in accordance with the International Catalogue of Small Planets and the numeration introduced in it [6]). This list of asteroids is updated every year. The largest and well-known asteroids are denoted by numbers: 1—Ceres; 2—Pallada; 3—Juno; 4—Vesta; etc.;
- for comets $N_c=1,\ldots,K_c$; the numeration of the objects in the set is accepted in accordance with the list suggested in [24].

The numeration of well-known natural planetary satellites is proposed to be kept in accordance with the corresponding planets:

$$N_s^p, \quad p=1,\ldots 9;\ s=1,\ldots K_s^p.$$

- for the Earth, 1—the Moon;
- for Mars, 1—Phobos, 2—Deimos;
- for Jupiter, 1—Io, 2—Europa, 3—Ganymede, 4—Callisto, 5—Amaltea, 6—VI, 7—VII,...,;
- for Saturn, 1—Janus, 2—Mimas, 3—Enceladus, 4—Tethys, 5—Diona, 6—Rhea, 7—Titan,...,;
- for Uranus, 1—Miranda, 2—Arial,...,;
- for Neptune, 1—Triton, 2—Nereida,...,;
- for Pluto, 1—Charon.

It is evident that newly discovered objects (asteroids, comets, natural planetary satellites) can be incorporated into the lists by increasing the final numbers N_{ak}, N_{ck}, N_{pk} in the appropriate object classes.

The lists of artificial celestial bodies can be formed by the researcher in accordance with the scope of the problems to be solved.

Thus, it would seem convenient to represent any object in the Solar System by a five-digit integer number. The first digit denotes the object class, and the digits from the second to the fifth denote the object number.

All the objects in the 1st, 2nd, and 3rd classes have their orbits in the gravitational field of the Sun.

Examples:

- 10003—Earth
- 20004—Asteroid no. 4 (Vesta)
- 30001—comet no. 1 (Enke)
- 00000—Sun.

The 2nd digit in the index of a 4th class object denotes the number of the planet of which the object is a natural satellite.

Examples:

- 43001—Moon
- 44002—Phobos
- 45012—XII satellite of Jupiter

The 2nd digit in the index of a 5th class object also denotes the number of the planet, and the 3rd and 4th digits denote the number of the central body (natural planetary satellite) in whose field the orbit of the artificial object is determined. The 5th digit denotes the number of the artificial object.

Examples:

- 50005—artificial planet no. 5
- 53009—Earth's artificial satellite no. 9

- 53102—artificial satellite of the Moon no. 2
- 55403—artificial satellite of Callisto no. 3.

Broadly speaking, the classification could be extended to include artificial satellites of asteroids and comets (by introducing additional classes); however, no problems concerning flights between artificial objects associated with these bodies are considered here.

2.3.2 Classification of maneuvers and algorithmic representation of spacecraft paths in the Solar System

The algorithm being developed is intended for the study of spacecraft flights in the Solar System; therefore the powered maneuvers near celestial bodies can be classified as well (for implementation of the multi-purpose interplanetary mission).

The basic spacecraft maneuvers can be enumerated as follows:

1. acceleration of the spacecraft from the given starting orbit;
2. powered-gravitational maneuver to transfer from the incoming to the outgoing hyperbola (flyby gravity assist maneuver);
3. sounding flight near the celestial body;
4. deceleration (spacecraft entering the given orbit around the celestial body);
5. spacecraft landing on a small celestial body (equalization of velocities).

In general, the list of the maneuver classes can be extended by incorporating more complicated spacecraft maneuvers; however, the five maneuvers above are basic and allow virtually any path in the Solar System to be constructed.

2.3.3 Algorithmic representation of spacecraft paths in space

The algorithmic representation based on the above classification is not a particular problem for any morphological chain determining the path. Indeed, this chain can be specified as a sequence of bodies and maneuvers near them (which obviously should not be in contradiction with the physical meaning of the problem).

In this case, the proposed code representation is quite convenient because it can be represented in the form of a chain of real numbers, where the integral part of each of the numbers specifies the body of the Solar System, and the fractional part specifies the maneuver to be implemented near the body.

Thus, any object included in the path can be represented as a certain number

class of the object	number of the object	number of the maneuver

For example, the flight Earth–Venus–Halley's Comet can be represented as 10003.1–10002.2–30071.3.

This record means that the spacecraft starts from the given orbit of an artificial Earth satellite, performs a gravity assist maneuver near Venus, and performs a sounding flight near Halley's Comet.

> Earth–Venus–Earth–Jupiter path:
> 10003.1–10002.2–10003.2–10005.3
>
> Earth–Venus–Mars–Earth path with a parking orbit near Mars and subsequent start toward Earth:
> 10003.1–10002.2–10004.4–10004.1–10003.3
>
> The path of a flight between the natural satellites of Jupiter: Callisto–Ganymede–Europa:
> 45004.1–45003.2–45002.3

It is evident that any of the spacecraft paths in the Solar System can be represented in this form, and the criteria function (e.g. the total characteristic velocity) will be determined given the specific maneuvers of the spacecraft to be performed within the path.

2.3.4 Elemental-computational basis of the algorithm

When developing a universal algorithm for the synthesis of multi-purpose spacecraft missions in the Solar System using the top-down modular design approach, one should define the principal computational link of such an algorithm—the basic module. This term is taken to mean the largest program-algorithmic unit which is autonomous in terms of computations, and at the same time is a common part of all the problems under consideration. This kind of unit can be used without any adjustment to determine the necessary characteristics of a specified segment of the flight and to utilize them in the formation of any complex multipurpose spacecraft mission.

The choice of such a basic unit will to a great extent determine the logical pattern and the structure of the algorithm.

A complex multi-purpose spacecraft mission in the Solar System can be represented as a sequence of N spacecraft orbital arcs in the field of a single central body; the arcs link $N+1$ bodies included in the path. Any of the segments, specified by the ordinal numbers of bodies and maneuvers in their vicinity as well as by the appropriate dates of start and arrival of the spacecraft, are characterized by the energy expenditure for the implementation of the necessary spacecraft maneuvers.

Thus, we can take a well-known problem of celestial mechanics (construction of the orbit based on two positions (Lambert's problem) with necessary algorithmic additions) as the basis of the algorithmic unit (basic module) [1, 3, 9].

Let us determine the structure of this unit.

The *input data* for any ith flight segment is determined by the codes of bodies and by the dates of the start and arrival $N_i, N_{i+1}, T_i, T_{i+1}$ (or Δt_i—the flight time), respectively.

The input characteristic of the module must also include the vector of hyperbolic excess of the spacecraft incoming velocity to the starting body (if the segment is not the initial one); the restriction on the number of revolutions on the trajectory segment and the gravitational constant of the central body in whose field the orbit is formed (note that this parameter must be constant throughout the trajectory, because the schemes of spacecraft flight between bodies in different central fields were not considered).

Analysis of the codes of bodies in the program must yield the classes and numbers of the bodies between which the flight is to be implemented and the numbers of the maneuvers to be performed near the bodies. The number of the body class will refer to the data set from which the necessary characteristics of the body and its orbit can be obtained (based on the number of the celestial body).

Once the radius-vectors and velocities of the bodies have been evaluated, the flight segment can be determined together with the hyperbolic excesses of the spacecraft velocities at the starting and final bodies.

Then, the energy expenditure for the segment is calculated (impulses of the spacecraft velocity alteration) in accordance with the maneuver numbers at the starting and final bodies. (Note that running the module can include the regime of checking the expediency of an impulse within the segment and determining its value if necessary.)

The following determination of energy expenditure requires the solution of internal problems of interplanetary flight. They can be determined either based on simple relationships from impulse approximation of the active segments of the flight or by using more complex models to study the segments of the spacecraft maneuver near the bodies [5, 18, 30, 63, 64, 140].

In the case under consideration, where the criteria function is the path characteristic velocity, we can assume that the spacecraft maneuvers are approximated by single-impulse transfers:

- circular (elliptic) orbit–hyperbola (in the case of the spacecraft starting from an orbit of artificial satellite);
- hyperbola–circular (or elliptic) orbit (in the case of spacecraft orbiting as a planetary artificial satellite);
- hyperbola–hyperbola (in the case of a pericentral powered-gravitational maneuver of the spacecraft near the flyby body).

The flowchart (Figure 2.15) reflects the structure and logic of this basic module.

The *output data* of this basic module will be the characteristic velocity at the given segment. It is determined as the sum of expenditure for the spacecraft maneuvering near the start and end bodies. In parallel output data is auxiliary: the central flight angle and radius-vectors of the initial and end points of the segment; the hyperbolic excesses of velocities near the bodies in the beginning and end of the segment; the altitude of the impulse application near the start body or during flyby; the angle between the asymptotes of the incoming and outgoing branches of the hyperbola at the flyby segment.

Thus, the basic module makes it possible to determine both the combined criteria function at the given segment and the additional data required for the formation of the following segment of the multi-purpose path.

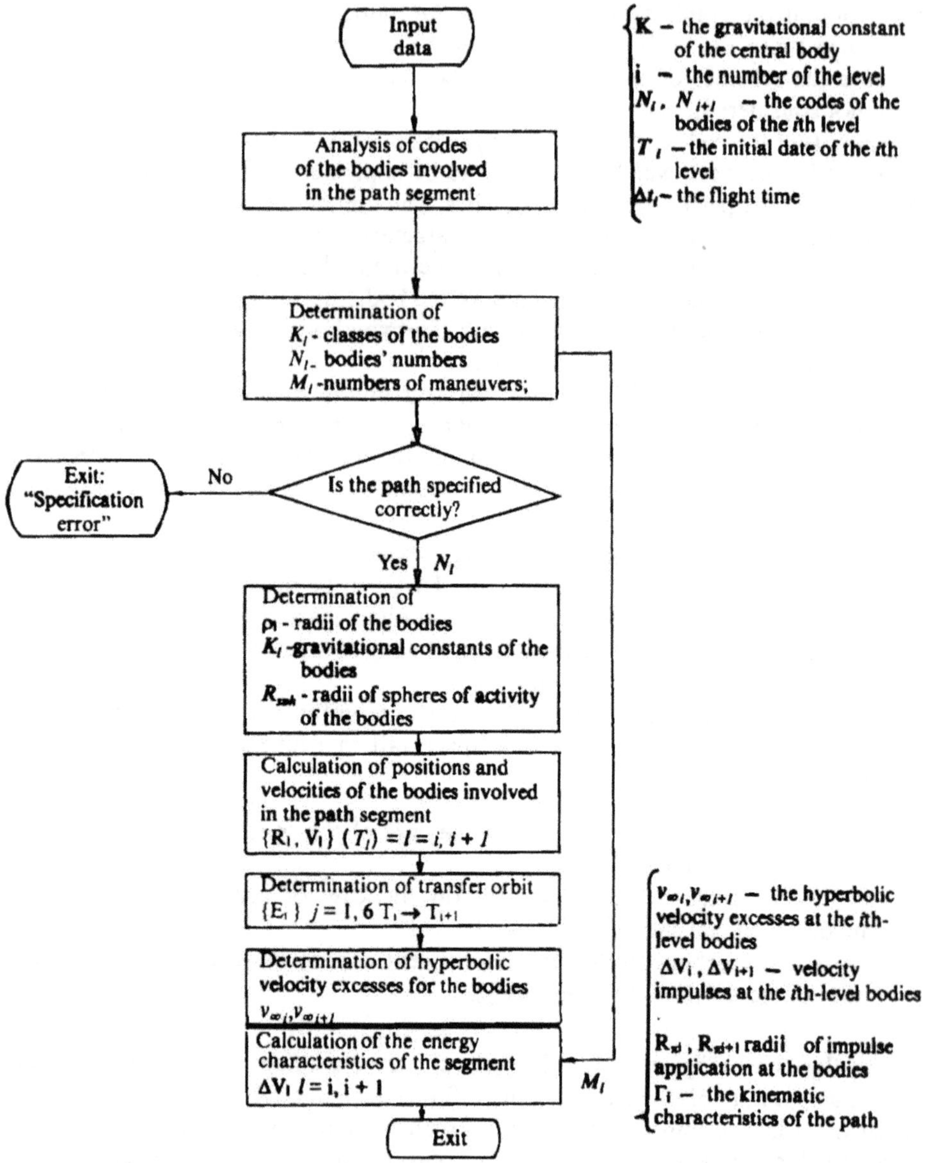

Figure 2.15 Flowchart of the basic module for the formation of multi-purpose spacecraft paths.

The module is autonomous and can be used as many times as required by the logic of the control program (depending on the number of the path segment, the start date, and the spacecraft flight time within the segment).

2.3.5 Control algorithm for the synthesis of complex multi-purpose spacecraft paths

When developing the control algorithm for the formation of optimal multi-purpose spacecraft paths, one should take into account a series of specific design and ballistic restrictions that to a great extent determine the logic and structure of the numerical algorithm.

The computational aspect of the optimization problems faced when dealing with multi-purpose spacecraft paths is rather complicated because of the nonlinearity and high dimensionality of these problems. This circumstance obviously hampers the application of conventional methods to the search for optimal solutions. Therefore, the researcher sometimes has to use various procedures to reduce the involved multiparameter problem, e.g. subdividing the problem into separate subproblems, searching for the optimal solution with part of parameters being fixed, etc.

These simplifications are often inefficient, because changes in the requirements, for instance, the need to fly by a further given body (to say nothing of the selection of this body from a list), result in a significant complication of the computations. This circumstance often leads the researcher to abandon the convenient programs and algorithms and search for new approaches to the problems arising.

At the same time, it should be noted that the problems of synthesising complex multi-purpose spacecraft trajectories implies a series of conditions and restrictions, which, if neglected, make it impossible to form the path [30]. These conditions can consist both in requirements for the spacecraft mission as a whole (e.g. the mission is total energy expenditure or mission time) and in requirements to separate flight segments (e.g. restrictions on the energy expenditure at the starting stage of the flight based on the need to use a regular accelerating device; reserve of the spacecraft velocity during maneuvering; requirements regarding the possibility of radio communication with the spacecraft during the mission, etc.).

In this connection it would appear expedient to develop a recursive-directed approach for the formation of such spacecraft paths considering all restrictions when optimizing the selected cost function F (e.g. minimization of the flight energy expenditure with restrictions on the mission time or minimization of some other path characteristic with the mission's given energy resource).

This can be defined as an approach with successive addition of the path segments with all the requirements met both for the given segment and for the path as a whole. This procedure should result in selection of the best variant, ensuring a minimum cost function.

This control algorithm should obviously use fork logical constructions including recursive calls.

The need to universalize such a flexible algorithm intended for a wide range of problems means that different solution optimization regimes have to be provided. These regimes can be as follows:

1. Search for the optimal solution for the given path in the case where all the bodies are specified.
2. Formation of an optimal path in the case where only some of the bodies within the path are specified, and in other breakpoints, only the classes of objects are specified from which the specific bodies are to be selected.
3. Formation of an optimal spacecraft path in the case where the number of flybys is unknown, and only the class (or classes) of the objects are known, from which the bodies are selected. The path is formed based on the condition of the given spacecraft energy resource $V_{\Sigma L}$ or total mission time $T_{\Sigma L}$.

The first regime of the algorithm's function can be applied to study any given path in the Solar System (a particular case is the study of a direct flight where only two celestial bodies are specified).

The two other processes of the algorithm's function are intended for the synthesis of complex multi-purpose paths including flybys of the planets and small celestial bodies (comets and asteroids). The inclusion of such objects in the path extends the scientific potential of the mission by allowing a larger number of objects of the Solar System to be investigated from a single spacecraft.

Figure 2.16 presents a general flowchart of the algorithm for the formation of this kind of spacecraft multi-purpose path. Within each of the segments, a multiple call of the basic computational block is performed, and a check is made (if it is specified in the solution regime) as to whether it is expedient to include an intermediate impulse within the segment. The formation of the path segments is made so as to ensure minimum costs and to meet the specified conditions and requirements both on individual segments and through the entire path.

In order to ensure the absolute minimum cost, the algorithm allows recursive calls with the depth limited by a maximum specified level. These consist in passing to a lower (preceding) level in the case where the flight time within the segment reaches a boundary value. In this case, the search is continued from the preceding level, starting from the last variant considered.

Thus, the exit from the regime of searching for the solution in the algorithm is implemented from the initial level after a recursive-directed search has been made through all the path levels (up to the maximum level), and the problem solution corresponds to the absolute minimum cost for the given interval of restrictions.

If the problem includes levels where only the class of objects is specified, and the search for such objects is to be implemented providing the minimization of a certain parameter of the path (e.g. the velocity of approaching the object), the exit from the search procedure will be made from the last of the levels formed. It is evident that all the restrictions imposed on the path should be satisfied.

Adjustment of the algorithm is quite simple. Based on the analysis of the problem and initial data, the designer specifies:

(a) the process of problem solution;
(b) mission coding;

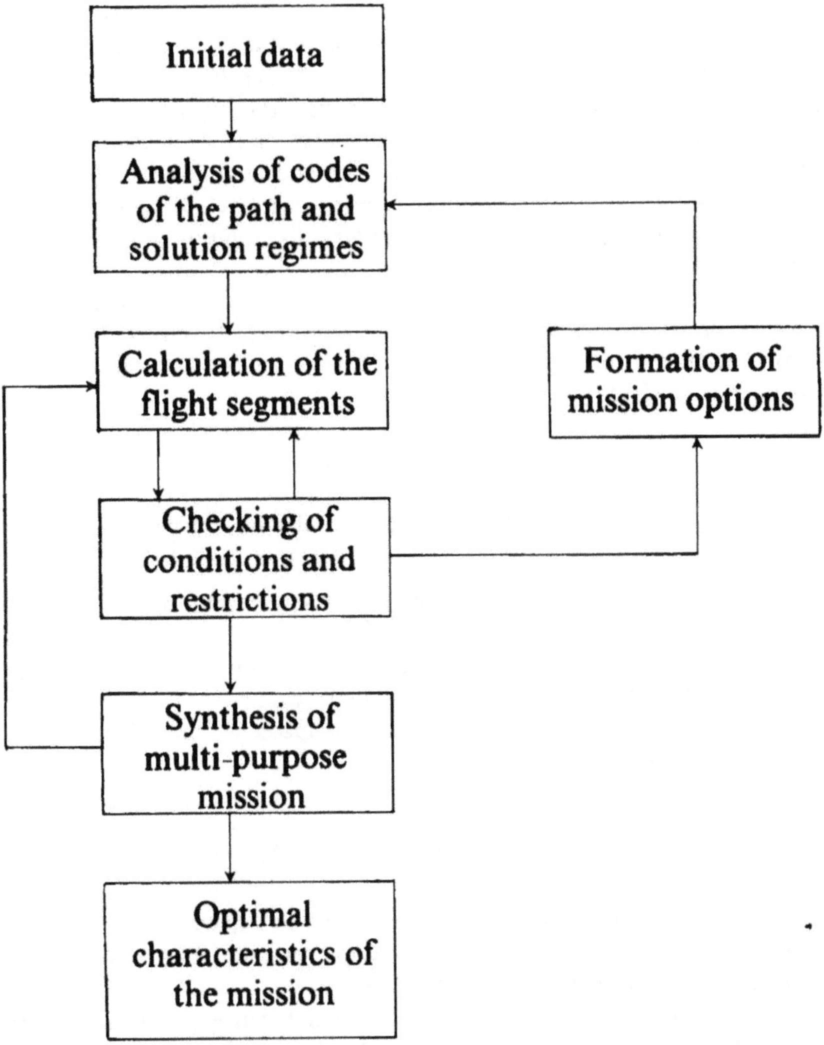

Figure 2.16 Flowchart of the procedure to form multi-purpose spacecraft path.

(c) conditions and restrictions (local and integral);
(d) steps for changing the flight times within the path segments and the step of changing the spacecraft start date.

Next is a step-by-step realization of the algorithm:

1. Based on the analysis of the process of solving the problem and the mission code, the two celestial bodies of the first ($i=1$) trajectory segment and the maneuver types are specified.
2. The basic module is used to calculate variants of the initial ($i=1$) segment of the trajectory, checking the specified conditions and restrictions.

3. If the restrictions are met, the next $(i+1)$ segment of the trajectory is calculated in a similar manner to steps 1 and 2; after this step the restrictions are checked, and if satisfied, the next $(i+2)$ segment of the trajectory is considered; if the restrictions are not met through the flight time, the algorithm returns to the previous $(i+1)$ trajectory segment, and the search for the solution is continued.
4. Step 3 is repeatedly performed to calculate subsequent trajectory segments while $i \leqslant N$, where N is the number of the specified last level. If $i = N$, the solution regime = 1, and the algorithm is completed with the solution obtained providing minimal flight energy expenditure *min* V_Σ.

   If process 2 or 3 takes place, the trajectory formation will be continued from the lists of proper classes of celestial bodies. Steps 1, 2, 3 are repeated.
5. The algorithm from processes 2 or 3 is left, when one of the following integral restrictions is reached

$$V_\Sigma = V_{\Sigma L} \quad \text{or} \quad T_\Sigma = T_{\Sigma L}.$$

Thus, the suggested algorithm for the synthesis of multi-purpose missions in the Solar System:

- allows one to find the solutions providing absolute minimum cost function within the given interval of restrictions;
- allows the designer to actively participate in the solution of the problem, making necessary adjustments in the restrictions at some levels;
- is a universal algorithm allowing a wide range of spacecraft interplanetary mission problems to be solved.

2.4 Formation of multi-purpose spacecraft missions in the solar system

1. Suppose that we are to investigate a given scheme of a spacecraft flight to Jupiter using the gravitational fields of Venus and Earth (mission Earth–Venus–Earth–Jupiter, E–V–E–J) with the starting date in 1994.

Suppose that restrictions are imposed on the mission concerning its energy expenditure (the total characteristic velocity $V_\Sigma \leqslant 6$ km/s and the mission duration $T_\Sigma \leqslant 4$ years).

Let us form the input data for this study.

Trajectory coding. The flight path is completely determined, i.e. the celestial bodies included in the path as well as the maneuvers near these bodies are known; therefore, according to the accepted classification, we can write a code chain denoting the E–V–E–J path:

10003.1–10002.2–10003.2–10005.3

This line means that the spacecraft:

- starts from the given circular orbit of an Earth artificial satellite;
- flies to Venus;

- performs a gravity assist maneuver at Venus;
- flies back to Earth;
- performs a gravity assist maneuver at Earth;
- starts off for Jupiter, where it performs a sounding flight over the planet.

The number of levels specified in the scheme $N=3$.
The number of levels to be formed $NNN=0$.

The range of the start date: $T_{so}=1994.50$
$T_{sk}=1994.80$

Step of variation $\Delta T_s=0.01$ year
Requirements to the mission $V_{\Sigma L}=6$ km/s, $T_{\Sigma L}=4$ years
Restrictions on the flight segments:
The number of complete revolutions within the segments $m_i=0$, $i=1,2,3$.
$\Delta V_1 \leqslant \Delta V_{1L}=4.2$ km/s (for given acceleration device)
$\sum_i \Delta V_i \leqslant 2$ km/s $i=2,3$ at the sites of maneuvering

$$\sum_i \Delta V_i \leqslant V_{\Sigma L}, \quad \sum_i \Delta T_i \leqslant T_{\Sigma L}, \quad i=1,2,3.$$

The program yielded a solution corresponding to the minimum cost function based on the integral characteristic velocity within the given range of spacecraft start dates.

Output data

$T_{st}=1994.76$	– start date
$\Delta t_1=0.47$ year	– flight time at the Earth–Venus segment
$\Delta t_2=0.87$ year	– flight time at the Venus–Earth segment
$\Delta t_3=2.44$ year	– flight time at the Earth–Jupiter segment
$\Delta V_1=3.85$ km/s	– starting impulse from the orbit of an artificial satellite of the Earth; H = 200 km
$\Delta V_2=0.010$ km/s	– intermediate velocity impulse during the flyby of Venus
$\Delta V_3=1.44$ km/s	– intermediate velocity impulse during the flyby of Earth
$V_{Ja}=6.275$ km/s	– approaching velocity (hyperbolic velocity excess near Jupiter)
$R_{\pi 2}=13646$ km	– Venus flight altitude
$R_{\pi 3}=6570$ km	– Earth flight altitude
$T_k=1998.54$	– date of arrival at Jupiter
$V_\Sigma=5.30$ km/s	– total characteristic velocity of the mission
$T_\Sigma=3.78$ years	– total mission time

2. Suppose that we are to investigate a scheme of flight to asteroids. Mars is selected as a flyby planet. After the flyby of Mars, flybys of two asteroids are supposed (mission E–M–A1–A2). The numbers of asteroids are not specified, so they are to be selected from the small planet list so as to meet the requirement of minimum total velocity of the mission for start dates in late 1994. Restrictions on

mission energy expenditure are assumed to be imposed (the total characteristic velocity $V_\Sigma \leqslant 6$ km/s and the mission time $T_\Sigma \leqslant 4$ years).

Let us form the input data for designing the optimal path under these conditions.

Path coding. The path is not completely specified. It is known that the first segment is Earth–Mars; as for the two other segments, it is known that they must consist of flybys of two asteroids; therefore, in accordance with the developed classification we can write: Earth–Mars–asteroid 1–asteroid 2:

10003.1–10004.2–20000.2–20000.3

This means that the spacecraft starts from the orbit of an artificial satellite of the Earth, flies to Mars, performs a perturbation maneuver in its gravitational field, and performs successive flybys of two asteroids.

The number of specified levels $N=1$

The number of levels after the scheme has been formed $NNN=3$

The range of the start date: $T_{s0}=1994.75$
$T_{sk}=1994.90$

The step of variation $\Delta T_s=0.01$ year

Requirements to the mission $V_{\Sigma L}=6$ km/s, $T_{\Sigma L}=4$ years

Restrictions on the flight segments:

The number of complete revolutions within the sites $m_i=0 \quad i=1,2,3.$

$\Delta V_1 \leqslant \Delta V_{1L}=4.2$ km/s (for given acceleration unit)

$\sum_i \Delta V_i \leqslant 2$ km/s, $i=2,3$ at the sites of maneuvering

$$\sum_i \Delta V_i \leqslant V_{\Sigma L}, \quad \sum_i \Delta T_i \leqslant T_{\Sigma L}, \quad i=1,2,3.$$

The program yielded a solution corresponding to the minimum value of the total characteristic velocity of the mission within the range considered (with estimates also obtained and displayed for other characteristics of the mission for each start date considered in the range under study).*

Output data

The start date of the spacecraft $T_{st}=1994.85$

The path formed

10003.1–10004.2–20341.2–20433.3

This means that after its start from the orbit of an artificial satellite of Earth, the spacecraft flies to Mars, performs a powered-gravitational flyby of Mars, flies to asteroids no. 341 (California) and no. 433 (Eros) (the numbers and names of

*Note. In the case where the solution cannot be obtained for some start date and with given restrictions, the program prints the date of start T_{st}, the message "no solution found", and the level number at which the calculation program failed.

asteroids are presented in accordance with the classification of the International Catalogue of Small Planets [6]).

Δt_1 = 0.63 year – flight time at the Earth–Mars segment
Δt_2 = 0.74 year – flight time at the Mars–no. 341 segment
Δt_3 = 1.60 year – flight time at the California–no. 433 segment
ΔV_1 = 3.86 km/s – impulse of the spacecraft to start from the circular orbit of the artificial satellite of the Earth H = 200 km
ΔV_2 = 0.020 km/s – velocity impulse during the flyby of Mars
ΔV_3 = 0.95 km/s – velocity impulse during the flyby of asteroid no. 341
V_{a1} = 2.98 km/s – velocity of approach to asteroid no. 341
V_{a2} = 4.26 km/s – velocity of approach to asteroid no. 433
$R_{\kappa 2}$ = 3685 km – spacecraft flyby altitude at Mars
V_{Σ} = 4.820 km/s – total characteristic velocity of the mission
T_{Σ} = 2.97 years – total duration of the mission
T_k = 1997.82 – the date of the mission completion

Thus, the optimal path for studying two asteroids with a flyby of Mars is formed for a given interval of the spacecraft start dates; this path can be used for further ballistic-design studies of the spacecraft and its multi-purpose trajectory.

In summary, it should be noted that the algorithm is very easy to adapt to various studies of multi-purpose paths (both for different classes of celestial bodies and different schemes of spacecraft flight in the Solar System).

Thus, we can set a reasonably wide range of studies, synthesize a number of different paths and determine for each of them the number and type of flybys of celestial bodies, and establish the time intervals for the spacecraft launch etc. And then the options, selected by their scientific, technical, and economic features, public interest in the projects, and their feasibility in the foreseeable future, can be analyzed in more detail, taking into account realistic conditions and requirements.

3. BALLISTIC METHODS IN MULTI-PURPOSE INTERPLANETARY SPACECRAFT DESIGN

In the previous chapter, we presented methods of pre-design studies using simple models to form a multi-purpose scheme of the flight together with the reference optimal trajectory of the spacecraft. These solutions can be used not only as initial approximations for more accurate calculation of the interplanetary trajectory or numerical integration, but also for estimating the design and ballistic characteristics of the spacecraft. This makes it possible to pass to the first stage of designing the spacecraft for the multi-purpose mission.

This stage of mission analysis is a component of spacecraft design, and therefore, the methods and models used in computations must meet certain requirements, because the output data must reflect all the relationships between the ballistic characteristics and the spacecraft design parameters.

The methods of mission analysis are based on a multi-level approach, which makes it possible to subdivide the problem into more simple subprocesses, thus reducing the number of parameters. Moreover, the results of analysis should represent not only a single solution, even if it is optimal, but also the characteristics of the entire domain of admissible trajectories. The design must deal with admissible domains rather than trajectories.

Experiments in mission design for interplanetary spacecraft show that the well-known activity spheres method (sometimes referred to as the method of piecewise conical approximation) meets these requirements (Chapter 1). According to this method, the interplanetary trajectory between two celestial bodies is approximated by three conic sections.

It is significant that the orbital elements of all the three conic sections, as well as any parameter of these trajectories q, can be represented as a function of four independent parameters:

$$q = q(T_1, T_2, \xi, \eta),$$

where T_1, T_2 are the starting and arrival times; ξ, η, the coordinates of the target point in the B-plane, i.e. the plane perpendicular to the velocity vector at infinity with respect to the target planet.

The values of T_1, T_2 completely determine the heliocentric orbit of the flight and the velocity vectors at infinity of the spacecraft moving to and from the planet. Once the parameters of the artificial earth satellite orbit have been specified, the hyperbolic orbit in the Earth's activity sphere is unambiguously determined, together with all energy and kinematic conditions of entering the interplanetary trajectory [2, 9].

For fixed T_1, T_2, parameters ξ, η uniquely determine the orbit within the activity sphere of the planet. The targeting parameters make it possible:

- to uniquely determine the parameters of the post-flyby heliocentric orbit for flyby trajectories;
- to determine the point of contact with the target planet surface for hitting trajectories;
- to determine the energy and kinematic conditions of transfer to a specified orbit of an artificial satellite of the planet or, given the energy parameters, to determine the class of possible orbits of the artificial satellite.

Methods for orbit analysis in the sphere of activity of a planet will be considered in detail in Chapter 7.

This approach to trajectory analysis allows a multi-parameter problem to be reduced to a series of two-parameter problems with independent variables. All the necessary characteristics of the interplanetary trajectory segments are represented at the plane of dates determining the position of planets (e.g., the start date to the date of arrival at the target planet).

After that, domains of possible trajectories of the spacecraft or its components (lander, artificial satellite of the planet) within the planetocentric flight segments are determined using a mapping to the plane of targeting parameters (Chapter 7).

The experiment shows that the entire set of requirements and restrictions can also be subdivided into the criteria depending on the heliocentric flight segments, the near-planetary flight segment, the orbits of artificial planetary satellites, and the conditions of contact and activity on the surface of the planet.

3.1 Isoline method

In a general case, to select a trajectory one should solve a nonlinear programming problem to choose an optimal or compromise trajectory, meeting the restrictions

$$q_i\{\leqslant, =, \geqslant\}q_i^*$$

where q_i^* are the specified values of design parameters.

Note that the number of design parameters is as a rule far greater than that of the independent variables.

Typical criteria and restrictions q_i that one should take into account when selecting the interplanetary segments of the flight and, therefore, should be able to represent in the date plane include:

- minimum admissible intervals of the dates of start and arrival at the planet (in the case of a multi-start flight scheme);
- providing the maximum weight of the spacecraft or its components, meeting the restrictions imposed on other parameters (in the case of combined flight schemes);

- restrictions on the inclination of velocity vectors at infinity with respect to the planetary equators; the restrictions relate to the spacecraft visibility when flying from the Earth, the possibility of the spacecraft landing in the specified region of the planetary surface, the possibility of creating an artificial satellite of the planet with specified parameters, etc.;
- restrictions on the velocity of entering the planetary atmosphere for the schemes implying direct landing of a lander;
- restrictions on the rate of acceleration from the Earth and the rate of deceleration entering the orbit of an artificial satellite of the planet; these restrictions are associated with the capacity of fuel tanks;
- restrictions on the Sun–planet–Earth angle at the moment where the spacecraft is approaching the planet; these afe associated with the feasibility of radio communication and operation of the spacecraft's celestial navigation system;
- restrictions on the flyby altitude at the planet; these are associated with both safety considerations and the efficiency of scientific studies, including delivery of the spacecraft components;
- restrictions on shading of the flyby trajectory around the planet from the Sun and Earth.

These and some other restrictions are rather complicated functions of the planetary positions. Therefore, a universal method for constructing level curves $q_i = const$ is needed to find out the admissible domains [87].

Thus, the study of interplanetary flight segments is the first stage of the comprehensive analysis of interplanetary trajectories. The main goal of this stage is the selection of the reference orbits meeting all the design restrictions and requirements. After that, the design and ballistic optimization of the reference trajectories is made, generally using variational methods or numerical solution of complete differential equations of the flight (Section 3.2).

3.1.1 Perturbation Equations

As shown in the previous section, any parameter depending on the trajectory between two planets can be expressed as a function depending on the dates and the position of the planets:

$$q = q(T_1, T_2). \tag{3.1}$$

Then the level curve of the parameter q is determined by the equation

$$q - q^* = q(T_1, T_2) = 0. \tag{3.2}$$

Geometrically, Expression (3.2) represents the projection of the curve in the space q, T_1, T_2 on the plane T_1, T_2 constructed as a section of the surface (3.1) by the plane $q = q^*$.

Let us derive the general equations of gravity assist maneuvers. Positions of the planets in the space will be represented by the set $T_i, i = 1, 2, \ldots, n$, where T_1 and T_n

are the launch dates from the starting planet and arrival at the target planet respectively, and $T_2, \ldots, T_{n-1}$ are the flyby dates of the planets. Let us use $V^i_{\infty 1}$ and $V^i_{\infty 2}$ to denote the velocities at infinity of the approach to and departure from the ith planet, respectively. Introduce the function:

$$F_i = V^i_{\infty 2} - V^i_{\infty 1}$$

and considering that

$$V^i_{\infty 1} = V^i_{\infty 1}(T_{i-1}, T_i), \quad V^i_{\infty 2} = V^i_{\infty 2}(T_i, T_{i+1})$$

we obtain the necessary conditions of existence of a trajectory with gravity assist flybys of the planets:

$$F_i(T_{i-1}, T_i, T_{i+1}) = 0, \quad i = 2, \ldots, (n-1). \tag{3.3}$$

Relationships (3.3) determine a two-dimensional manifold in the space $T_1, \ldots, T_n$. Let $q = q(T_1, \ldots, T_n)$ be a characteristic of an interplanetary trajectory, and q^* a given value of this characteristic. Then the equality

$$f(T_1, \ldots, T_n) = 0 \tag{3.4}$$

where $f = q - q^*$, determines a set of trajectories with $q = q^*$.

Combining (3.3) and (3.4) we obtain

$$F_i = 0, \quad f = 0, \quad i = 2, \ldots, (n-1). \tag{3.5}$$

Equations (3.5) correspond to a curve in n-dimensional space; this curve is formed as the intersection of the surface $f = 0$, corresponding to the specified parameter q^*, with $(n-2)$ surfaces $F_i = 0$.

Thus, the problem of constructing the isoline fields was reduced to constructing a series of curves in the n-dimensional space corresponding to different values of q^* and projecting them on the coordinate plane of a pair of coordinates among T_i.

Note that in terms of physical meaning and dimensions, Equations (3.3), (3.5) coincide with (3.1), (3.2). However, mathematically, the problem is reformulated for a multidimensional space.

It should be noted that in most cases, q is a function of no more than two or three variables. For instance, the values of V^i_∞ depend on two dates, and the pericenter altitude of a flyby planetocentric trajectory is a function of three successive dates.

As shown below, the application of the isoline method requires the perturbation equations to be represented in the linear form.

Let the sequence of dates T_i^0 be an approximate solution to Equations (3.5). Then, in order to linearize the constraint conditions, let us construct a Tailor expansion of F_i in the vicinity of T_i^0 up to the first terms and impose the conditions (3.3):

$$\Delta F_i + \frac{\partial F_i}{\partial T_{i-1}} \Delta T_{i-1} + \frac{\partial F_i}{\partial T_i} \Delta T_i + \frac{\partial F_i}{\partial T_{i+1}} \Delta T_{i+1} = 0$$

where

$$\Delta F_i = F_i(T_i^0), \quad \Delta T_i = T_i - T_i^0.$$

Let us find the derivatives:

$$\frac{\partial F_i}{\partial T_{i-1}} = -\frac{\partial V^i_{\infty 1}}{\partial T_{i-1}}, \quad \frac{\partial F_i}{\partial T_i} = \frac{\partial V^i_{\infty 2}}{\partial T_i} - \frac{\partial V^i_{\infty 1}}{\partial T_i}, \quad \frac{\partial F_i}{\partial T_{i+1}} = \frac{\partial V^i_{\infty 2}}{\partial T_{i+1}}.$$

We obtain the linear system with respect to ΔT_i:

$$-\frac{\partial V^i_{\infty 1}}{\partial T_{i-1}} \Delta T_{i-1} + \left(\frac{\partial V^i_{\infty 2}}{\partial T_i} - \frac{\partial V^i_{\infty 1}}{\partial T_i} \right) \Delta T_i + \frac{\partial V^i_{\infty 2}}{\partial T_{i+1}} \Delta T_{i+1} + \Delta F_i = 0 \quad i = 2, \ldots, (n-1) \tag{3.6}$$

The matrix of system (3.6) has dimensions $(n-2)n$.

Thus, if the values of any pair ΔT from $\Delta T_1, \ldots, \Delta T_n$ are specified, system (3.6) will be determined and the rest of $n-2$ values of ΔT_i can be found. Note that if the values of ΔT_1 and ΔT_n are specified, the system matrix becomes tridiagonal, i.e. elements of the matrix $a_{ij}=0$ for $|i-j| \geqslant 2$ [23].

If variables ΔT_1 and ΔT_2 are chosen as independent variables, it is expedient to represent expression (3.6) in the form:

$$\Delta T_{i+1} = \left(\frac{\partial V^i_{\infty 2}}{\partial T_{i+1}} \right)^{-1} \left[\frac{\partial V^i_{\infty 1}}{\partial T_{i-1}} \Delta T_{i-1} - \left(\frac{\partial V^i_{\infty 2}}{\partial T_i} - \frac{\partial V^i_{\infty 1}}{\partial T_i} \right) \Delta T_i - \Delta F_i \right] \tag{3.7}$$

i.e. in this case the values of $\Delta T_3, \ldots, \Delta T_n$ can be determined from the linear recurrent relationship (3.7).

Physically, this means that if a small displacement is specified for the first two planets along their orbits, then the recurrent relationship (3.7) allows successive determination of the displacements of all other planets necessary to satisfy the conditions of the perturbation maneuver (3.3).

The above expressions are convenient because the partial derivatives are evaluated independently for each successive pair of planets $T_1 - T_2, T_2 - T_3, \ldots T_{n-1} - T_n$.

The derivatives can be calculated numerically using the Euler–Lambert Equation.

Consider the linearization of the condition $q = q^*$. In the general case, linear representation of Equation (3.4) in the vicinity of T_i^0 will have the form:

$$\Delta f_0 + \frac{\partial f}{\partial T_1} \Delta T_1 + \frac{\partial f}{\partial T_2} \Delta T_2 + \cdots + \frac{\partial f}{\partial T_n} \Delta T_n = 0 \tag{3.8}$$

where

$$\Delta f_0 = f(T_i^0), \quad \Delta T_i = T_i - T_i^0.$$

System (3.6) in combination with Equation (3.8) yields linearization of system (3.5), i.e. it determines a tangent to the curve in n-dimensional space in the point T_i^0.

In general, the determination of the domain of admissible trajectories requires the construction of $(K \times P)$ level curves for orbital parameters on the date plane that are selected as independent variables:

$$f_{KP} = q_{KP} - q_{KP}^{*} = 0 \tag{3.9}$$

where K is the number of design parameters of the trajectories; P is the number of isolines of the Kth parameter.

Most commonly, the start date T_1 and the date of arrival at the last planet T_n are chosen as the independent variables.

Note that when analyzing trajectories with perturbation maneuvers, one should first reveal the domain where the value of the pericenter of the flyby planetocentral orbit ρ_π is greater than the planetary radius R_p, which implies the feasibility of the flight:

$$\rho_\pi = \frac{K_p(1-\sin(\psi/2))}{V_\infty^2 \sin(\psi/2)} > R_p + h$$

where: h is the specified minimum flyby altitude;

$$\cos\psi = V^0_{\infty 1} \cdot V^0_{\infty 2}, \quad |V_{\infty 1}| = |V_{\infty 2}| = V_\infty$$

K_p is the gravitational parameter of the planet.

The plane of the planetocentric orbit is determined by a unit vector C^0, parallel to the vector of the kinetic moment of the flyby

$$C_0 = \frac{V^0_{\infty 1} \times V^0_{\infty 2}}{\sin\psi}$$

All the above values are functions of the positions of the planets.

3.1.2 Method for the direct construction of level curves

Many applied problems require parallel analysis of a number of parameters that can be represented as functions of two independent variables. In this case, it is expedient to perform the study by mapping the level curves onto the plane of independent variables.

A number of techniques can be used to construct isolines. One is the method of sections, in which one of the variables is fixed, and a directed search is made along the other variable with interpolation in the points between which the given parameter value is found to lie. After that, the fixed variable is assigned a new value, and the process of search and interpolation is repeated, etc. Another technique is called the grid method, and consists in calculating tables of the values of the function of two variables, these values being stored in computer memory. The isolines are determined by searching the table with subsequent interpolation. There are some other methods that are based, as a rule, on particular features of the functions.

In the present section, we present a technique for constructing level curves based on following along the isoline in a given vicinity [87]. This technique has the following advantages:

- the technique is universal, i.e. it can be applied to any differentiable function of two variables $f=f(X_1, X_2)$;
- the technique allows a generalization to a two-dimensional manifold in an n-dimensional space;
- the given isoline can be constructed without searching for the values of the variables over the entire domain;
- the technique does not require large amounts of data to be kept in the computer memory;
- the technique can be applied to a series of other problems: constructing the curves of surface intersection, solving equations, searching for a conditional extremum, etc.

Consider first the algorithm of the technique for a two-dimensional case, and then pass to an n-dimensional generalization.

Suppose that we are to construct a level curve:

$$f(X_1, X_2)=q(X_1, X_2)-q^*=0 \tag{3.10}$$

on the plane X_1, X_2 within the domain enclosed by the rectangle:

$$X_{1\,\min} \leqslant X_1 \leqslant X_{1\,\max}; \quad X_{2\,\min} \leqslant X_2 \leqslant X_{2\,\max} \tag{3.11}$$

with an admissible accuracy of $|f| \leqslant \varepsilon_f$.

Introduce the notations:

$$\begin{gathered}
\bar{X}=\begin{Vmatrix} X_1 \\ X_2 \end{Vmatrix} \qquad \Delta X=\begin{Vmatrix} \Delta X_1 \\ \Delta X_2 \end{Vmatrix}; \\
\bar{g}_h=\begin{Vmatrix} \partial f/\partial X_1 \\ \partial f/\partial X_2 \end{Vmatrix} \qquad \bar{g}_K=\begin{Vmatrix} -\partial f/\partial X_2 \\ \partial f/\partial X_1 \end{Vmatrix}; \\
\bar{g}_h^0=\bar{g}_h/g; \quad \bar{g}_K^0=\bar{g}_K/g; \quad \rho=f/g; \\
g=\sqrt{(\partial f/\partial X_1)^2+(\partial f/\partial X_2)^2}.
\end{gathered} \tag{3.12}$$

Here, g_h^0, g_h^K are unit vectors of the gradient and tangent to the level curve in the point X_1, X_2; $\bar{g}_h^0 \cdot \bar{g}_K^0=0$; g is the module of the gradient; and ρ is the distance from the point X_1, X_2 to the level curve along the vector $\bar{g}_h^0$.

We suppose that the point X_1, X_2 lies sufficiently close to the level curve, i.e. $|f(X_1, X_2)| \leqslant \varepsilon_f$.

Suppose that in the point $\bar{X}^p$ we have $|f(\bar{X}^p)| \leqslant \varepsilon_f$, then to pass from the Pth point to the $(P+1)$th we use the relationships:

$$\begin{aligned} \bar{X}^{p+1} &= \bar{X}^p + \Delta\bar{X}^p, \quad \Delta\bar{X}^p = \bar{K}^p - \bar{h}^p \\ \bar{K}^p &= \beta_K^p g_K^{0p}, \quad \bar{h}^p = \beta_h \rho^p \bar{g}_h^{0p}. \end{aligned} \tag{3.13}$$

Expression (3.13) determines a discrete algorithm of movement along the isoline. The transfer into the $(P+1)$th point takes place along two directions:

- along the tangent to the level curve the vector $\bar{K}$ (this is the movement itself);
- against the gradient along vector $\bar{h}$ (descent to the isoline).

Coefficients β_K and β_h control the rate and stability of the process in the following way.

The value of β_h is responsible mainly for the condition $|f| \leqslant \varepsilon_f$, therefore, it is expedient to vary it within $1 \leqslant \beta_h \leqslant 2$. When $\beta_h = 1$, an exact descent to the isoline in the pth point takes place, which is convenient for rectilinear segments of the level curve. In the case of isolines with greater curvature, this value will lead to a one-directional movement with respect to the isoline, and therefore, to a decrease in the admissible step along the tangent. At $\beta_h = 2$, the reproduction of high-curvature segments of the isoline is good, but excessive fluctuations occur within rectilinear segments. Extension of the algorithm by introducing a block for determining the curvature makes it too involved; therefore, it appears reasonable to keep the value of β_h constant within the specified range.

The coefficient β_K is variable and controls the rate of movement along the isoline.

Let us introduce an additional, more narrow accuracy range $\tilde{\varepsilon}_f$ (Figure 3.1). The new range accounts for the fact that if $f(\bar{X}^p) \leqslant \tilde{\varepsilon}_f$, then the movement is too accurate, and the step can be increased. If $\tilde{\varepsilon}_f < f(\bar{X}^p) \leqslant \varepsilon_f$, the movement can be continued with the same step as before. And if a step yielded the point lying beyond the accuracy range ε_f, the step must be repeated from the same point, but it must be shorter along the tangent. In accordance with this:

$$\begin{aligned} \alpha_1 \beta_K^{p-1}, \alpha_1 > 1, &\quad \text{if } |f(\bar{X}^p)| \leqslant \bar{\varepsilon}_f \\ \alpha \beta_K^{p-1}, \alpha = 1, &\quad \text{if } \tilde{\varepsilon}_f < |f(\bar{X}^p)| \leqslant \bar{\varepsilon}_f \\ \alpha_2 \beta_K^{p-1}, \alpha_2 < 1, &\quad \text{if } |f(\bar{X}^{p+1})| > \varepsilon_f. \end{aligned} \tag{3.14}$$

This approach to selecting the length of the step along the tangent results in controlling the rate of motion along the isoline in a manner such that the rate (the step length) increases within rectilinear sites of the isoline and decreases within the sites with a significant curvature. The stability of the algorithm is guaranteed by the condition that the motion takes place only within the vicinity of the isoline determined by the specified accuracy.

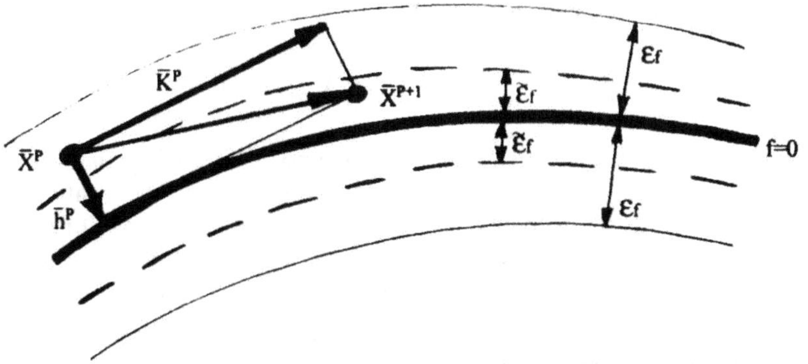

Figure 3.1 Motion along the level line.

Many experiments on the application of the algorithm have led to the conclusion that it is advantageous to choose:

$$\beta_h \approx 1{,}5 \quad \tilde{\varepsilon}_f = (0{,}5 \div 0{,}7)\varepsilon_f$$
$$\alpha_1 \approx 1{,}5 \quad \alpha_2 \approx 0{,}7.$$

The isoline from a given initial point $\mathcal{X}^0$ is constructed by the gradient method based on the same formulae (3.13), but with $\beta_K = 0, \beta_h = 1$

$$\mathcal{X}^{p+1} = \mathcal{X}^p - \rho^p \bar{g}_h^{0p}. \tag{3.15}$$

The iteration procedure is completed in a point S where

$$|f(\mathcal{X}^s)| \leq \varepsilon_f.$$

Vector $\mathcal{X}^s$ is the initial point for movement along the isoline using Relationships (3.13).

Note that the sign of β_K determines the direction of movement along the isoline. The process terminates in the following cases:

1. Subdivision of the step.
   In this case, at a certain step $\beta_K^p < \varepsilon_\beta$, where ε_β is the specified value. This can take place as a result of a repeated return into the previous point (the last condition in (3.14)). This generally occurs in places of dramatic convergence of the isolines; in this case, the accuracy range ε_f should be increased.
2. Closing.
   This is the most natural case; it means that the isoline construction is completed, and we returned into an ε_s neighborhood of the initial point $\mathcal{X}^s$. This suggests also that the closed isoline lies entirely in the domain under consideration.

3. Going beyond the domain boundaries

$$X_{i\min} \leqslant X_i \leqslant X_{i\max}.$$

When the isoline being constructed goes beyond the domain boundaries for the first time, the process returns to the initial point and then continues in the opposite direction by changing the sign, that is, Relationships (3.14) hold, but $\beta_K < 0$. When the isoline goes beyond the boundaries for the second time (it can be found out by the sign of β_K), calculation of the current isoline terminates.

After completing the process, a new isoline is selected from Expression (3.15), and the movement along it always starts with $\beta_K > 0$.

Figure 3.2 presents a typical example of search for the motion and passing to new level curves for two isolines with the conditions of closeness and going beyond the boundaries.

The proposed algorithm is used to construct isoline fields for the parameters of trajectories between two planets.

Consider a generalized n-dimensional algorithm for application to the trajectories containing gravity assist maneuvers. The number of flybys equals $n-2$, that is, the starting and target planets are excluded.

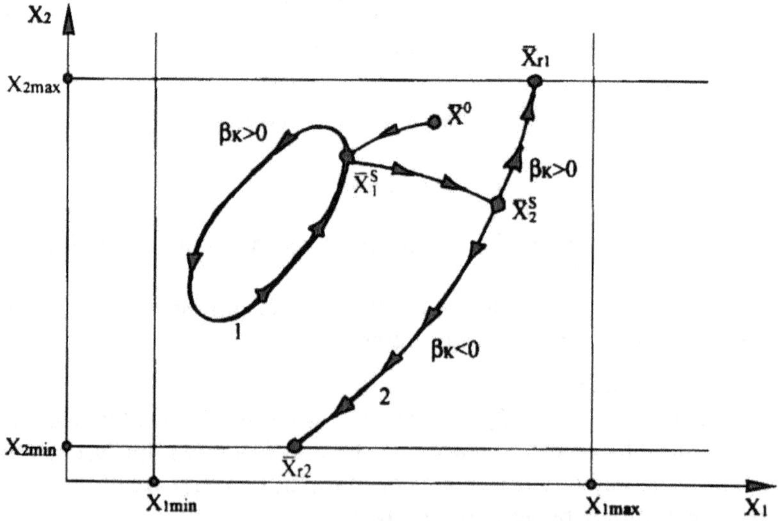

1, 2—Line of a level; $\bar{X}^0$—initial point; $\bar{X}^0$, $\bar{X}_1^s$—Trajectory of search of isoline 1; $\bar{X}_1^s$—Initial and final point of isoline 1 (locking); $\bar{X}_1^s$, $\bar{X}_2^s$—Trajectory of search of isoline 2; $\bar{X}_{r2}$, $\bar{X}_{r1}$—Point of the first and second infringement of borders on isoline 2.

Figure 3.2 Allowance for restrictions in the construction of isolines.

Suppose that we have a system of nonlinear Equations:

$$f_i(X_1, X_2, \ldots, X_n) = f_i^*, \quad i = 1, \ldots, m.$$

Let us designate $\Phi_i = f_i - f_i^*$, then the system can be rewritten as:

$$\Phi(X) = 0, \tag{3.16}$$

where Φ is a vector-function with values in an m-dimensional space, and $\bar{X}$ is a vector in an n-dimensional space, $m < n$. It is necessary to develop an algorithm for describing variations in the vector $\bar{X}$ so as to provide movement in the direction of an n-dimensional vector $\bar{Y}$ under a condition that the curve must not go beyond the ε-neighborhood of the constraints:

$$|\Phi_i| \leqslant \varepsilon_i, \quad i = 1, 2, \ldots, m \tag{3.17}$$

where ε_i are given values.

Let us linearize the system (3.16) with respect to the point $\bar{X}^0$:

$$G\Delta\bar{X} = \Delta\Phi \tag{3.18}$$

where

$$\Delta\bar{X} = \bar{X} - \bar{X}^0, \quad \Delta\Phi = -\Phi(\bar{X}^0),$$

$$G = \begin{Vmatrix} \overline{\text{grad}}^T \Phi_1 \\ \overline{\text{grad}}^T \Phi_2 \\ \cdots \\ \overline{\text{grad}}^T \Phi_m \end{Vmatrix}$$

The superscript T denotes transposition, and the matrix dimensionality is $m \times n$.

Let us find a vector $\Delta\bar{X}^*$ for which

$$|\bar{Y} - \Delta\bar{X}^*| = \min|\bar{Y} - \Delta\bar{X}|, \quad \Delta\bar{X} \in N.$$

N is the linear manifold of the solutions to system (3.16).

Let us form the function

$$\mathcal{F} = (\bar{Y} - \Delta\bar{X})^T(\bar{Y} - \Delta\bar{X}) + \bar{\Lambda}^T(G\Delta\bar{X} - \Delta\bar{\Phi}),$$

where $\bar{\Lambda}$ is the m-dimensional vector of Lagrange multipliers. Differentiating n times with respect to the components of vector $\Delta\bar{X}$ yields the system of equations

$$\left.\begin{aligned} &-2(\bar{Y} - \Delta\bar{X}^*) + G^T\bar{\Lambda} = 0 \\ &G\Delta\bar{X}^* = \Delta\bar{\Phi} \end{aligned}\right\},$$

whence it follows that

$$\Delta \bar{X}^* = \frac{1}{2} G^T \bar{\Lambda} - \bar{Y}$$

$$\bar{\Lambda} = 2(GG^T)^{-1}(G\bar{Y} + \Delta\bar{\Phi}),$$

and finally

$$\Delta \bar{X}^* = \bar{K} + \bar{H}$$

$$\bar{K} = [G^T(GG^T)^{-1}G - \mathrm{I}]\bar{Y}, \quad \bar{H} = G^T(GG^T)^{-1}\Delta\Phi. \tag{3.19}$$

Expressions (3.19) form the basis of the algorithm and allow a simple geometric interpretation. It can be shown that $\bar{K} \cdot \bar{H} = 0$, i.e. the vector $\Delta \bar{X}^*$ can be represented as the sum of two orthogonal vectors, the first of which $\bar{H}$ is the perpendicular from the point $\bar{X}^0$ to the linear manifold N, and $\bar{K} \in N$ and coincides with the projection of vector to $\bar{Y}$ to N.

If $m = n - 1$, then the constraints (3.16) determine a curve, and the system (3.18) determines a straight line, then the vector collinear to the vector $\bar{K}$ can be expressed in terms of the vector product of $(n-1)$ gradients of functions Φ_i in n-dimensional space:

$$\bar{g}_K = \overline{\mathrm{grad}}\,\Phi_1 \times \overline{\mathrm{grad}}\,\Phi_2 \times \cdots \times \overline{\mathrm{grad}}\,\Phi_{n-1}$$

or

$$g_K = \begin{vmatrix} e_1, e_2, \ldots, e_n \\ G \end{vmatrix}, \tag{3.20}$$

where $e_1, \ldots, e_n$ are unit basis vectors of the n-dimensional space of vectors $\bar{X}$.

The movement takes place in accordance with the following relationships

$$\bar{X}^{p+1} = \bar{X}^p + \Delta \bar{X}^p$$

$$\Delta \bar{X}^p = \beta_K^p \bar{K}^{0p} + \beta_h \bar{H}^p. \tag{3.21}$$

Here $\bar{X}^p$ is a vector meeting condition (3.17), $\bar{K}^{0p}$ is the unit vector of the direction $\bar{K}^p$; and $\bar{K}^p, \bar{H}^p$ are evaluated from (3.19) in the point $\bar{X}^p$; in the case of $m = n - 1$, vector $\bar{K}^0$ can be determined also from (3.20).

The values of β_K^p and β_h are determined in the same way as in the two-dimensional case. However, the accuracy conditions (3.17) are checked simultaneously for all m. In order to reach the given ε-neighborhood from the initial point, the iteration procedure

$$\bar{X}^{r+1} = \bar{X}^r + \bar{H}^r \tag{3.22}$$

is used, which terminates at some $r = s$, for which conditions (3.17) are satisfied, and the vector $\bar{X}^s$ is the initial point for movement along the constraints.

Termination of the movement is controlled through the same indices as in the two-dimensional algorithm, considering that the movement is restricted by the n-dimensional parallelepiped.

$$X_j^{\min} \leqslant X_j \leqslant X_j^{\max}, \quad j=1,2,\ldots,n.$$

The generalized algorithm is applied to construct the parameters of trajectories with perturbation planetary flybys. In this case, the general systems (3.16) and (3.18) are equivalent to the perturbation Equations (3.5), (3.6), and (3.7); and $m=n-1$.

When $m=1$, $n=2$, we obtain a two-dimensional variant of the above method. More simple relationships (3.13) are more advantageous in this case.

It is significant that the generalized method opens interesting new possibilities for solving other applied problems of mathematical programming:

1. Motion along a curve in an n-dimensional space.
   In this case, $m=n-1$, and vector $\bar{Y}$ determines only the sign of $\bar{K}^0$, i.e., the direction of motion. Therefore, it can be selected arbitrarily, e.g. for the sake of argument, we can set $\bar{Y}=\{\pm 1, 0, \ldots, 0\}$. Expressions (3.21) are used to calculate the motion along the curve with the possibility of constructing its projection onto any plane X_K, X_e in the space $X_1, X_2, \ldots, X_n$.
2. Solution to the problem of conditional extremum.
   We are to find extr $F(\bar{x})$ under the condition $\Phi(\bar{x})=0$. In this case, the motion within the $(n-m)$-dimensional surface takes place in the direction of the vector $\bar{Y}=\pm\overline{\text{grad}}\,F$. It is significant that in this process, the constraints must not be violated.
3. Nonlinear programming problem.
   We are to find min $F(\bar{x})$ under the condition $\Phi(\bar{x})\{<;>\}0$. The motion algorithm (3.21) can be used in the gradient projection method [11]. In this case, the motion takes place along the boundaries of the admissible domain with a variable matrix of projection

   $$G^T(GG^T)^{-1}G-I,$$

   into the vector

   $$Y=-\overline{\text{grad}}\,F.$$

4. Search for the value of a function $F(\bar{x})=F^*$ on $(n-m)$-dimensional surface determined by constraints $\Phi(\bar{x})=0$.
   In this case, vector Y will be equal to

   $$\bar{Y}=-\text{sgn}[F(\bar{x})-F^*]\,\overline{\text{grad}}\,F.$$

5. Search for the $(n-m)$-surface.
   The motion towards the surface $\Phi(\bar{x})=0$ to be found is implemented from the point $\bar{X}^0$ in accordance with (3.22); this motion coincides with that in the method of motion into the ε-neighborhood of the constraints.

Note that the above-considered problems are rather common in space-flight mechanics. For example, various problems of searching for optimal trajectories under conditions specified in the form of equalities or inequalities can be reduced to cases 2 and 3; and providing the given value to a trajectory parameter with specified values of other parameters reduces to case 4. Condition 5 determines the domain of the space restricted by the specified parameter values.

Case 1 represents the method for the direct construction of level curves.

In particular, the method was repeatedly used in this form not only for mission analysis, but for other problems as well. For example, in design work, it is sometimes necessary to construct the line of intersection of two surfaces in a three-dimensional space:

$$f_1(X,Y,Z)=0, \quad f_2(X,Y,Z)=0.$$

In this case, the vector tangent to the level curve is determined from (3.20) which reduce to a common vector product.

The method proposed here has been used for the analysis and selection of reference trajectories in designing the interplanetary orbits of unmanned interplanetary probes "Venera-9" to "Venera-16" and "Phobos" [111–113], in analysis of the trajectories Earth–Venus–Halley's Comet in the "Vega" project [205, 210], and in a number of design projects dealing with flights to distant planets, asteroids, and comets [140, 187, 201].

3.2 Optimization of design and ballistic characteristics of multi-purpose spacecraft

The isoline method presented in 3.1 is an initial stage of a multi-purpose spacecraft mission analysis. It allows one to reduce a multiparameter mission design problem to a series of two-parameter problems with independent variables, to select acceptable domains for the realization of the spacecraft trajectory, and to choose the trajectory parameters. This kind of reduction is quite acceptable for the preliminary designing and analysis of the mission. However, more detailed studies require an approach that can take into account the interaction of the factors of a multi-purpose expedition [30, 31, 82].

Analysis shows that the interaction of external and internal problems of a direct interplanetary flight exerts a considerable effect on the design and ballistic characteristics of the spacecraft [33].

This effect will be even greater in the case of analysis of multi-purpose missions including multiple planetary flybys, and will be very sensitive to changes in the reference trajectory.

The subject of this section is the development of the method and algorithm for the optimization of design and ballistic characteristics of multi-purpose interplanetary spacecraft.

3.2.1 Mathematical model of the study

Below we consider the design and ballistic problems of an interplanetary spacecraft flight designed to deliver a certain payload into a specified region of the Solar System using one or several planetary flybys.

The search for optimal design and ballistic characteristics of a spacecraft comprising a given number of stages and performing an interplanetary flight suggests that in addition to the gravitational effect exerted on the spacecraft by the sun, the gravitational fields of planets (within the framework of the principle of the activity spheres) should also be taken into account. Thus, we assume that the interplanetary spacecraft, having started from the given near-earth orbit, moves along the design trajectory, being at each moment subject to the effect of only one attracting body—the Earth, the Sun, or any other planet of the Solar System (in the respective activity spheres), and within powered flight segments, the effect of the force of thrust of the engine. It is also assumed that all the spacecraft maneuvers are performed at a sufficient distance from the planets to allow the effect of the atmosphere to be neglected. A schematic diagram of this kind of flight is presented in Figure 3.3.

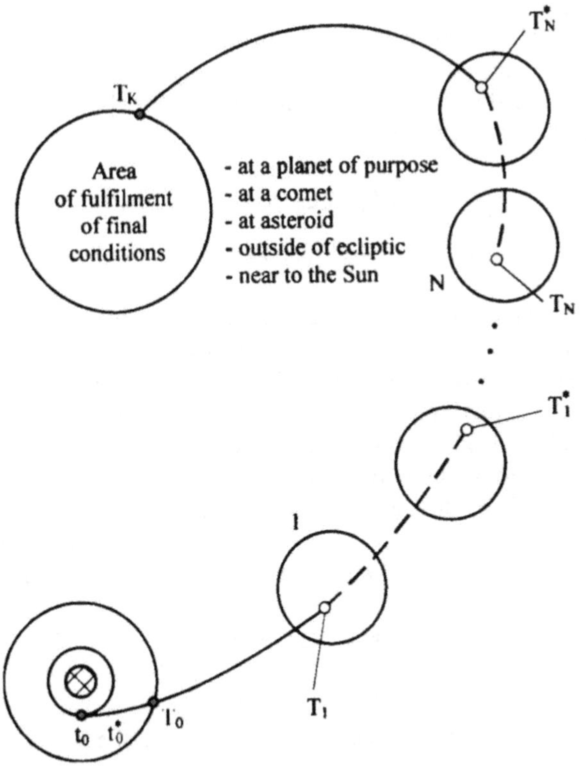

Figure 3.3 Schematic diagram of forming multi-purpose interplanetary mission.

The objective of the interplanetary spacecraft mission analysis is to determine the reference interplanetary trajectory, control regime, and principal design parameters of the spacecraft, which will allow the mission to be performed with extremal values of the accepted optimal weight criterion: minimum spacecraft takeoff mass G_0 given the payload G_p or maximum payload G_p at given G_0. These interplanetary flight problems are investigated within the framework of the activity spheres with the length of powered flight segments taken into account in a combined solution of the internal and external problems; therefore, the solutions will be sought on the set of the differential and algebraic constraints describing the spacecraft motion throughout its flight. Below, these constraints are written in the geocentric, heliocentric, and planetocentric coordinate systems.

The spacecraft position and velocity $\mathscr{H}\{\bar{r}_0, \bar{v}_0\}$ at the parking circular orbit specified by its elements can be determined for the moment of start t_0 from the relationships of the following form

$$f_{j0} = \mathscr{H}_{j0} - \mathscr{H}_{j0}(r_0, i_0, \Omega_0, M_0, t_0, K_0) = 0, \quad j = 1, \ldots, 6, \tag{3.23}$$

where r_0, i_0, Ω_0 are the radius, inclination, and the longitude of the ascending node of the starting orbit, $M_0 = M(t_0)$ is the average anomaly of the starting point, K_0 is the gravitation constant of the starting planet (the Earth). Within the powered flight segments (the initial and other ones) the constraints on the phase coordinates of the spacecraft velocity and location are described by the system of six differential equations of controlled spatial motion together with the characteristic of mass losses due to the engine of the spacecraft mth stage:

$$\varphi_{jN}(\mathscr{H}_{jN}, \dot{\mathscr{H}}_{jN}, \bar{u}, \bar{a}_m, \mu, K_N) = 0, \quad j = 1, \ldots, 6. \tag{3.24}$$
$$\varphi_{7m} = \dot{\mu} = -n_{0m}/P_{sm} = \text{const},$$

where $\bar{u}\{\alpha, \varphi\}$ are the angles of the spacecraft thrust vector; μ is the current relative mass of the spacecraft; n_{0m}, P_{sm} are the thrust-to-weight ratio and specific thrust of the spacecraft mth stage; $\bar{a}_m$ is the vector of design parameters of the mth stage of the spacecraft; K_N is the gravitational constant of the Nth central body, $N = 0, \ldots, K$.

Within the ballistic-flight segments in the heliosphere and the activity spheres of planets, the spacecraft flight is determined by the gravitational forces of a single central body. Therefore, the velocity components and location coordinates of the spacecraft in the beginning and endpoint of the ith ballistic-flight segment are correlated by the equations for the undisturbed Keplerian motion

$$f_{jN}(\mathscr{H}_{jN0}, \mathscr{H}_{jNK}, \Delta t_i, K_N) = 0, \quad j = 1, \ldots, 6. \tag{3.25}$$

At the given time moments T, the conditions of the spacecraft entering the activity sphere of a planet with the given radius R_{sp} can be specified:

$$\Psi_p(T_N) = R_{sp} - |\bar{R}_p(T_N) - \bar{R}(T_N)| = 0 \tag{3.26}$$

where $\bar{R}_p(T_N)$ and $\bar{R}(T_N)$ are the heliocentric coordinates of the planet and spacecraft locations. On the boundaries of the planetary activity spheres, the spacecraft coordinates must be converted from the planetocentric $\bar{r}_N, \bar{v}_N$ into heliocentric $\bar{R}_N, \bar{V}_N$ in the points of exit from the zone and vice versa, in the points of entering it

$$R_{jN}(\bar{r}_N, \bar{v}_N, \bar{R}_N, \bar{V}_N, T_N, K_N)=0, \quad j=1,\ldots,6. \tag{3.27}$$

In the planetary flyby segments, allowance must be made for the displacement of the planet during the time the spacecraft is moving within its activity sphere

$$F_{jN}(\bar{R}_{pN}, \bar{V}_{pN}, \bar{R}^*_{pN}, \bar{V}^*_{pN}, \Delta t_N)=0, \quad j=1,\ldots,6 \tag{3.28}$$

where $\Delta t_N = T^*_N - T_N = f(\mathscr{H}_{jN}, R_{spN}, K_N)$ is the time it takes the spacecraft to pass through the activity sphere of the planet.

Boundary values in the right end of the trajectory and in inner points of the multi-purpose path (if it is found necessary) can be written in the general form as

$$f_{jK}(\mathscr{H}_{jK}, T_N, K_N)=0, \quad j\leqslant 6, \quad N=1,\ldots,K. \tag{3.29}$$

The specific form of the conditions (3.29) is determined immediately by the objective of the mission. It can be a flyby—sounding trajectory at the given object of the Solar System (the target planet, comet, or asteroid), a trajectory which will enter the orbit of an artificial satellite, or a landing trajectory. It is evident that these conditions to a great extent determine the interplanetary trajectory formation and the values of design and ballistic characteristics of the spacecraft.

And finally, the design-weight equations [25] correlating the principal weight parameters of the spacecraft G_0 and G_p, initial and all-burnt weights of its stages G_{0m} and G_{Km}, the engine thrusts of the stages P_m:

$$\Pi_m(G_0, G_{0m}, P_m, \mu_p, n_{0m}, \mu_{Km}, K_s)=0. \tag{3.30}$$

Here μ_p is the relative payload delivered by the spacecraft to the target planet; $n_{0m}=(P_m/G_{0m})$ is the thrust-to-weight ratio of the mth stage; $\mu_{Km}=(G_{km}/G_{0m})$ is the relative all-burnt weight of the mth stage; K_s are specific design-mass characteristics of the spacecraft stages, $K_s=const$.

The specific form of the relationships (3.23)–(3.30) can be different depending on the mission objective and the path of the flight under consideration.

Thus, combined consideration of weight and dynamical problems of the spacecraft interplanetary flight leads to the following problem of mission analysis.

Given the spacecraft path, we are to find the spacecraft trajectory, the program of orientation of the engine thrust vector in the powered flight segments, the moments at which the engine turns on and off, and the values of the spacecraft principal design parameters so as to implement the mission with the selected weight criterion taking extremal values: either G_p is maximum or G_0 is minimum.

3.2.2 Variational problem of spacecraft mission analysis

Mathematically, the problem of minimizing the spacecraft initial mass considered as a functional criterion to select the design and ballistic parameters and control functions for powered-flight segments can be formulated and studied as a variational problem using conventional approaches to optimization of design decisions.

In the problem under consideration, application of the variational method of optimal craft design [33] allows one to determine all the conditions necessary for the optimization of the design and ballistic characteristics of the interplanetary spacecraft flying along the specified path with flybys of a series of planets.

These conditions can be derived from the stationary condition

$$d\Phi = 0 \tag{3.31}$$

of the unconditional functional

$$\Phi = G_0 + \Pi + P + \sum_i \int_{t_i}^{t_i^*} F_i \, dt, \quad 0 < i \leqslant K$$

where $\Pi = \sum_m e_{nm} \Pi_m$ is the Lagrangian of design constraints,

$$P = \sum_J e_{J0} f_{J0} + \sum_{JN} e_{JN} f_{JN} + \sum_N e_N \Psi_{PN} + \sum_{J,N} e_{JN} R_{JN} + \sum_{J,N} e_{JN} F_{JN} + \sum_i e_{JK} f_{JK}$$

is the Lagrangian of boundary and intermediate conditions, $F_i = \sum_{J=1}^{7} (\dot{\mathscr{X}}_{JN} - \varphi_{JN}) \times \lambda_{JN} \, dt$ is the Lagrangian of differential constraints. Here, the multipliers of the form $e_{j,0,K}$ and λ_{JN} are the constant and variable Lagrange multipliers, respectively.

Thus, the necessary conditions for optimization of the trajectory and design parameters of the multi-purpose spacecraft comprise:

- Euler–Lagrange equations, determining the conditions of spacecraft optimal control within the ith powered flight segment

$$\dot{\lambda}_{JN} = -\frac{\partial H}{\partial \mathscr{X}_{JN}}, \quad \frac{\partial H_i}{\partial u_s} = 0 \tag{3.32}$$

where

$$H = \sum_{J=1}^{7} \varphi_{JN} \lambda_{JN}, \quad N = 0, \ldots K, \; S = 1, 2$$

- Weierstrass condition, determining additional information on the control of spacecraft engine at powered flight segments

$$E = H_i - \tilde{H}_i \geqslant 0. \tag{3.33}$$

H_i is defined by the values φ_{JN} given the admissible control functions α^*, φ^*.

• Conditions of discontinuity of Lagrange coefficients in the points of staging t_q.

$$\lambda_{jN}^{+}-f_{jN}\{\lambda_{jN}^{-},\mu(t_q)\}=0,\quad j=1,\ldots,7,\ q=1,\ldots,m-1. \tag{3.34}$$

• Conditions of transversality at the ends of the spacecraft trajectory

$$\left.\frac{\partial p}{\partial \mathcal{H}_l}\right|_{t_0,t_K}=0 \tag{3.35}$$

$\mathcal{H}_l$ are the parameters that can be varied freely.

In a similar way, the conditions at intermediate points of the multi-purpose spacecraft trajectory can be written as well as the conditions for the transformation of Lagrange coefficients λ_{jN} in the ballistic flight segments.

The analysis of the problem shows that the algorithm of transformation of the Lagrange coefficients within any ballistic segment of the multi-purpose path can be represented in the form of a matrix, which allows us to avoid a numerical solution of the system of differential equations and to use instead multiplication of matrices (in accordance with the succession of the spacecraft flight segments in the gravispheres of the bodies involved in the interplanetary path being formed):

$$\bar{\lambda}_{jK}^{(i)}=\mathcal{J}\left(\frac{\partial f_{jK}}{\partial \mathcal{H}_{jK}}\right)\cdot\mathcal{J}^{-1}\left(\frac{\partial f_{jK-1}}{\partial \mathcal{H}_{jK-1}}\right)\cdots\mathcal{J}\left(\frac{\partial f_{j1}}{\partial \mathcal{H}_{j1}}\right)\cdot\mathcal{J}^{-1}\left(\frac{\partial f_{j0}}{\partial \mathcal{H}_{j0}}\right)\cdot\bar{\lambda}_{j0}^{(i)},\quad j=1,\ldots,6.$$

This allows the values of the Lagrange coefficients to be obtained at the starting point of the following powered segment. In this case, the main connecting link is a matrix of the form

$$\mathcal{J}\left(\frac{\partial f_j}{\partial \mathcal{H}_j}\right)=\begin{vmatrix}(\partial f_1/\partial v_x)\cdots(\partial f_6/\partial v_x)\\ \cdots\cdots\cdots\cdots\cdots\cdots\\ (\partial f_1/\partial z)\cdots(\partial f_6/\partial z)\end{vmatrix}.$$

Finally, the conditions of optimization of the basic design parameters of the spacecraft must be considered

$$\frac{\partial \Pi}{\partial a_{t,m}}=0. \tag{3.36}$$

$a_{t,m}$ are the basic design parameters of the spacecraft stages being varied.

Thus, meeting all the requirements necessary for optimization together with the boundary conditions will yield the optimal solution to the problem, i.e. the optimal multi-purpose trajectory, control regime, and spacecraft basic design parameters that will make the spacecraft's initial mass on the near-earth orbit minimum. In terms of computations, the problem to be solved is a multi-point boundary value problem with boundary conditions in the intermediate and terminal points of the trajectory:

$$P_n(x_n)=0, \quad x_n=\{\mathscr{H}_e, a_{em}, \lambda_{ji}, e_0, \ldots, e_K\} \tag{3.37}$$

satisfying the constraints and conditions (3.23–3.36).

The computational difficulties associated with solving multi-point boundary value problems that are due to the selection of the unknown variables greatly increase when nonlinear multi-parametric objects are involved. Therefore, at the conceptual design stage of the interplanetary spacecraft, where a number of variants of the spacecraft are to be assessed with different parameter combinations, and different start and arrival dates, it appears reasonable to use approximate methods of solution.

Studying the problems of direct interplanetary flights based on similar models and the analysis of the individual segments of spacecraft motion shows that the computational aspect of the problem can be greatly simplified by making some assumptions (the essence of the problem being unaffected). Thus, we assume that the profile of the interplanetary flight is specified: the powered segments of the spacecraft trajectory can be located only in the domains of near-planetary motion. It is also assumed that during flybys, the spacecraft engine is turned on (if necessary) in the pericenters of the flyby segments. In addition, we will use the approximate tangential control of the thrust vector

$$[\bar{p}(t)\times\bar{v}(t)]=0 \tag{3.38}$$

in the powered segments of the spacecraft flight. Analysis of model problems show this control law to be very close to the optimal regime. The unknown undetermined Lagrange coefficients λ_{j0} obtained by solving the boundary problem with tangential control can be considered a good approximation in the subsequent solution of the boundary-value problem and selection of the spacecraft optimal control.

The assumptions made allow the variational problem of minimizing the spacecraft launch weight to be reduced to a more simple problem of finding the conditional extremum of a function of several variables:

$$G_0(\mathscr{H}_{j0}, n_{0m}, \mu_{Km})\Rightarrow\min \tag{3.39}$$

under the conditions (3.23–3.30) and considering (3.38), which is sufficient to determine the spacecraft trajectory $\mathscr{H}_j(t), t_0\leqslant t\leqslant t_K$ for the given multi-purpose path.

The procedure of solving the reduced problem of optimizing the trajectory and principal design parameters of the spacecraft should be organized in such a way that, given the path, the dates of its implementation T_N $(N=0,\ldots,K)$, and boundary conditions, one will find the initial conditions of leaving the parking orbit $\mathscr{H}_{j0}$, the initial thrust-to-weight ratios n_{0m} and the relative all-burnt weights of the stages μ_{Km} that make G_0 minimum given the payload G_{pl} or G_{pl} maximum given the launch weight G_0. In either case, the problem can be reduced to the maximization of the spacecraft relative payload $\mu_p\Rightarrow\max$.

3.3 Optimization of design solutions based on the generalized criterion method

Before turning to the problems of constructing the optimization algorithm, we note that the design of multi-purpose interplanetary flights with gravity assist maneuvers has a specific feature; if the condition of the spacecraft entering the activity sphere of a planet to be flown by is satisfied exactly, this determines to a great extent the meeting of the subsequent boundary conditions. Moreover, in order to make it possible to accomplish the subsequent segment of the path (intermediate or terminal, depending on the number N of flybys), it is necessary to direct the trajectory into a certain domain in the activity zone of the planet.

For the sake of argument we will consider the problem of interplanetary flight of a two-stage spacecraft with a flyby of a planet and subsequent arrival into the neighborhood of a target celestial body. As an initial approximation, we can use the solution of the similar problem in the point-impulse formulation. Indeed, the use of the universal algorithm for the synthesis of spacecraft multi-purpose interplanetary trajectory (see Chapter 2) allows us to find the initial approximations for all the necessary characteristics of this path: reference times of launching, planetary flyby, arrival at the target planet, energy expenditure of the flight in the form of the total characteristic velocity, and impulses to alter the velocity within the necessary segments (they can also allow an assessment of the relative final weights of the spacecraft stages μ_{Ki}), and, finally, the orientation of the flight plane and the flyby altitude.

Thus, the above condition of the spacecraft falling within a certain near-planet zone can be written as the condition for the required aiming point distance β_p in B-plane at the moment when the spacecraft enters the activity sphere of the flyby planet

$$\beta-\beta_p=0.$$

The required and actual values of β can be determined from the following relationships for the moment the spacecraft enters the activity sphere

$$\beta=\Gamma_p\sqrt{1+\frac{2K_p}{\Gamma_p(v^2-2K_p/R_{sp})^{1/2}}},\quad \beta_p=R_p\sqrt{1+\frac{2K_p}{R_p v_\infty^2}}. \tag{3.40}$$

The pericentral distance of the flyby hyperbola is

$$\Gamma_p=\frac{K_p}{v^2-(2K_p/R_{sp})}\left(\sqrt{1+\frac{C^2}{K_p^2}h}-1\right).$$

Here, v is the planetocentric velocity of the spacecraft in the activity sphere of the flyby planet; c, h are the values of constants of the integral of area and the energy of the spacecraft incoming hyperbola; v_∞, R_p is the hyperbolic excess of velocity and the flyby altitude (from the impulse approximation solution).

Considering that for the spacecraft with high-thrust engines considered here, the duration of the active powered deceleration is small as compared with the segment of ballistic flight within the activity sphere of the flyby planet, we assume that as a first approximation, the required deviation in the B-plane at the moment T_1 can be found from (3.40) (indeed, investigation of internal problems of escape and deceleration of the spacecraft showed that the error in determining β_p from the relationship amounts to about 0.5%).

Thus, a reasonable strategy for searching for the conditional extremum of the multivariate function appears to be a subdivision of the procedure into stages.

At the first stage, a reference interplanetary flight trajectory is determined, which allows the spacecraft to enter the activity sphere of the flyby planet at given moment T_1 with the necessary requirements being met. To implement this, one can select the lacking parameters $n_{01}, \mu_{K1}, M_0, \Omega_0$ (or i_0) and then use these approximations to solve the problem of minimizing the spacecraft launch weight and meeting the target condition of the flight (reaching the target planet, near-solar zones, or zones beyond the ecliptic, the specified asteroid or comet).

Consider a segment of the spacecraft path from the parking orbit to the flyby planet. Generally speaking, the longitude of the ascending node Ω_0 in the parking orbit can be selected arbitrarily, because Ω_0 is determined by the time of the spacecraft launch to the near-earth orbit, whereas the inclination of the parking orbit i_0 cannot be selected arbitrarily with no allowance made for the conditions of the spacecraft entering the near-earth parking orbit, which are determined primarily by the launch point location (latitude and longitude). Therefore, we assume that the parking orbit inclination i_0 is specified and we will meet the condition of the spacecraft entering the activity sphere of the target planet through the selection of parameters $n_{01}, \mu_{K1}, M_0, \Omega_0$.

If the above condition is not satisfied for the given inclination of the parking orbit, it becomes necessary to scan the value of i_0 within a certain given interval $i_{0\min} \leqslant i_0 \leqslant i_{0\max}$ determined by conditions of the spacecraft entering the activity sphere of the flyby planet with a specified mass in B-plane β_p.

Thus, at the first stage, the problem of searching for the reference trajectory from the Earth to the flyby planet can be formulated as the problem of minimizing the spacecraft launch weight $G_0(\mathscr{X}_m), \mathscr{X}_m\{n_{01}, \mu_{K1}, M_0, \Omega_0\}$ with the following conditions in the point I:

$$P_1 = R_{sp1} - \sqrt{(X_1 - X_{p1})^2 + (Y - Y_{p1})^2 + (Z_1 - Z_{p1})^2} = 0 \tag{3.41}$$

$$P_2 = 1 - \beta/\beta_p = 0. \tag{3.42}$$

The selected solutions must provide the spacecraft entering the descending branch of the incoming hyperbola to the flyby planet; therefore, inequality $\bar{r}_1, \bar{v}_1 \leqslant 0$ must hold at each iteration; here $\bar{r}_1, \bar{v}_1$ are the radius vector of the escape point and the spacecraft velocity vector on the activity sphere of the target planet.

The conditional minimum of the function under equality-type restrictions can be found by methods of nonlinear programming. Considering the computational peculiarities of the problem in question, the most suitable is the method of generalized criterion (also called penalty function method), which allows the problem in question to be reduced to the problem of unconditional minimization of the target function of the form

$$\Phi(\mathcal{H}_m)=G_0(\mathcal{H}_m)+\sum_{i=1,2}K_i|P(\mathcal{H}_m)|, \tag{3.43}$$

where $G_0(\mathcal{H}_m)$ is the criterion function to be minimized, $P_i(\mathcal{H}_m)$ are the flight boundary conditions, and K_i are the penalty scale factors.

It is evident that the greater the selected values of K_i, the narrower the zone on the restriction surface in which the minimum of the function $\Phi(\mathcal{H}_m)$ will be sought. In the case where the restrictions $P_i(\mathcal{H}_m)$ are met exactly, the function (3.43) coincides with the function $G_0(\mathcal{H}_m)$; therefore, the values of K_i being sufficiently large, the minimum of $\Phi(\mathcal{H}_m)$ will coincide with the minimum of $G_0(\mathcal{H}_m)$.

It should be mentioned, however, that as the values of K_i grow, the "ravine" of the target function (3.43) located along the surface of restrictions becomes more narrow, which can hamper the convergence of the process of searching for the minimum.

Once the minimum value of the target function

$$\Phi(\mathcal{H}_m)=\min$$

has been found, we obtain the missing values, providing the trajectory entering the activity sphere of the flyby planet with a necessary mass, the spacecraft launch weight being minimum.

The obtained values of $n_{01}, \mu_{K1}, M_0, \Omega_0$ (or i_0) can be used as a first approximation for the second stage of the problem solution.

The aim of the second stage will be to find the unconditional minimum of the target function of the form

$$Q(\mathcal{H}_j)=\frac{K_0}{1+\mu_p}+\sum_{i=1,n}K_i|P_i(\mathcal{H}_j)|,\quad \mu_p=\mu_{01}\cdot\mu_{02}=\frac{G_p}{G_0} \tag{3.44}$$

where

$$\mathcal{H}_j=\{\Omega_0, M_o, \mu_{K1}, n_{01}, \mu_{K2}, n_{02}\}.$$

In this case, the condition P_1 of the spacecraft entering the activity sphere of the flyby planet are determined as (3.41), whereas the restrictions $P_2,\ldots,P_n$ being boundary conditions will be determined by the peculiarities of the interplanetary flight problem under consideration. For example, these can be for example, the conditions on the flyby altitude of the target planet or the conditions of the spacecraft entering the orbit of an artificial satellite of the target planet with specified parameters.

Introducing the penalty terms into the target function to be minimized brings about the question of selecting the coefficients of these terms. This question should be solved taking into account the specific conditions and restrictions of the problem in question. In the general case, the coefficients with the loading terms must be positive and sufficiently large to provide the condition

$$K_i\left|\frac{\partial P(\mathscr{X}_j)}{\partial \mathscr{X}_j}\right| \gg \left|\frac{\partial G_0(\mathscr{X}_j)}{\partial \mathscr{X}_j}\right| \tag{3.45}$$

throughout the range of the variable parameters, but for a certain δ-neighborhood of the restrictions determined by the condition $P(\mathscr{X}_j)<\delta$.

In the δ-neighborhood of the boundary conditions, the inequality (3.45) becomes weaker, and in this case the gradient of the original function to be minimized acquires a significant role in determining the direction of movement to the optimum.

The obtained minimum of function (3.44) will correspond to the minimal start weight and the optimal design ballistic characteristics of the spacecraft to perform the flight along the specified multi-purpose path with flybys. The solution of the problem as a whole will also allow one to construct the entire trajectory of the spacecraft by successively forming it in the gravity fields of the celestial bodies of the path being studied.

Figure 3.4 presents the flow-chart of the algorithm with stage-by-stage search strategy for the optimal solution to the interplanetary flight problem, taking into account the length of powered segments of the trajectory and the activity spheres of the planets.

Numerical examination of the algorithm [83, 172] showed that the solutions it yields (with simplification) are rather close to the optimal solutions to the general variational problem. Thus, the discrepancy between the values of design ballistic characteristics following from these two solutions never exceeds 0.5%. At the same time, the assumptions made to simplify the computational algorithm extend the range of its application in various problems of interplanetary flights. To apply the algorithm, it is sufficient to construct the target function to be minimized in accordance with the conditions of the problem under consideration.

Let us consider some of the most interesting problems.

3.3.1 Probe for investigating planets

One of the simplest variants of the above path is the scheme of sounding the region near the target planet. Suppose that the spacecraft implements a sounding flight in the activity sphere of the target planet with a planetary flyby. In this case, no restrictions are imposed on the orientation of the part of the spacecraft orbit near the target planet, and the target function to be minimized can be represented as

$$Q(\mathscr{X})=\frac{K_0}{1+\mu_p}+P_{p1}(T_1)+P_{p2}(T_2) \tag{3.46}$$

$$R_{pj}=R_{\mathrm{sp}j}-\sqrt{(X_j-X_{pj})^2+(Y_j-Y_{pj})^2+(Z_j-Z_{pj})^2}; \quad j=1,2$$

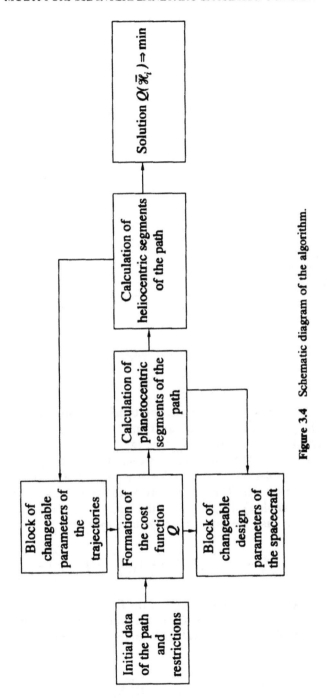

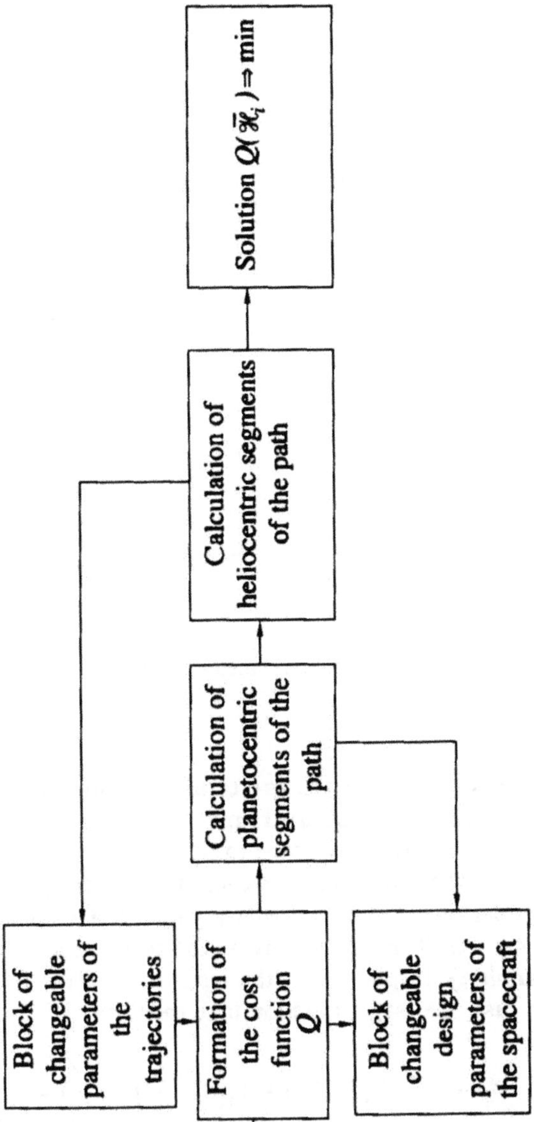

Figure 3.4 Schematic diagram of the algorithm.

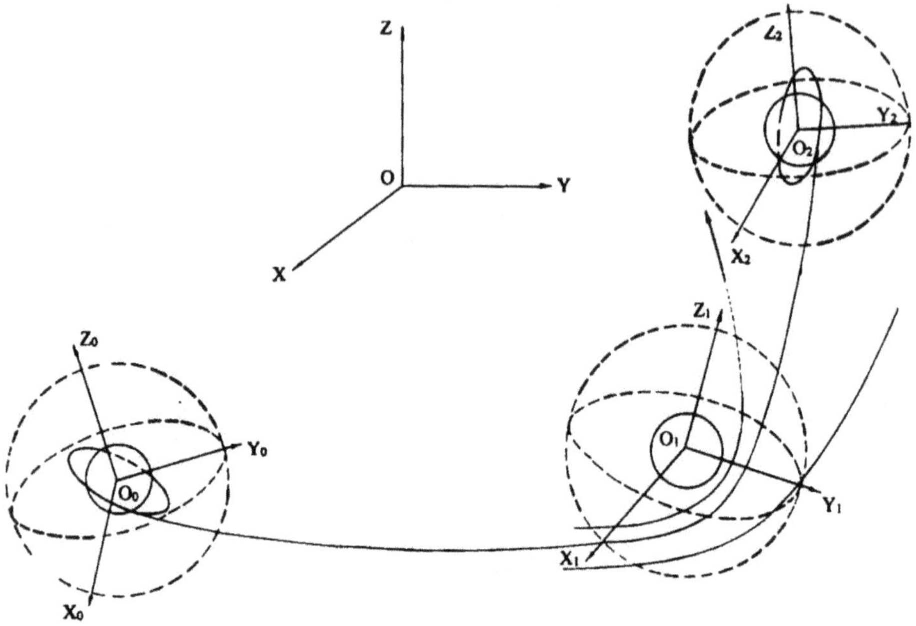

Figure 3.5 Diagram of interplanetary flight with a flyby.

where P_{p1}, P_{p2} are the conditions of the spacecraft entering the activity spheres of the flyby and target planets in the specified time.

The requirements for the spacecraft-probe to fly at a given altitude over specified areas of the planetary surface under investigation, the feasibility of radio communication with the interplanetary probe and the feasibility of data transfer from the spacecraft to Earth can result in restrictions on the orientation of the flyby orbital plane (partial or complete). In particular, the inclination of the flyby orbital plane can be fixed. In this case, the target function take the form

$$Q=\frac{K_0}{1+\mu_p}+\sum_{j=1,2} K_j P_{pj}(T_j)+K_\beta\left(1-\frac{\beta}{\beta_p}\right)+K_i\left(1-\frac{i}{i^*}\right) \tag{3.47}$$

where i^* is the specified inclination angle of the flyby orbit.

It should be noted that i^* must lie within the range of realizable inclination angles $i_{min} \leqslant i^* \leqslant \frac{\pi}{2}$, where the minimum realizable angle of the orbital inclination can be determined as

$$i_{min}=\operatorname{arctg}\frac{v_z(T_2)}{v(T_2)}. \tag{3.48}$$

If the probe enters polar flyby orbit, the target function to be minimized will take the form

$$Q=\frac{K_0}{1+\mu_p}+\sum_{j=1,2}K_jP_{pj}(T_j)+K_\beta\left(1-\frac{\beta}{\beta_p}\right)+K_c|C_3|. \quad (3.49)$$

If the spacecraft flies over low-latitude regions of the planet under study, the target function will be

$$Q=\frac{K_0}{1+\mu_p}+\sum_{j=1,2}K_jP_{pj}(T_j)+K_\beta\left(1-\frac{\beta}{\beta_p}\right)+K_c\left|\frac{1}{C_3}\right|. \quad (3.50)$$

Here, $C_3=X(T_2)V_y(T_2)-Y(T_2)V_x(T_2)$ is the component of the vector of the integral of areas along z-axis in the equatorial planetocentric coordinate system (this component is determined by the planetocentric coordinates of the spacecraft location and velocity at the moment it enters the activity sphere of the target planet). Condition $C_3=0$ corresponds to the spacecraft entering the polar flyby orbit, the condition $C_3=max$ provides the spacecraft flying within the plane of minimum inclination with respect to the equatorial plane of the target planet.

The approach proposed here can be extended to more complex problems of interplanetary flight, e.g. multiple flybys, or segments where a spacecraft module is to be orbited as an artificial satellite of a flyby planet or the target planet (the orbital orientation can be free or fixed). The form of the target function to be minimized will be determined by the peculiarities of the problem in question.

Thus, the solution to the problem of optimization of a multi-purpose flight of a probe with several planetary flybys will require minimization of the following target function

$$Q=K_0M_0(\mathscr{H}_0)+\sum_i K_iP_{pi}(T_i)+\sum_l K_lF_l \quad (3.51)$$

where $M_0(\mathscr{H}_0)$ is the target function being minimized; K_0, K_i, K_l are the coefficients of the penalty functions; $P_{pi}(T_i)$ are the conditions of entering the activity spheres of the flyby planets at the given times; and T_i; F_l are additional boundary conditions.

The scale coefficients for the penalty terms in the intermediate and end points of the trajectories should satisfy the condition $K_1, \ldots \leqslant K_i \leqslant \ldots K_N$ (and increase in the ordinal number of the coefficient corresponds to more distant intermediate conditions). Minimization of the target function (3.51) implies the search for a multi-purpose interplanetary trajectory, the entry and exit points at the activity spheres of the planets, and the spacecraft characteristics so as to provide a minimum cost function for the path with flybys of several planets.

3.3.2 A probe for studying the Sun and the space beyond the ecliptic

Currently the paths of flights towards the Sun has attracted ever-growing attention from both purely scientific and practical view points, within the framework of the problem of radioactive waste removal.

The flight of a spacecraft to the Sun immediately from the orbit of Earth is known to require very high energy expenditures, because the spacecraft velocity required to escape the activity sphere of the Earth must be comparable with the orbital velocity of the Earth. Flying by a heavy planet (like Jupiter) allows this operation to be implemented with far less energy expended. The flight can have various aims: to reach the Sun or to fly at a given altitude over its surface. More complicated aims can be also considered, e.g. the requirement for the spacecraft to fly over high-altitude regions of the solar system.

A condition for the probe to fly directly to the Sun after leaving the activity sphere of a planet is $V_\tau(T^*_{N})=0$, where V_τ is the transversal component of the spacecraft heliocentric velocity after the flyby of the planet. Thus, the target function in the problem of optimizing the design ballistic characteristics of the solar probe takes the form

$$\Phi=\frac{K_0}{1+\mu_p}+P_{p1}(T_1)+K_1V_\tau(T^*_1). \tag{3.52}$$

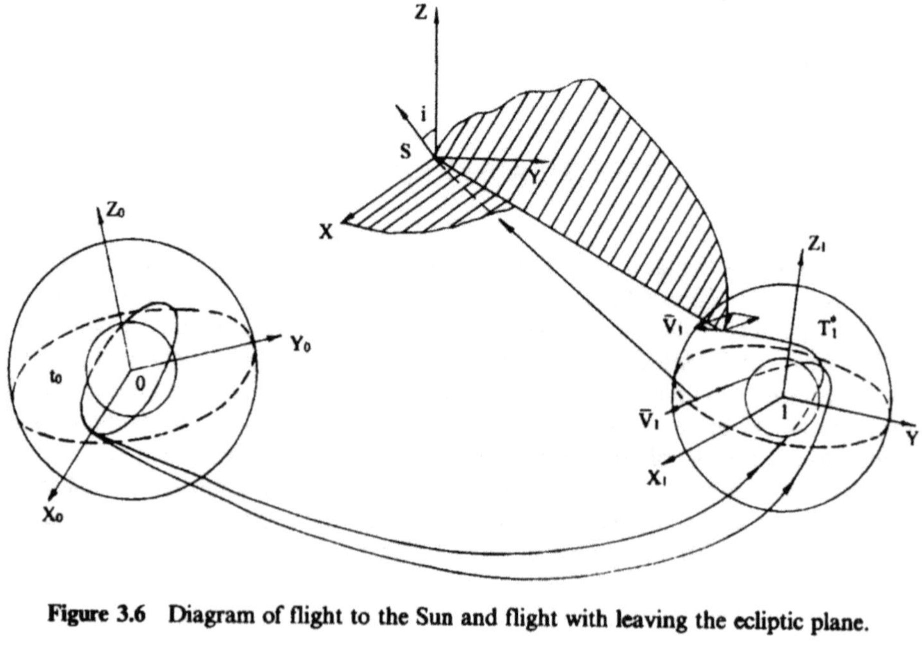

Figure 3.6 Diagram of flight to the Sun and flight with leaving the ecliptic plane.

In the case where the probe is to fly at a given altitude of R_{ps} over the surface of the Sun, the target function is

$$\Phi=\frac{K_0}{1+\mu_p}+P_{p1}(T_1)+K_1\left(1-\frac{R_p}{R_{ps}}\right). \tag{3.53}$$

The design-weight equation of the booster orbiting the probe to the trajectory towards the Sun can be written as [25]

$$\mu_p=[\mu_{K1}(1+a_{fs})-a_{fs}-\gamma_e n_0-\mu_r]\cdot(1-\mu_r)^{-1}$$

where the design coefficients for the blocks of this type can be chosen as: $a_{fs}=(G_{fs}/G_0)=0.075$ is the relative weight of the fuel section; $\gamma_e=G_e/P=0.03$ is the specific weight of the engine; $\mu_r=(G_r/G_0)=0.05$ is the relative weight of other elements.

Minimization of the target functions (3.52) and (3.53) yields the basic design parameters of the spacecraft and its trajectory with a flyby of Jupiter so as to provide maximum μ_p.

Below we present the solution to the problem of spacecraft flight along the Earth–Jupiter–Sun path with a flyby of the Sun at a distance of $R_{ps}=0.7\cdot10^6$ km from the surface of the Sun.

Table 3.1 presents successive values of the spacecraft phase coordinates in basic points of a complete optimal trajectory formed successively in the fields of Earth, the Sun, and Jupiter.

Columns 2–7 show the velocity and location coordinates of the spacecraft: from the given launch time t_0 to the time of entering the activity sphere of Earth T_0^- (in the geocentric orbital coordinate system), within the segment $T_0^+-T_1^-$ (in the heliocentric ecliptic coordinate system; and within the segment $T_1^+-T_2^-$ (in the planetocentric ecliptic coordinate system in the activity sphere of Jupiter). All the coordinate systems are rectangular, are located in the plane of the ecliptic, and are oriented as vernal.

Columns 1 and 8 of the table present the current time in the Julian's years (except for the moment at which the engine turns off t_0^*) and the spacecraft relative weight μ.

T_i^-, T_i^+ are the moments of leaving the activity spheres of the planets and transformation of the spacecraft coordinates, $i=0,1,2$.

The solution obtained shows that the Earth–Jupiter–Sun flight with the spacecraft flying through the near-solar zone can be implemented using a booster with a minimum relative fuel reserve onboard of $\mu_T=1-\mu_{K1}=0.8109$ and optimal initial overload (thrust-to-weight ratio) $n_{01}=$ 0.5342 for $\mu_{p\max}=0.062$.

So, the spacecraft starts from a circular orbit of an artificial satellite of Earth ($H=200$ km), goes through powered acceleration within 698.5 sec, and reaches the activity sphere of Earth within 1.2 days. Continuing its motion in the heliosphere, the spacecraft will reach the activity sphere of Jupiter within 538.8 days and will continue its movement along a flyby hyperbola within 80.2 days. After a gravity

Table 3.1

Current time (JD)	v_x (km/s)	v_y (km/s)	v_z (km/s)	X (km)	Y (km)	Z (km)	μ
$t_0 = 244.13770$	-6.883	-3.689	0	-3087.6	5760.9	0	1
$t_0^x = t_0 + 698.5\,\text{s}$	-6.679	-12.924	0	-7699.0	324.6	0	0.1891
$T_0^- = 244.13782$	-2.403	-10.154	0	-239790	-970824	0	—
$T_0^+ = 244.13782$	-13.124	-38.194	-0.669	-139.391×10^6	50.004	0.0637×10^6	—
$T_1^- = 244.19170$	12.341	-4.420	-0.5252	4691×10^6	-652.597×10^6	-7.245×10^6	—
$T_1^+ = 244.19170$	2.086	-12.955	0.2681	8.0102×10^6	47.518×10^6	1.027×10^6	—
$T_2^- = 244.19972$	10.756	7.959	0.9811	-25.223×10^6	39750×10^6	-11.205×10^6	—
$T_2^+ = 244.19972$	-0.534	-0.423	0.01214	495.311×10^6	-606.012×10^6	-48.561×10^6	0.1891

assist maneuver at Jupiter, the spacecraft will leave its activity sphere and pass to a trajectory towards the Sun with a pericentral altitude $R_{p*}=0.7 \cdot 10^6$ km.

The problem of sounding the space beyond the ecliptic is also of considerable interest.

It is known that when launched from the Earth surface, the spacecraft will require excessive energy to reach the space regions beyond the ecliptic (the reasons are the same as for the spacecraft flight to the Sun).

However, using a planetary flyby with a gravity assist maneuver allows the spacecraft flight plane to be rotated through the given angle with respect to the ecliptic plane with far lower energy expenditure [171–173]. The energy requirements in this case are determined by the conditions of the spacecraft reaching the activity sphere of the flyby planet, and the spacecraft trajectory beyond the ecliptic is determined by the choice of the point at which the spacecraft enters the activity sphere of the flyby planet and by a proper gravity assist maneuver of the spacecraft.

In the case where after the flyby the spacecraft enters the plane with a given inclination i^* with respect to the ecliptic plane, the target function to be minimized can be written as

$$\Phi=\frac{K_0}{1+\mu_p}+P_{p1}(T_1)+K_i\left|1-\frac{i}{i^*}\right| \tag{3.54}$$

where

$$i=\mathrm{arctg}(C_1^2+C_2^2)^{1/2}/C_3 \quad 0\leqslant i\leqslant\pi$$

$$C_1=Y_3V_{z3}-Z_3V_{y3},\quad C_2=Z_3V_{x3}-X_3V_{z3},\quad C_3=X_3V_{y3}-Y_3V_{x3}.$$

If, after the gravity assist maneuver, the spacecraft should fly within a plane perpendicular to the ecliptic plane, which will allow it to fly over the near-solar regions of space beyond the ecliptic, the target function will be

$$\Phi=\frac{K_0}{1+\mu_p}+P_{p1}(T_1)+K_1|C_3|. \tag{3.55}$$

Also of interest are the trajectories allowing the sound to fly at maximum distance from the ecliptic plane. This can be accomplished by maximizing the component of the heliocentric velocity $V(T_1^*)$, normal to the ecliptic plane, which the spacecraft will have after the planetary flyby. These trajectories can be investigated using the proposed algorithm through minimization of the target function of the form

$$\Phi=\frac{K_0}{1+\mu_p}+P_{p1}(T_1)+K_1\left|\frac{1}{V(T_1^*)}\right|. \tag{3.56}$$

Proper choice of the coefficients K_0 and K_1 allows one to find a compromise solution, namely, the trajectory and design ballistic characteristics of the probe providing the greatest distance of the spacecraft from the ecliptic plane [172].

Table 3.2

Solution	$v_{\infty 0}$ (km/s)	$v_{\infty 1}$ (km/s)	$\beta \cdot 10^6$ (km)	μ_{k1}	n_{01}	μ_p
Separate	9147	8148	1.910	0.2301	0.5267	0.02271
Combined	8966	6613	2.892	0.2353	0.5251	0.02923

Table 3.2 presents the design ballistic solutions of one such problem concerning the spacecraft leaving the ecliptic plane. After starting from the Earth and flying by Jupiter ($T_0 = 244.13850, \Delta t = T_1 - T_0 = 669$ days), the spacecraft is to enter a research heliocentric orbit inclined at $i = 20°$ with respect to the ecliptic plane, under the condition of maximum relative payload. For reference, the table also presents a solution based on the conventional model where the external and internal flight problems are considered separately.

Experiments involving the application of the suggested algorithm to interplanetary flight problems show that the search for the optimal solution requires not more than 20–25 iterations. The approach allows us to assess the basic design parameters of the spacecraft as early as the conceptual design stage, taking into account the combined effect of the factors forming the multi-purpose mission.

4. ANALYSIS OF MULTI-PURPOSE SCHEMES OF INTERPLANETARY FLIGHT

This chapter presents the results of studies on various schemes of interplanetary flights with gravity assist maneuvers, based on the methods and algorithms developed in Chapters 2 and 3.

Before going into an analysis of multi-purpose spacecraft paths to the planets of the solar system, we note that a number of such paths have already been analyzed by numerical methods, and some (flight to Mercury via Venus, Grand Tour Programme, and others) have been successfully accomplished. At the same time, the search for new variants which are optimal in terms of energy expenditure will open new possibilities for future programs of solar system study including low cost interplanetary missions.

The first part of the chapter deals with the application of the graphic-analytical method developed in Chapter 2 to the analysis of various multi-purpose methods of reaching the planets of the solar system and the sun. This allows us to correlate different paths within the framework of a single model and to find the basic characteristics of the schemes with no reference to specific dates. This kind of comprehensive analysis gives an insight into the general picture of the multi-purpose interplanetary paths and allows their correlation in terms of their energy characteristics.

The second part of the section is concerned with numerical investigations of some promising multi-purpose schemes of interplanetary spacecrafts and their design ballistic characteristics. The studies were based on the numerical methods and algorithms developed in Chapter 3 using initial estimates from the graphic-analytical analysis of the paths.

4.1 Nomographic analysis of spacecraft multi-purpose paths in the solar system

The technique developed in Chapter 2 allows us to avoid intricate and time-consuming computational procedures in analyzing a wide range of interplanetary spacecraft paths. To accomplish the analysis it is sufficient to use the nomogram of combined isolines of equal relative velocity for the orbits of the solar system planets, to select the path to be analyzed, and to assess its basic orbital and energy characteristics.

Again, when analyzing a mission option, the researcher can obtain from the nomogram some additional data on the path, namely, if a planetary flyby is made before or behind the planet, what are the conditions whereby the flyby and target planets can be observed from the spacecraft and Earth?

This kind of analysis provides the designer with a general notion of the interplanetary path selected. Once these nomographic data have been obtained, the researcher can use them in numerical studies of the path based on more rigorous computational models, thus restricting the range of search for the optimal solution among the flight paths with gravity assist maneuvers.

The present chapter presents the analysis of possible paths of the spacecraft flight in the solar system with flyby of a single planet or several (usually two) planets. In each case, the results are compared with the nomographic estimates of a direct flight, which for any target planet, allows one to find out whether the gravity assist maneuvers can be included in the path.

Figure 4.1 shows the generalized nomogram of flights to each of the nine planets of the solar system and to the asteroid belt presented in the coordinates of orbital parameters: R_π—pericentral altitude and T—the flight orbital period (the scale of the nomogram axes is logarithmic). The nomogram represents a map of interplanetary flights, and its domains form a set of all periodic orbits along which spacecraft flights in the solar system can be accomplished.

Isoperiodic curves of post-perturbation changes in the spacecraft orbital parameters are also presented for the case where the spacecraft energy either increases or decreases (in accordance with whether the acceleration or deceleration gravity assist maneuver is performed at the planet Figures 4.2–4.9).

Before proceeding to the flight analysis, we will consider the assumptions and peculiarities of the model used in the nomographic analysis.

The orbits of all the solar system planets are assumed coplanar, circular, according to their average radii. The spacecraft motion is represented by segments of unperturbed Keplerian movement in the central field of the sun.

Gravity assist maneuver of the spacecraft during the planetary flyby is considered within the framework of the model of point-impulse approximation (see Chapter 1), i.e. the spacecraft velocity vector is assumed to instantaneously turn in the gravitational center of the flyby planet with the appropriate changes in the spacecraft orbit heliocentric parameters.

We consider only gravitational acceleration or deceleration maneuvers (the engine is not turned on during the planetary flyby). The paths including gravity assist maneuvers are usually assessed in terms of the spacecraft maximum approaching the flyby planet ($m = 1$).

The paths of flights to planets and the sun are estimated in terms of hyperbolic excess of velocity at the start, during flyby, and at the target planets. The lowest-energy solutions for the flights are presented in tables (these paths, as a rule, include segments tangent to the orbits of start and target planets). At the same time, the estimate neglects the planetary phases and therefore presents a lower bound for flight energy, taking no account of the dates of start, flyby, and arrival of the spacecraft at the planets. The corresponding dates can be determined later based on more rigorous computational models of a comprehensive interplanetary flight.

All the results of the analysis are summarized in tables presenting the solutions for variants of flights to all the solar system planets and the sun in accordance with the variants of forming the multi-purpose path (flight variants are as a rule designated

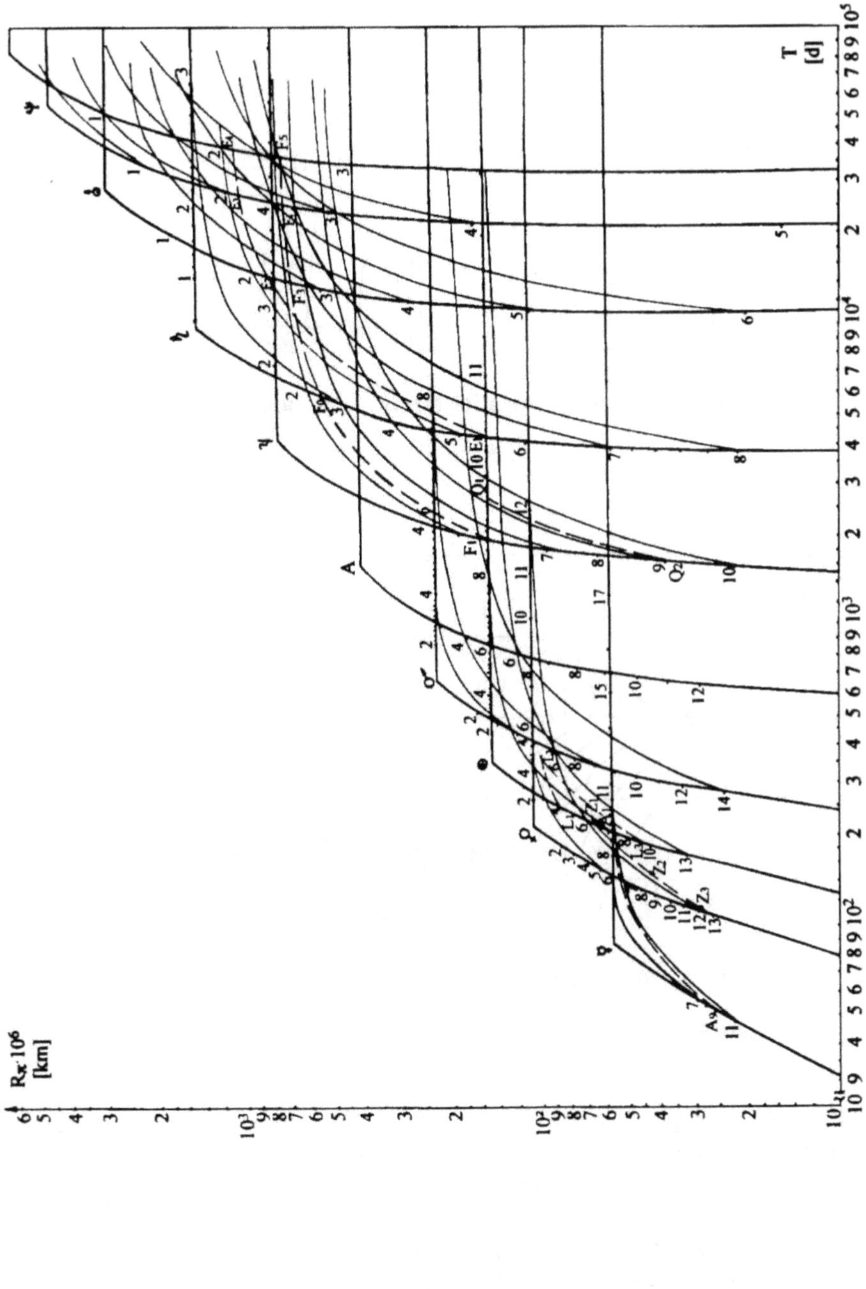

Figure 4.1 Nomogram for the examination of interplanetary spacecraft flights.

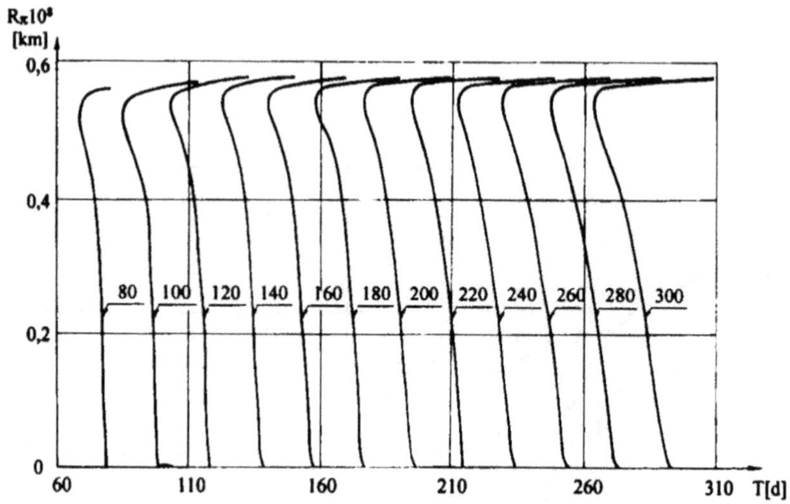

Figure 4.2 Isolines $T^* = const$ for Mercury (increasing energy).

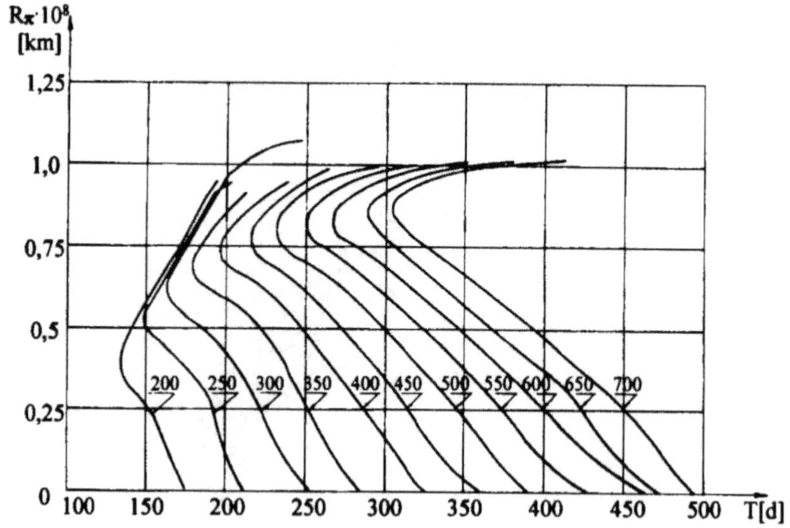

Figure 4.3 Isolines $T^* = const$ for Venus (increasing energy).

by an abbreviation composed of the initial letters of the planets involved in the path). Each variant is supplied with the type of transfer determining the position of the corresponding segments of the path with respect to the orbits of the start, flyby, and target planets.

Each segment of the flight is represented in the diagram by the corresponding point; therefore, the entire transfer is determined by a combination of points found

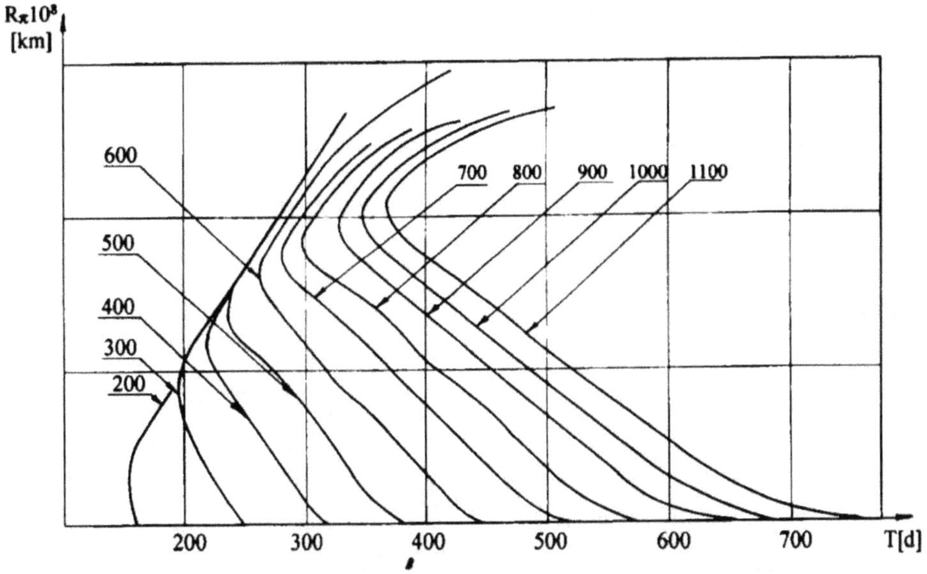

Figure 4.4 Isolines $T^* = const$ for Earth (increasing energy).

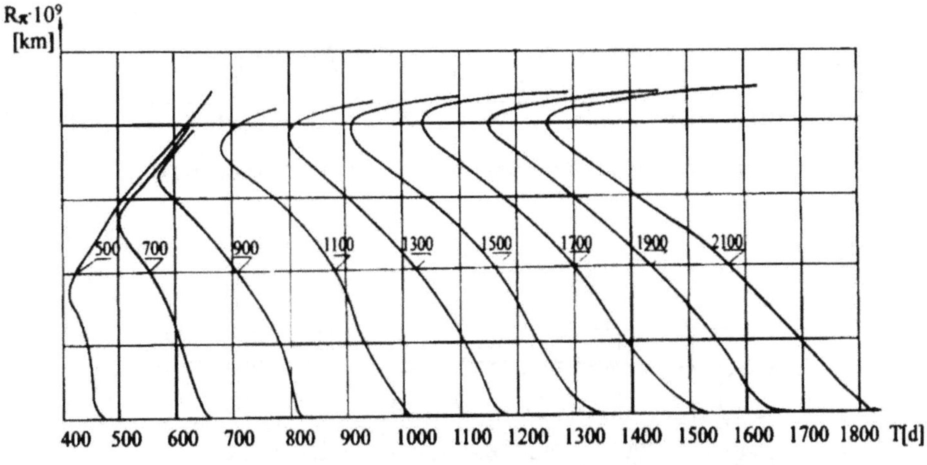

Figure 4.5 Isolines $T^* = const$ for Mars (increasing energy).

in different parts of the nomograms. The type of transfer is specified by a set of subscripts determining the succession of the segments of the interplanetary trajectory being formed, where

- α denotes a segment of the interplanetary trajectory with an apocentral tangency to the planetary orbit;
- π denotes a segment of the interplanetary trajectory with a pericentral tangency to the planetary orbit;

- N denotes a segment of the interplanetary trajectory with no tangency points with the planetary orbits;
- Π^+ denotes a gravity assist flyby maneuver with an increase in the spacecraft orbital energy;
- Π^- denotes a gravity assist flyby maneuver with a decrease in the spacecraft orbital energy;
- N_π denotes the pericentral point of the trajectory segment.

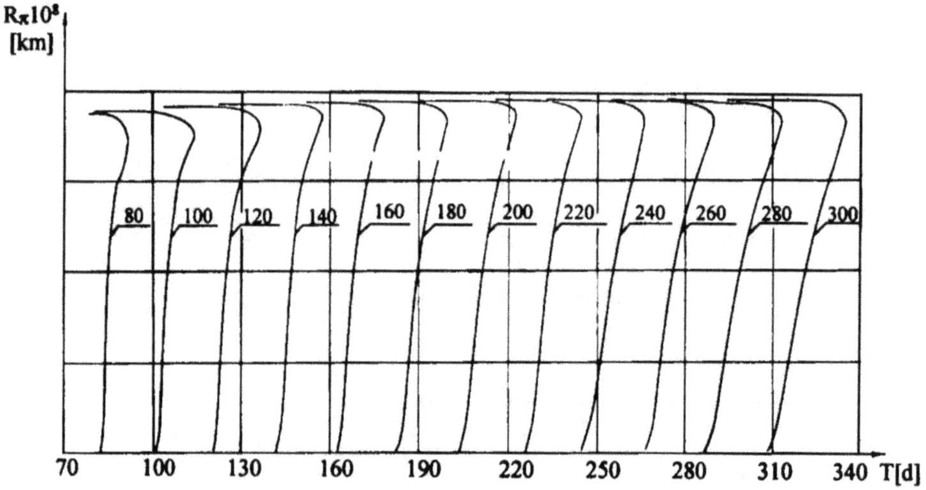

Figure 4.6 Isolines $T^* = const$ for Mercury (decreasing energy).

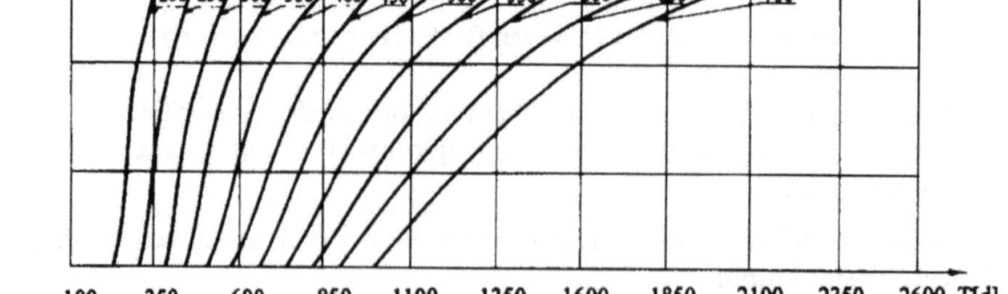

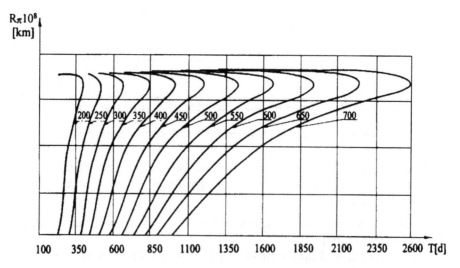

Figure 4.7 Isolines $T^* = const$ for Venus (decreasing energy).

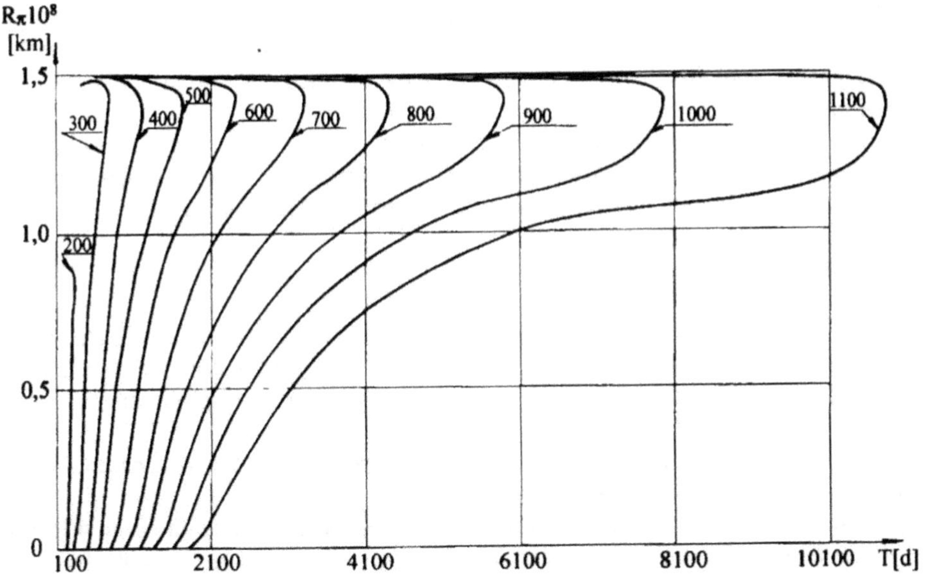

Figure 4.8 Isolines $T^* = const$ for Earth (decreasing energy).

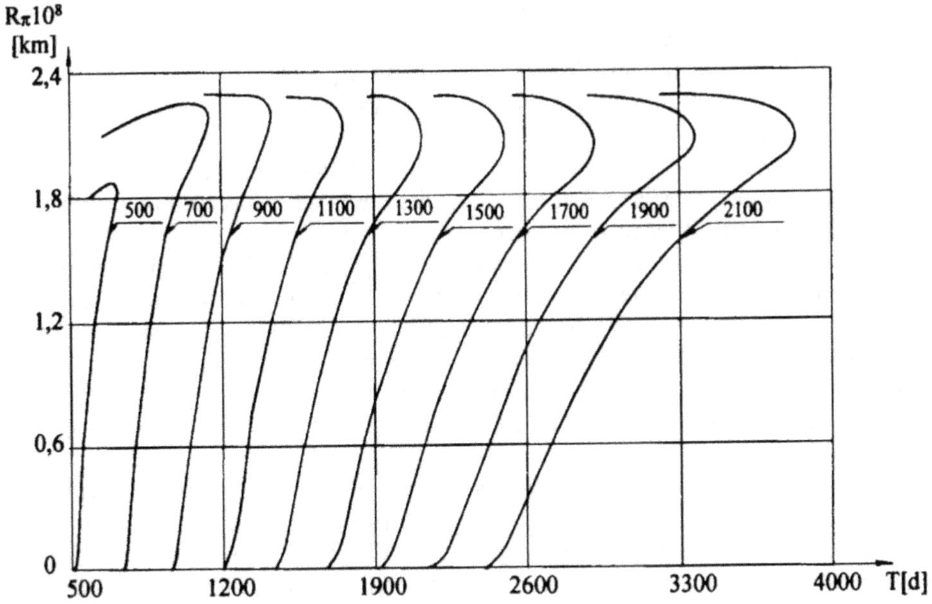

Figure 4.9 Isolines $T^* = const$ for Mars (decreasing energy).

The paths denoted by the subscript N are as a rule distinguished by high energy expenditures, because no cotangent interplanetary transfer is possible in this case.

The paths of flights to the sun are supplied in the table by the value of the perihelion altitude at the final segment of the path.

In the cases where it is found to be expedient, solutions for paths with multiple flybys of a planet are included in the tables. Their formation in accordance with the condition of a return to the flyby planet is illustrated in the nomogram.

4.1.1 Paths to Mars and Venus

Multiple-flyby missions to Venus and Mars with flybys of planets of the Earth group and their correlation with direct flights have repeatedly attracted the attention of researchers [36–46]. The suggested method allows an initial (preliminary) analysis of such schemes under similar model assumptions.

Consider in more detail the procedure of nomographic analysis based on an example of the Earth–Venus–Mars path.

As can be seen from the nomogram (Figure 4.10, fragment of nomogram 4.1), the interorbital transfer along this path can be accomplished with a single flyby of Venus. The minimum energy expenditure for this path can be easily estimated using the presented nomogram (transfer $B_1 \rightarrow M$).

The minimum relative velocities in the orbits of the start and target planets amount to $v_{\infty 1} = 3.3$ km/s and $v_{\infty 2} = 4.8$ km/s, respectively. In this case, the flyby of Venus must follow a flyby hyperbola with $v_{\infty} = 5.72$ km/s with a relative flyby altitude of $m = 1$. The flyby trajectory consists of two successive segments—pre-flyby and post-flyby with characteristics $R_{\pi 1} = 96 \cdot 10^6$ km, $T_1 = 270$ days, $R_{\pi 2} = 108 \cdot 10^6$ km, $T_2 = 430$ days, where R_π and T are the pericentral altitude and the period of rotation of the flight segment.

The suggested nomogram makes it possible not only to obtain the minimum possible energy characteristics but also to assess the range of acceptable realizations of the selected path in terms of both energy and time.

Suppose that a range of starting velocities $v_{\infty E}^{min} \leqslant v_{\infty E} \leqslant v_{\infty E}^{max}$ acceptable for the implementation of the given mission in terms of energy expenditure is specified (e.g., $v_{\infty E}^{max} = 5$ km/s), and there are no restrictions on the arrival velocity at Mars (a flyby-sounding variant of the scheme is considered).

Now we can find the domain of acceptable orbits of the initial segment of the path (Earth–Venus) as the domain D limited by the lines α_3, $v_{\infty E}^{max}$, $v_{\infty E}^{min}$ (see the nomogram in Figure 4.11).

The gravity assist maneuver at Venus (with a relative flyby altitude $m = 1$) will result in the set of initial orbits being mapped into a set of post-perturbation orbits leading to the solution of the problem g: $D \rightarrow D'$. Considering that the mission under consideration can be accomplished at $m > 1$ as well, the set D' can be extended by including an additional domain D'' (the set of orbits providing the accomplishment of the path at $m > 1$) $D_1 \subset D' \cup D''$.

Thus, the mission under consideration can be accomplished if the mapping g_m: $D \rightarrow D_1$ is valid in the plane of the nomogram orbital parameters.

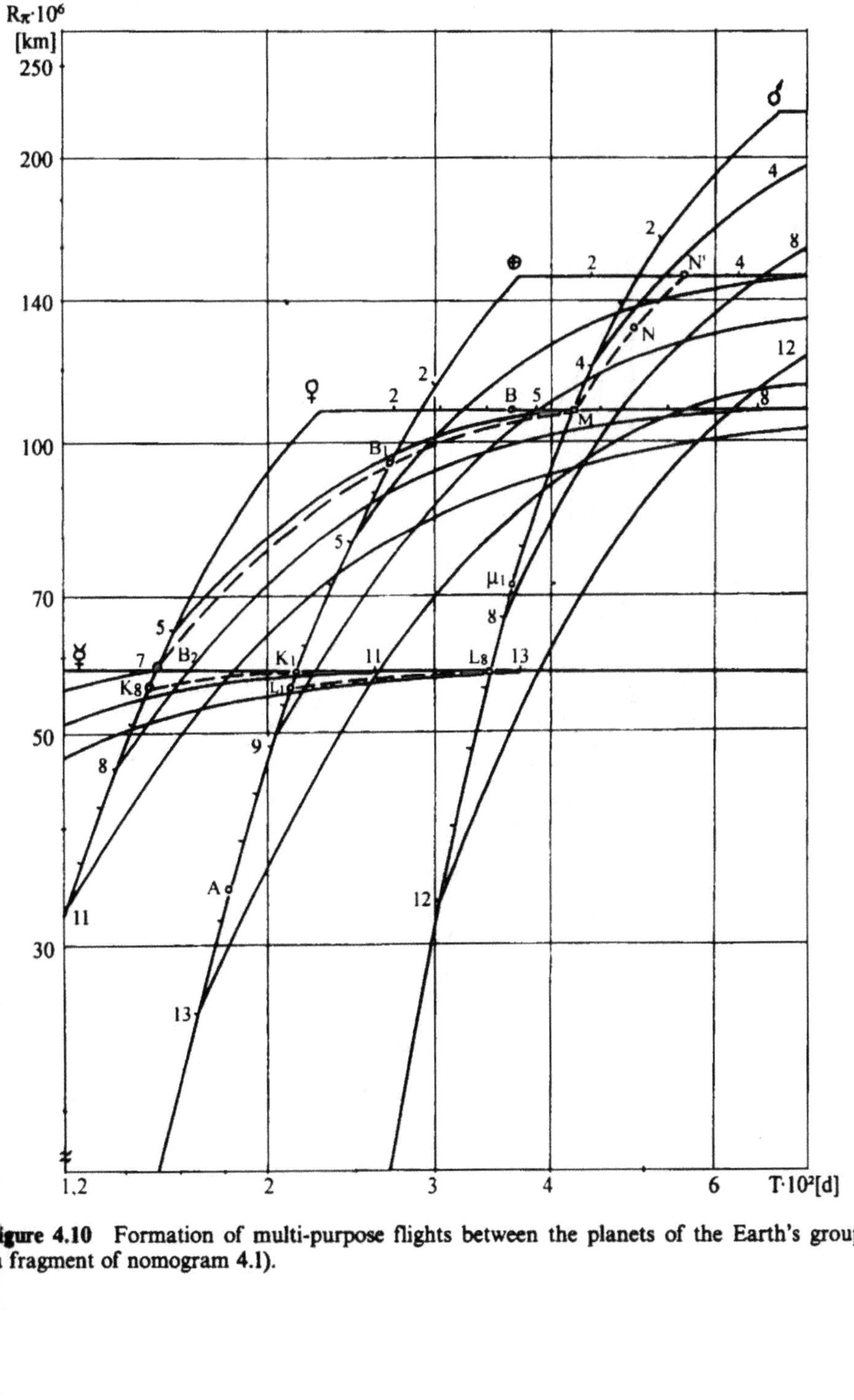

Figure 4.10 Formation of multi-purpose flights between the planets of the Earth's group (a fragment of nomogram 4.1).

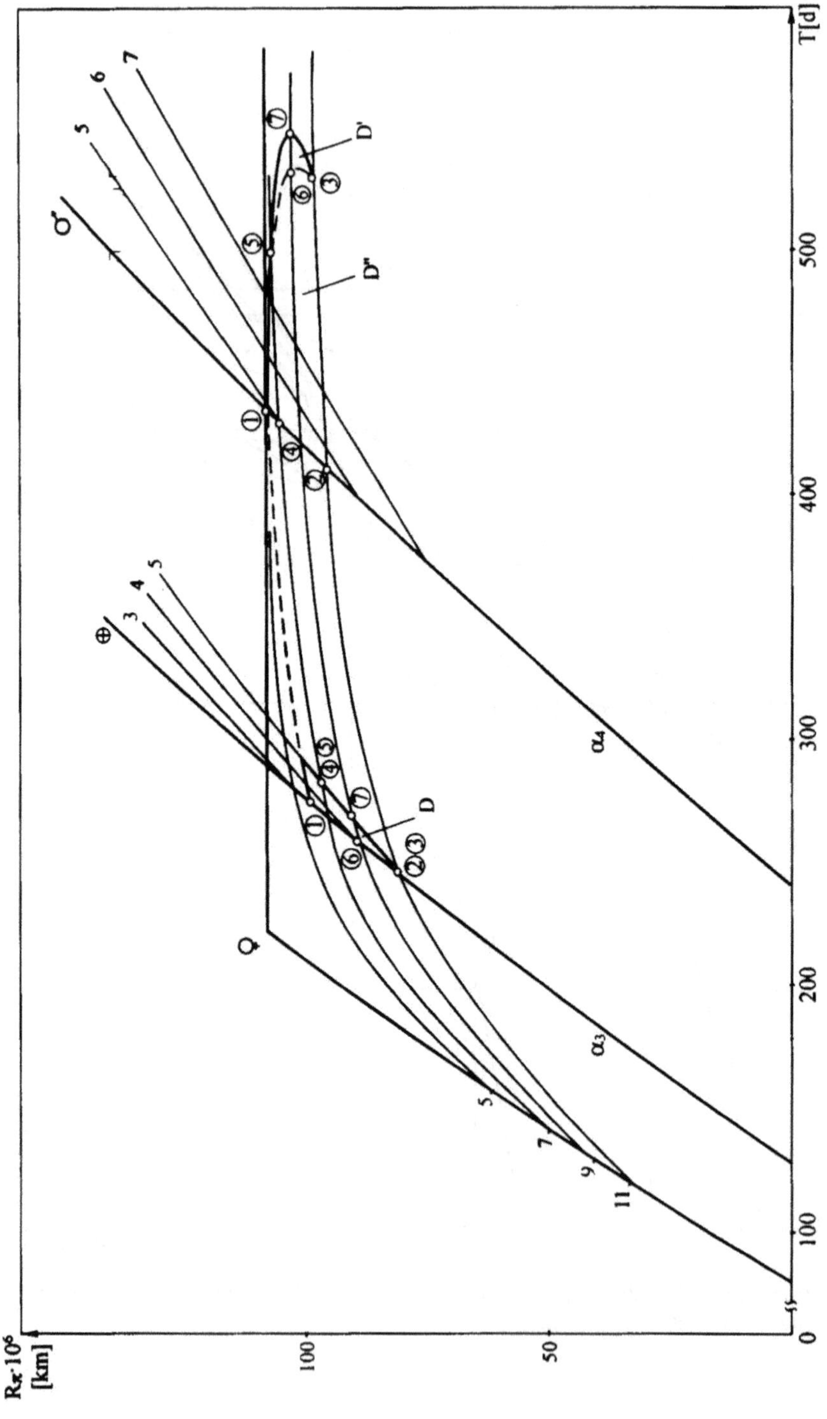

Figure 4.11 Domains of orbital and energy characteristics for Earth–Venus–Mars flights.

Various types of transfer are possible in this case (they are determined by the position of the images of the initial and end points M and N in the domains D and D_1):

(a) cotangent transfer $M \in \alpha_3(D)$, $N \in \alpha_4(D_1)$;
(b) partially cotangent transfer $M \notin \alpha_3(D)$, $N \in \alpha_4(D_1)$ or $M \in \alpha_3(D)$, $N \notin \alpha_4(D_1)$;
(c) non-cotangent transfer $M \notin \alpha_3(D)$, $N \notin \alpha_4(D_1)$.

Consider the range of variations in the flight scheme characteristics for the trajectory being studied. To do that, let us choose the extreme points in the domains D and D_1, corresponding to the pre- and post-perturbation flight segments. Based on the procedure developed for determining the transfer scheme parameters, we can obtain the characteristics of the "extremal" variants of the mission under consideration (see Table 4.1).

The type of transfer is determined by the combination of the initial and end segments of the trajectory and is presented in the table, where α is a cotangent flight segment, M_k or N_k is a non-cotangent flight segment with respect to the corresponding orbits of the launch and target planets (all the solutions are considered for the first half-revolution).

As can be seen from the table, the extreme left and top position of the mapping points in the domains D and D_1 yield a variant that is best in terms of energy expenditure (variant 1) though it has the maximum flight time. Other possible positions of the mapping points in the domains result in faster flight schemes, but their energy indices are less satisfactory.

Note also that it is not always expedient to maximize the use of the flyby body's perturbation capability. For example, if the image point $N_k \notin \alpha_4(D_1)$, the arrival velocity at the target planet will increase (compare the transfer variants 2 and 3).

Thus, analysis of the correspondence between points within domains D and D', D'' in the nomogram 4.11 allows an estimate of some kinematic characteristics of pre- and post-perturbation segments of the Earth–Venus–Mars trajectory.

This kind of nomographic analysis of a trajectory can provide a comprehensive notion about the orbital and energy characteristics at the initial stage of studies.

Tables 4.2 and 4.3 present the results of analysis of several variants of the multi-purpose schemes of flights to Venus and Mars.

Table 4.1

No.	Type of transfer	$v_{\infty E}$, km/s	$v_{\infty V}$, km/s	$v_{\infty M}$, km/s	m
1	$\alpha-\alpha$	3.3	5.72	4.8	1
2	$\alpha-\alpha$	5.0	11.0	5.58	>1
3	$\alpha-N_k$	5.0	11.0	10.3	1
4	$M_k-\alpha$	5.0	7.0	5.0	>1
5	M_k-N_k	5.0	7.0	8.9	1
6	$\alpha-N_k$	4.0	9.0	10.5	1
7	M_k-N_k	5.0	9.0	11.4	1

Table 4.2 Flights to Venus.

1st flyby planet	2nd flyby planet	$v_{\infty E}$, km/s	$v_{\infty 1}$, km/s	$v_{\infty 2}$, km/s	$v_{\infty V}$, km/s	Transfer type
—	—	2.50	—	—	2.70	$\alpha-\pi$
Mars	—	7.50	5.00	—	5.72	$N-\Pi^{-}-\pi$
	Mars	3.50	5.00	5.0	5.72	$\pi-\Pi^{-}-\Pi^{-}-\pi$
	Mercury	14.50	8.85	12.55	21.00	$N-\Pi^{-}-\Pi^{-}-\pi$
	Earth (2 y)	3.45	4.5	3.8	3.05	$\pi-\Pi^{-}-\Pi^{-}-\pi$
	Earth (3 y)	5.1	9.1	5.5	3.70	—
Mercury	—	7.65	9.55	—	15.00	$\alpha-\Pi^{-}-N$
	Mars	17.00	12.55	8.85	14.00	$N-\Pi^{-}-\Pi^{+}-N$
	Earth	7.85	9.95	9.0	4.95	$\alpha-\Pi^{+}-\Pi^{+}-\pi$
Jupiter	—	9.05	6.60	—	11.40	$\pi-\Pi^{-}-\pi$
	Mars	9.00	6.30	17.50	10.80	$N-\Pi^{-}-\Pi^{-}-\pi$
	Earth	8.95	5.91	12.50	6.70	$\pi-\Pi^{-}-\Pi^{-}-\pi$

Table 4.3 Flights to Mars.

1st flyby planet	2nd flyby planet	$v_{\infty E}$, km/s	$v_{\infty 1}$, km/s	$v_{\infty 2}$, km/s	$v_{\infty M}$, km/s	Transfer type
—	—	2.90	—	—	2.62	$\pi-\alpha$
Venus	—	3.30	5.72	—	4.80	$\alpha-\Pi^{+}-\alpha$
	Earth	3.20	5.80	8.60	4.00	$\alpha-\Pi^{+}-\Pi^{+}-\alpha$
	Mercury	13.50	21.00	12.60	8.85	$N-\Pi^{+}-\Pi^{+}-\alpha$
Mercury	—	17.50	12.60	—	8.85	$N-\Pi^{+}-\alpha$
	Venus	7.68	10.1	14.5	6.20	$\alpha-\Pi^{+}-\Pi^{+}-\alpha$
	Earth	7.75	10.37	10.5	4.51	$\alpha-\Pi^{+}-\Pi^{+}-\alpha$
Jupiter	—	8.85	5.61	—	6.30	$\pi-\Pi^{+}-\alpha$
	Venus	9.05	6.60	11.5	16.2	$\pi-\Pi^{-}-\Pi^{-}-N$
	Earth	8.85	5.62	8.85	12.0	$\pi-\Pi^{-}-\Pi^{-}-N$

Contrary to the Earth–Venus–Mars scheme considered above, the scheme Earth–Mars–Venus implies significantly higher energy expenditures. Indeed, as can be seen from the nomogram, a starting trajectory tangent to the Earth's orbit is impossible in this case. One has to look for the initial segment of the trajectory which will make it possible to deliver the spacecraft to Venus' orbit after the gravity assist maneuver at Mars. This trajectory is represented on the nomogram in Figure 4.10 by the transfer $N \rightarrow M$, whose heliocentric trajectory parameters are ($R_{\pi 1}=135\cdot 10^{6}$ km, $T_1=495$ days)$\rightarrow$($R_{\pi 2}=108\cdot 10^{6}$ km, $T_2=430$ days). Note that the position of the initial point of the trajectory $N \notin \pi_{\oplus}$ (Figure 4.10) implies a rather high energy expenditure for this flight: $v_{\infty E}=7.50$ km/s, $v_{\infty M}=5.00$ km/s, $v_{\infty V}=5.72$ km/s.

It should be noted that in the case where a single flyby of a planet is not efficient, the energy expenditure can be reduced by multiple flybys (the mission time will evidently increase). Thus, the nomogram shows that even two flybys of Mars make the Earth–Mars–Venus flight acceptable (the transfer $N' \rightarrow N \rightarrow M$). In this case, the location of the starting point of the trajectory E–M–M–V ($N' \in \pi_{\oplus}$ in the nomogram 4.10) corresponds to the condition of tangent start of the spacecraft from the Earth's orbit, and the start velocity can be reduced to $v_{\infty E} = 3.5$ km/s.

It is of interest to consider the nomographical analysis of the Earth–Mercury–Venus mission (transfer $K_1 \rightarrow K_8$). Two variants of the trajectory are possible for the flyby of Mercury, namely, before and behind the flyby planet. As the nomogram evidently shows, it is more expedient to fly before Mercury (deceleration gravity assist maneuver). In this case, the point K_1, representing the orbital parameters in the nomogram 4.10, shifts to the left, which results in a lower arrival velocity of the spacecraft at Venus. Multiple flybys of Mercury (transfer $K_1 \rightarrow K_8$ $n=8$) allow the arrival velocity at Venus to be reduced even more (down to $v_{\infty V} = 6.5$ km/s). The table also presents other examples of the application of multiple gravity assist maneuvers with respect to the flyby planet (including combinations of two flyby planets).

Different options of multi-purpose spacecraft trajectories to reach Mars are shown in Table 4.3. Just as with the flights to Venus, it can be seen that the gravity assists yield no energy savings. At the same time, these results may be of interest for the analysis of various combined exploration schemes using flybys of the Earth group planets.

The trajectory Earth–Venus–Mars (transfer $B_1 \rightarrow M$ in Figure 4.10) was studied in detail earlier. Its correlation with the trajectory Earth–Mars–Venus (transfer $N' \rightarrow N \rightarrow M$) demonstrates the potential of the nomographic analysis, allowing us to reveal the difference between the trajectories in terms of energy expenditure at a very early stage of the study.

Other flights also demonstrate this, for example, those with the assistance of Mercury—a planet with a weak gravitational field. The mission Earth–Mercury–Mars with a single flyby of Mercury requires high energy expenditure, and the effect of gravity assist is small (see Table 4.10). Multiple flybys of Mercury ($n=8$; according to nomographic estimate, this is the number of flybys of Mercury required for the orbit of Mars to be encountered in the apocenter of the spacecraft orbit) allow the energy characteristics of the trajectory to be improved. As can be seen from the nomogram, this kind of mission arrangement (it is represented by the transfer $L_1 \rightarrow L_8$ in Figure 4.10) allows the start velocity to be reduced to $v_{\infty E} = 7.85$ km/s. The drawback, however, is an increase in the mission time.

Of interest among the above results are the trajectories that use the Earth in a second flyby (Earth–Venus–Earth–Mars and Earth–Mars–Earth–Venus). The energy expenditures of these flights are near those for the direct flights from Earth to Venus and from Earth to Mars (see Table 4.2 and 4.3).

Let us assess these trajectories based on the nomogram (Figure 4.12), on which the lines of transformation of orbits E–V–E and E–M–E are depicted (they are constructed considering the orbital phasing of the planets).

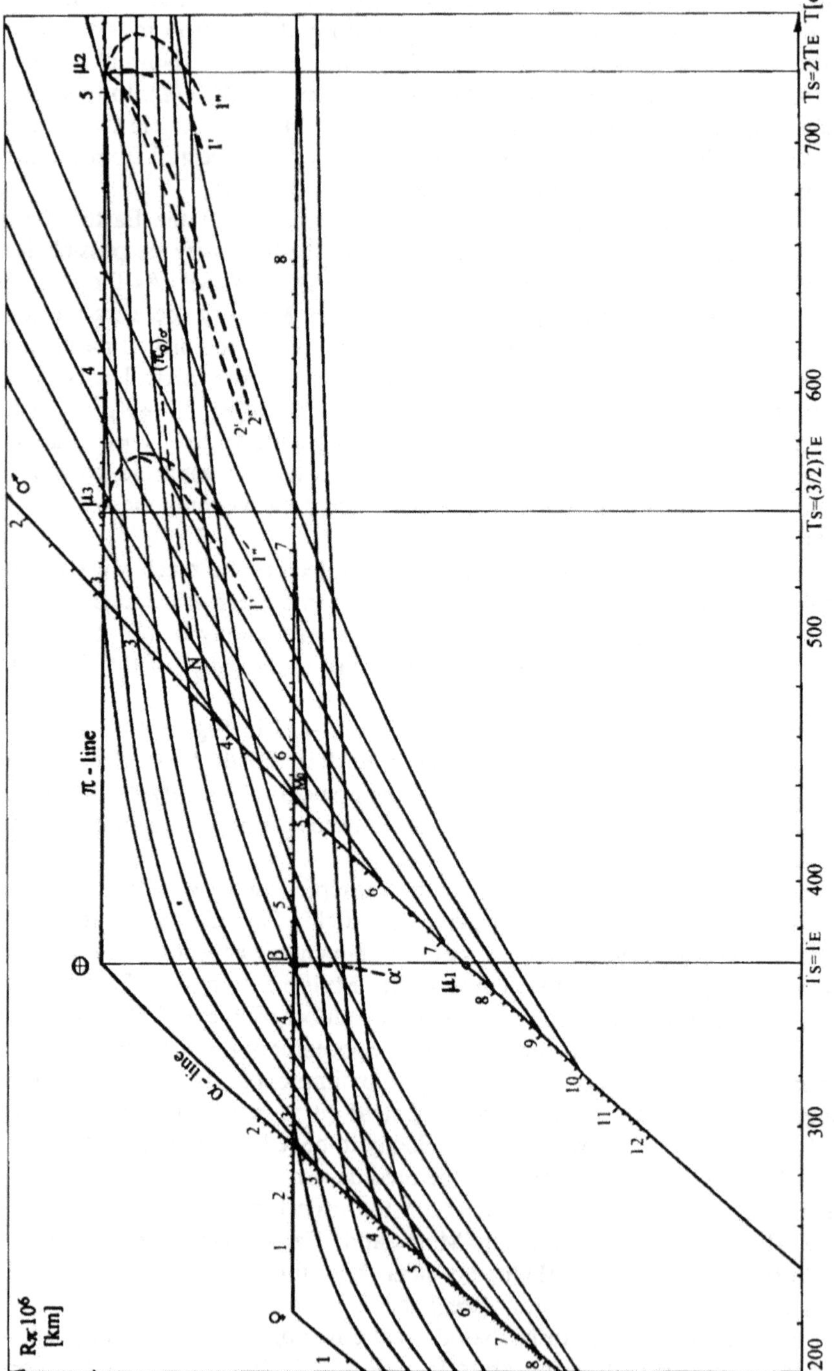

Figure 4.12 Formation of recurrent periodical orbits of the spacecraft.

The nomogram shows that the Earth–Venus–Earth–Mars trajectory, which is economical in terms of energy expenditure, can be formed based on the following prerequisites:

- the initial leg of the trajectory, Earth to Venus, should be a segment of an orbit with apocentral tangency with respect to the Earth's orbit (from the condition of minimum spacecraft start velocity);
- the gravity assist (acceleration) maneuver should result in the spacecraft returning to Earth (in approximately one year) with a higher velocity $v_{\infty E} > v_{\infty st}$, which allows a more efficient utilization of the Earth gravity assist (line β–α^1 in Figure 4.12);
- after a gravity assist maneuver at the Earth (it should also be an acceleration maneuver), the spacecraft can be set to Mars along a "quick" flight trajectory.

It is worth mentioning that when implementing this mission, it is important that the spacecraft approach velocity at Venus and Earth be as low as possible, for in this case, both the start velocity and the approach velocity at Mars tend to their minimum values (as can be seen from the nomogram).

The Earth–Mars–Earth–Venus flight requires greater energy expenditures. This trajectory can be arranged to include either two- or three-year return trajectories back to the Earth. The second option is better in terms of energy expenditure, because the second encounter with the Earth after the spacecraft has been launched to an orbit with the period of 1.5 years can take place only after two revolutions, i.e. within three years after launch.

As in the previous case, Mars plays the role of a perturbation-correction body for the spacecraft's subsequent encounter with the Earth to be performed with the most favorable approaching velocities (lines $\mu_3 - 1$ in Figure 4.12).

4.1.2 Recurrent and periodically-recurrent trajectories

Of great interest are the interplanetary flight trajectories with gravity assist maneuvers that allow a spacecraft to return to the launch planet [36, 37, 41, 43–50, 59].

With the model accepted, the nomogram allows one to explore some important features of such trajectories and to assess the range of energy expenditure for their accomplishment. A characteristic feature of the nomographic study of such trajectories is the formation of the pre- and post-perturbation orbits to allow the spacecraft to return to the launch planet. Solutions with one or two revolutions are preferable in this case. The formation of a recurrent trajectory to provide the encounter at subsequent revolutions reduces the practical value of such a trajectory, owing to the accordingly increased flight time.

Let us consider options for the accomplishment of the Earth–Mars–Earth trajectory. As can be seen from the nomogram 4.12, one-year trajectories to Mars are possible (point μ_1); however, the corresponding energy expenses are so high ($v_{\infty E}^{start} = 16$ km/s; $v_{\infty M} = 7.5$ km/s; $m \gg 1$; $v_{\infty E}^{ret} = 16$ km/s), that currently these trajectories cannot be accepted. As can be seen from the nomogram, a decrease in m—the flyby altitude at Mars—allows the return velocity at the Earth to be somewhat reduced (down to 12.5–13 km/s) at the expense of a small increase in the spacecraft start velocity. It appears that π-trajectories (with the start leg pericentrally tangent to the Earth's orbit) are more promising in terms of energy expenditure.

The nomogram shows that these trajectories can be two- or three-year. In the first case, the spacecraft should be launched to an orbit with a period twice as large as that of the Earth ($T_{sc}=2T_E$), (point μ_2 in the nomogram), whereas the three-year orbits can be of two types: with a period of $T_{sc}=3T_E$ or $T_{sc}=(3/2)T_E$ (point μ_3 in the nomogram). In the first case, the spacecraft will return to Earth after the first revolution, and in the second case, it makes two revolutions around the sun to return to the Earth within three years. These trajectories are equivalent in terms of flight time, but are different in terms of energy $v_{\infty E}=3.35$ km/s, $v_{\infty M}=4.5$ km/s for $T_{sc}=(3/2)T_E$, and $v_{\infty E}=$ 6.95 km/s, $v_{\infty M}=12$ km/s for $T_{sc}=3T_E$. The nomogram yields the following characteristics for the two-year trajectories: $v_{\infty E}=5.05$ km/s, $v_{\infty M}=9$ km/s, for $T_{sc}=2T_E$. The results refer to the flyby of Mars at $m \gg 1$ with "zero" gravity assist maneuver.

Consider now the effect of a gravity assist maneuver at Mars on the energy characteristics of the trajectory.

First, let us discuss some qualitative estimates of this effect. The nomogram shows that the flyby of Mars can be made only before the planet (i.e. the deceleration gravity assist maneuver should be performed). The behavior of the isolines $v_\infty=const$ at Mars shows that an acceleration gravity assist maneuver at Mars will result in a decrease in R_π of the spacecraft orbit, which implies that the condition of the spacecraft returning to the Earth will not be met. In addition, it can be seen that the implementation of any non-zero gravity assist maneuver at Mars will result in an increase in the spacecraft starting velocity, the flyby velocity of the planet, and the return velocity to Earth, which corresponds to a shift of the initial mapping point to the right from μ_2 (in the case of two-year recurrent orbits). The post-perturbation mapping point should be located so as to ensure the return of the spacecraft to Earth at the first revolution. Thus, the following condition should be satisfied within the framework of the simplified model under consideration:

$$2\pi n-\psi_{E-p}+\psi_{p-E}=\frac{2\pi}{T_E}(t_{E-p}+t_{p-E}). \tag{4.1}$$

Here: ψ_{E-p}, t_{E-p} are the central angle and the time of the spacecraft flight at the "Earth to flyby planet" leg; ψ_{p-E}, t_{p-E} are the central angle and the time of flight at the "flyby planet to Earth" leg; $T_E=1$ year is the period of revolution of the Earth; n is the number of complete revolutions of the Earth before the encounter with the spacecraft.

To check these conditions, the angular and temporal characteristics of both pre- and post-perturbation flight legs are to be determined. These calculations made it possible to construct the lines of mapped post-perturbation orbits ensuring the spacecraft's return to Earth after a flyby of Mars.

Within each isoline $v_{\infty M}=const$, the mapping point is chosen in such a way as to ensure the condition (4.1) of return to Earth after the corresponding gravity assist maneuver at the flyby planet.

The lines of mapping points for two-year orbits (with initial π-legs) are depicted in the nomogram (Figure 4.12) under the condition of flyby of Mars within the first half-revolution (line 1) and second half-revolution (line 2).

Here, the initial point μ_2 corresponds to the flyby of Mars "at infinity" (the parameters of the pre- and post-perturbation legs are identical, and the relevant mapping points coincide with μ_2). The shift of the point representing the initial part of the flight to the right from μ_2 yields the trajectories with "non-zero" gravity assist maneuvers at Mars. The point mapping the end part of the flight can be found at lines 1 or 2.

It should be noted that there are two possible return solutions—one meeting the Earth before ("short" return leg) and the other ("long" return leg) after the spacecraft passes the pericentral point of its orbit. The relevant solutions are represented by the lines 1′, 2′, and 1″, 2″.

It can be seen that the second option of the mission E–M–E is more favorable in terms of energy expenditure (for the 2nd half-revolution of the Earth-to-Mars leg).

Moreover, the nomogram allows one to reveal the relationship between the conditions of the mission accomplishment: a decrease in the flight altitude at Mars ($\infty > m \geqslant 1$) results in an increase in the spacecraft start velocity $v_{\infty E}^{start}$ (from 5.05 to 6.5 km/s), on increase in the Mars flyby velocity $v_{\infty M}$ (from 9 to 11 km/s), and an increase in the return velocity at the Earth $v_{\infty E}^{ret}$ (from 5.05 to 11 km/s).

Similar solutions are represented at the nomogram for a three-year recurrent trajectory (for $T_{sc} = (3/2)T_E$). As can be seen in this case, the start and flyby velocities are lower than in the case of two-year trajectories, $v_{\infty E}^{st} = 3.35$ to 4.5 km/s, $v_{\infty M} = 4.5$ to 8.2 km/s, whereas the return velocities to the Earth vary within a wider range $v_{\infty E}^{ret} = 3.35$ to 12 km/s. By and large, the three-year E–M–E trajectories are more favorable in terms of energy expenditures, though they require more time.

Now let us consider the Earth–Venus–Earth missions.

Flights via Venus to return to the Earth within 1 year can be accomplished through the orbits with $T_{sc} = (1/2)T_E$ and $T_{sc} = T_E$ (points A and B in the nomogram 4.10). However, the flight E–V–E in the first case will require rather high energy expenditure ($v_{\infty E}^{st} = 11.5$ km/s, $v_{\infty V} = 19$ km/s, $v_{\infty E}^{ret} = 11.5$ km/s). Gravity assist at Venus produces virtually no effect on the energy requirements of this flight. As can be seen from the behavior of the isolines, a gravity assist maneuver at Venus (which must be an acceleration maneuver) will result in an increase in the spacecraft return velocity.

More favorable are one-year recurrent orbits ($T_{sc} = T_E$). In this case, implementation of the E–V–E flight with a flyby of Venus "at infinity" will require, as the nomogram shows, rather high energy expenditures, see point B ($v_{\infty E}^{st} = 8$ km/s, $v_{\infty V} = 4.5$ km/s, $v_{\infty E}^{ret} = 8$ km/s). However, the energy parameters of this trajectory can be significantly improved through the gravity assist of Venus.

Indeed, we can select the initial point at the line $\alpha_\oplus$ and use an acceleration gravity assist maneuver at Venus to meet the condition (4.1) of the spacecraft's return to Earth. Nomogram (4.12) contains the line α' of perturbation-transformed α-orbits, meeting the condition of the spacecraft's return to Earth.

It is evident that the range of energy characteristics of this flight can be significantly improved by a gravity assist maneuver at Venus: $v_{\infty E}^{st} = 2.85$ to 4.6 km/s, $v_{\infty V} = 4.6$ to 10.5 km/s, $v_{\infty E}^{ret} = 8$ to 11 km/s. In this case, the gravity assist maneuver, allowing the use of relatively low-energy π-starting segments of the orbit, provides the necessary energy increment to the spacecraft and makes it possible to return it to the Earth within about a year (this maneuver is represented in the nomogram by the displacement of the mapping point into the higher-energy zone).

The Earth–Mars–Earth and Earth–Venus–Earth missions considered with the use of the nomogram can be of interest not only as recurrent exploration flights, but also as perturbation-acceleration segments that can provide high spacecraft velocities during the flyby of the Earth. They can be used for missions requiring high energy, i.e. to distant planets, the sun, comets, asteroids, or for the flights beyond the ecliptic plane [138, 141].

One of the promising ways of using the gravity assist maneuver effect is to form periodic exploration spacecraft orbits. We mean those orbits that will allow the spacecraft to periodically return to the neighborhood of the flyby planet with the aim of long-term space studies [20, 28, 48].

Note that sending spacecraft directly into such orbits requires considerable energy. For example, directly launching a spacecraft into periodical orbits (with respect to the Earth) implies the following energy expenditure:

T [years]	1/2	2/3	2	3
v_∞ [km/s]	−10.626	−5.045	5.045	6.950

The half-year recursive orbit, which is attractive for the purpose of exploration, requires rather high energy expenditure (more than the flight to Jupiter will require). Spacecraft entering other periodic orbits will also require large expenditure.

Application of schemes with gravity assists allows us to improve the performance of these paths.

Thus, in the Earth–Mars–Earth mission, a periodically recurrent orbit with a multiple return to the Earth every two years can be formed at the first encounter with the Earth.

In the case of spacecraft orbit Earth–Mars–Earth with a 1.5-year period (a swingby of Mars at the first half-revolution), the nomogram indicates that the sequence of flyby orbits can be as follows:

$$\begin{matrix} T \text{ [days]} \\ R_\pi\ 10^6 \text{ [km]} \end{matrix} \quad \left.\begin{matrix} 630 \\ 150 \end{matrix}\right\} \rightarrow \left.\begin{matrix} 518 \\ 118 \end{matrix}\right\} \rightarrow \left.\begin{matrix} 730 \\ 128 \end{matrix}\right\} \rightarrow \left.\begin{matrix} 730 \\ 128 \end{matrix}\right\} \rightarrow \cdots$$

In this case, the spacecraft starts from the Earth with $v_{\infty E} = 4.25$ km/s, flies by Mars ($v_{\infty M} = 7.4$ km/s), and returns to the Earth, where it performs an acceleration gravity assist maneuver $v_{\infty E} = 11.2$ km/s) to enter a two-year periodically returning orbit.

If a deceleration gravity assist maneuver is performed at the Earth, the spacecraft can enter a one-year recurrent orbit with respect to the Earth. In this case, the spacecraft can return to the Earth alternately every one and two years. This path can be represented by an orbit sequence:

$$\begin{matrix} T \text{ [days]} \\ R_\pi\ 10^6 \text{ [km]} \end{matrix} \quad \left.\begin{matrix} 630 \\ 150 \end{matrix}\right\} \rightarrow \left.\begin{matrix} 518 \\ 118 \end{matrix}\right\} \rightarrow \left.\begin{matrix} 365 \\ 95 \end{matrix}\right\} \rightarrow \left.\begin{matrix} 730 \\ 128 \end{matrix}\right\} \rightarrow \left.\begin{matrix} 365 \\ 95 \end{matrix}\right\} \rightarrow \cdots$$

Application of Venus gravity assist is even more efficient in this kind of mission. The start velocity in this case can be even lower.

The nomogram shows that a flyby of Venus along a one-year recurrent orbit can yield the succession of orbits

$$\begin{matrix} T\ [\text{days}] \\ R_\pi\ 10^6\ [\text{km}] \end{matrix} \quad \left.\begin{matrix} 282 \\ 103 \end{matrix}\right\} \rightarrow \left.\begin{matrix} 365 \\ 107 \end{matrix}\right\} \rightarrow \left.\begin{matrix} 730 \\ 140 \end{matrix}\right\} \rightarrow \left.\begin{matrix} 365 \\ 107 \end{matrix}\right\} \rightarrow \left.\begin{matrix} 730 \\ 140 \end{matrix}\right\} \rightarrow \cdots$$

In this case, the spacecraft starts from the Earth's orbit with $v^{st}_{\infty E}=2.95$ km/s, flies by Venus ($v_{\infty V}=4.55$ km/s), and returns to the Earth, whereupon it will perform a series of gravity assist maneuvers with successive returns to Earth alternately every 1 and 2 years (with $v_{\infty E}=3.0$ km/s). Generally speaking, once the spacecraft has entered the periodically returning orbit, any combination of 1-year and 2-year revolutions can be realized depending on the gravity assist maneuver performed.

It is evident that this kind of periodically returning exploration orbit can be formed with different spatial orientations of the spacecraft's orbital plane at the expense of a relevant gravity assist maneuver at the Earth; this will allow an exploration spacecraft to be sent to various zones in the solar system that lie beyond the ecliptic plane [117].

By and large, the paths using orbits with returns to Earth appear rather promising for the formation of various interplanetary flight schemes which are optimal in terms of energy expenditure.

4.1.3 Flights to Mercury

A direct flight to Mercury, the planet nearest to the Sun, requires significant energy expenditure, which makes the designers favor paths involving gravity assist maneuvers [54, 132–135].

The orbital characteristics of this planet agree to a lesser extent with the accepted nomographic model of the planetary orbits (coplanar and circular). The considerable ellipticity of Mercury's orbit and its inclination ($\approx 7°$) make the nomographic estimate of the flights under consideration more approximate than in the previous cases.

However, the nomographic analysis of the flight options appears reasonable, because it allows us to make a comparative estimate in terms of energy for different trajectories to Mercury under identical model assumptions.

Table 4.4, obtained from analysis of the nomograms (Figures 4.1 and 4.10), indicates that there are a number of possible paths to Mercury demonstrating different performances.

The direct flight from the Earth to Mercury, even along the optimal Hohmann's trajectory (point K_1), requires rather high energy expenditure: $v_{\infty E}=7.65$ km/s, $v_{\infty Me}=9.54$ km/s. Gravity assists used in the best options allows a considerable reduction in the energy expenditure.

Among the trajectories with a flyby of a single planet, the optimal one in terms of energy is that using gravity assist at Venus (transfer $B_1 \rightarrow B_2$). The energy benefit

Table 4.4 Flights to Mercury.

1st flyby planet	2nd flyby planet	$v_{\infty E}$, km/s	$v_{\infty 1}$, km/s	$v_{\infty 2}$, km/s	$v_{\infty Me}$, km/s	Transfer type
—	—	7.65	—	—	9.54	$\alpha-\pi$
Venus	—	3.20	5.95	—	6.7	$\alpha-\Pi^{-}-\pi$
	Earth	3.20	5.80	9.00	9.90	$\alpha-\Pi^{+}-\Pi^{-}-\pi$
	Mars	6.30	14.00	8.90	12.60	$\alpha-\Pi^{+}-\Pi^{-}-\pi$
Mars	—	14.50	8.90	—	12.60	$N-\Pi^{-}-\pi$
	Venus	13.50	7.20	18.00	11.25	$N-\Pi^{-}-\Pi^{-}-\pi$
	Earth	4.15	7.2	9.00	9.90	$N-\Pi^{-}-\Pi^{-}-\pi$
Jupiter	—	9.30	8.20	—	17.65	$\pi-\Pi^{-}-\pi$
	Venus	9.10	7.32	23.5	15.20	$\pi-\Pi^{-}-\Pi^{-}-\pi$
	Earth	9.17	7.45	25.0	15.35	$\pi-\Pi^{-}-\Pi^{-}-\pi$
	Mars	9.27	7.91	24.2	17.12	$\pi-\Pi^{-}-\Pi^{-}-\pi$

here can be obtained from the conditions of the spacecraft start and arrival at the planet (this can be seen from the behavior of the nomogram isolines).

Gravity assist at Mars gives no benefits: the spacecraft must first reach a more distant planet in order to reach Mercury; in addition, the gravitational field of Mars fails to produce significant perturbation effect.

Contrary to that, Jupiter, which possesses a powerful gravitational field, can return the spacecraft to the planet nearest to the sun. However, the velocities required for this flight (both start and arrival) are rather high, and the trip is long. These drawbacks are not typical of the trajectory Earth–Venus–Mercury.

As the nomogram indicates, the ranges of energy expenditure for this path are as follows: $v_{\infty E}=3.2$ to 6.0 km/s, $v_{\infty V}=5.95$ to 13 km/s, $v_{\infty Me}=6.7$ to 8.6 km/s depending on the conditions of the Venus flyby. As for the trip time, its value in the case of a single flyby differs by as little as about 15% from the duration of the direct path. The nomogram also shows that the repetition of the Venus flyby on the way to Mercury is not reasonable, for it will not improve the energy efficiency of the flight (the transfer $B_1 \rightarrow B_2$ can be realized with a single Venus flyby).

Thus, it appears most reasonable to use the Venus gravity assist, because it allows a more than twofold reduction in the spacecraft start velocity (as compared with the direct flight). A considerable reduction in the spacecraft arrival velocity at Mercury is also obtained. These solutions in their best values agree well with numerical investigations of the flights to Mercury via Venus.

One can expect that the use of one more planetary flyby in the flight to Mercury will further reduce the energy expenditure. Thus, a flyby of Earth before approaching Mercury is highly beneficial in terms of energy. This maneuver at the trajectories Earth–Venus–Earth–Mercury and Earth–Mars–Earth–Mercury makes it possible to reduce the start velocities to the values typical of direct flights to Venus or Mars. However, an analysis of the nomogram shows that one more flyby of Earth is necessary to accomplish this trajectory; this will significantly increase the total trip time to Mercury and evidently reduce the value of these paths. Moreover, in this case the arrival velocity at Mercury is higher than in the path Earth–Venus–Mercury.

As for the paths including flyby of Jupiter, the use of a second flyby of a planet from the Earth group will make it possible to reduce somewhat the energy expenditure for the mission. However, this reduction is not very large, because of the high flyby velocities of the spacecraft, which hampers the full-scale utilization of the gravitational capabilities of these planets.

Summing up the possibility of flight to Mercury, we can say that the best path is Earth–Venus–Mercury. These missions can be realized with relatively low energy expenditures at the trajectories using a flyby of one of the planets of the Earth group with a second flyby of Earth (Earth–Venus–Earth–Mercury and Earth–Mars–Earth–Mercury).

More complicated exploration trajectories, including recurrent ones, are evidently possible (with three or more planetary flybys). These paths can also be analyzed using the nomographic technique suggested.

4.1.4 Flights to Jupiter

Flights to Jupiter, one of the distant planets of the solar system, are known to require high energy expenditure, which makes the designers consider multiple-flyby schemes of reaching this planet [136–147]. According to the nomographic analysis, even minimum hyperbolic launch and arrival velocities of the spacecraft at the Earth and Jupiter amount to $v_{\infty E}=8.85$ km/s, $v_{\infty J}=5.61$ km/s, respectively. This makes it necessary to search for possible ways of reducing the velocities required for the flight to the planet. Consider the nomographic analysis of possible paths to Jupiter (the results are presented in Table 4.5).

Let us discuss first the flight path options, including a single planetary flyby.

Consider the possibility of using Mars for the flights to Jupiter.

Table 4.5 Flights to Jupiter.

1st flyby planet	2nd flyby planet	$v_{\infty E}$, km/s	$v_{\infty 1}$, km/s	$v_{\infty 2}$, km/s	$v_{\infty Me}$, km/s	Transfer type
—	—	8.85	—	—	5.61	$\pi-\alpha$
Venus	—	16.50	11.50	—	6.60	$N-\Pi^{+}-\alpha$
	Venus	9.95	11.50	11.50	6.60	$N-\Pi^{+}-\Pi^{+}-\alpha$
	Earth	3.40	5.80	8.85	—	$R_{\alpha}<R_{J}$
	Earth (2)	3.40	5.80	8.85	5.61	$\alpha-\Pi^{+}-\Pi^{+}-\Pi^{+}-\alpha$
	Mars	14.00	10.80	17.50	6.26	$N-\Pi^{+}-\Pi^{+}-\alpha$
Mars	—	8.05	13.80	—	5.40	$\pi-\Pi^{+}-\alpha$
	Venus	13.00	12.8	12.0	6.62	$N-\Pi^{-}-\Pi^{+}-\alpha$
	Earth M2	5.8	10.4	8.85	5.61	$\pi-\Pi^{-}-\Pi^{+}-\alpha$
	M3/2	4.5	8.1	8.85	5.61	—
	Mars	5.28	9.0	9.0	4.60	$\pi-\Pi^{+}-\Pi^{+}-\alpha$

Mars is the only planet located on the way to Jupiter. Its flyby improves all the energy parameters of the flight to Jupiter as compared with the scheme of direct flight [137, 139, 140]. The lower boundary of the energy expenditure for the flight to Jupiter via Mars, as the behavior of the isolines in the nomogram shows, is determined by the following values $v_{\infty E}=3.8$ km/s, $v_{\infty M}=5.9$ km/s, $v_{\infty J}=4.2$ km/s. However, a single Mars flyby fails to yield such characteristics (because of the limited perturbation capabilities of Mars' gravity field). For example, the path Earth–Mars–Jupiter (represented in the nomogram 4.13 by the transfer $J_1 \rightarrow J_2$) can be implemented only if the Mars flyby velocity $v_{\infty M} \geqslant 13.8$ km/s, which results in a reduction in the departure velocity by as little as 0.8 km/s (as compared with direct optimal flights Earth–Jupiter). A small reduction in the spacecraft arrival velocity at Jupiter (by 0.2 to 0.25 km/s) is also obtained owing to the perturbation effect of Mars (see Table 4.5).

Considerable gain cannot be expected from the trajectories with a single Mars flyby. As can be seen from the nomogram, two or three Mars flybys should be performed to reduce the energy expenditure of the flight (one of the solutions Earth–Mars–Mars–Earth is given in Table 4.5; it corresponds to the transfer $R_1 \rightarrow R_2 \rightarrow R_3$ in Figure 4.13). This will provide a certain improvement of the flight characteristics, though at the expense of the trip time (an increase of about 2 years). It is interesting that this path results in a reduction in both the initial spacecraft velocity (by 2.7 to 2.8 km/s) and the hyperbolic velocities of Mars flyby (by about 4.8 km/s) and arrival at Jupiter (by 0.8 km/s).

Let us consider the flights to Jupiter via Venus—a planet that is further from Jupiter than the Earth. The behavior of the nomogram lines indicates that the minimum possible characteristics of this flight can be $v_{\infty E}=5.31$ km/s, $v_{\infty J}=6.6$ km/s, $v_{\infty Me}=11.5$ km/s. However, flights with such characteristics are only possible along the paths with multiple flybys of Venus. Analysis of the nomogram shows that flights with one (or even two) Venus flybys require high energy expenditure (see Table 4.5), and only three Venus flybys allow the lower energy limit of the flight to be attained. It is represented in the nomogram Figure 4.13 by the transfer $C_1 \rightarrow C_4$. A drawback of this scheme (as is the case with several Mars flybys) is a significant increase in the mission time.

It appears more promising to use paths utilizing successive flybys of two planets. Most reasonable in this case can be various combinations of the Earth group planets (Earth, Mars, and Venus).

Let us use the nomogram to examine the feasibility of such paths. Consider first the gravitational capability of the Earth for the formation of trajectories to Jupiter. It is evident that the utilization of the Earth's gravitational field at the second flyby is useful only after a transformation of the initial heliocentric trajectory of the spacecraft. To accomplish this, gravity assist maneuvers of the spacecraft at Venus or Mars can be used. In this case, the conditions for the spacecraft to return to the Earth after Venus or Mars flyby must be provided.

The trajectory Earth–Venus–Earth–Jupiter meets these conditions. Orbits returning to the Earth every one or two years are possible (note that this path was mentioned as a promising one in [138]).

The behavior of the isolines indicates that the minimum Earth flyby velocities must be $v^{fl}_{\infty E} \geqslant 8.85$ km/s (it follows from the condition to reach Jupiter's orbit). This

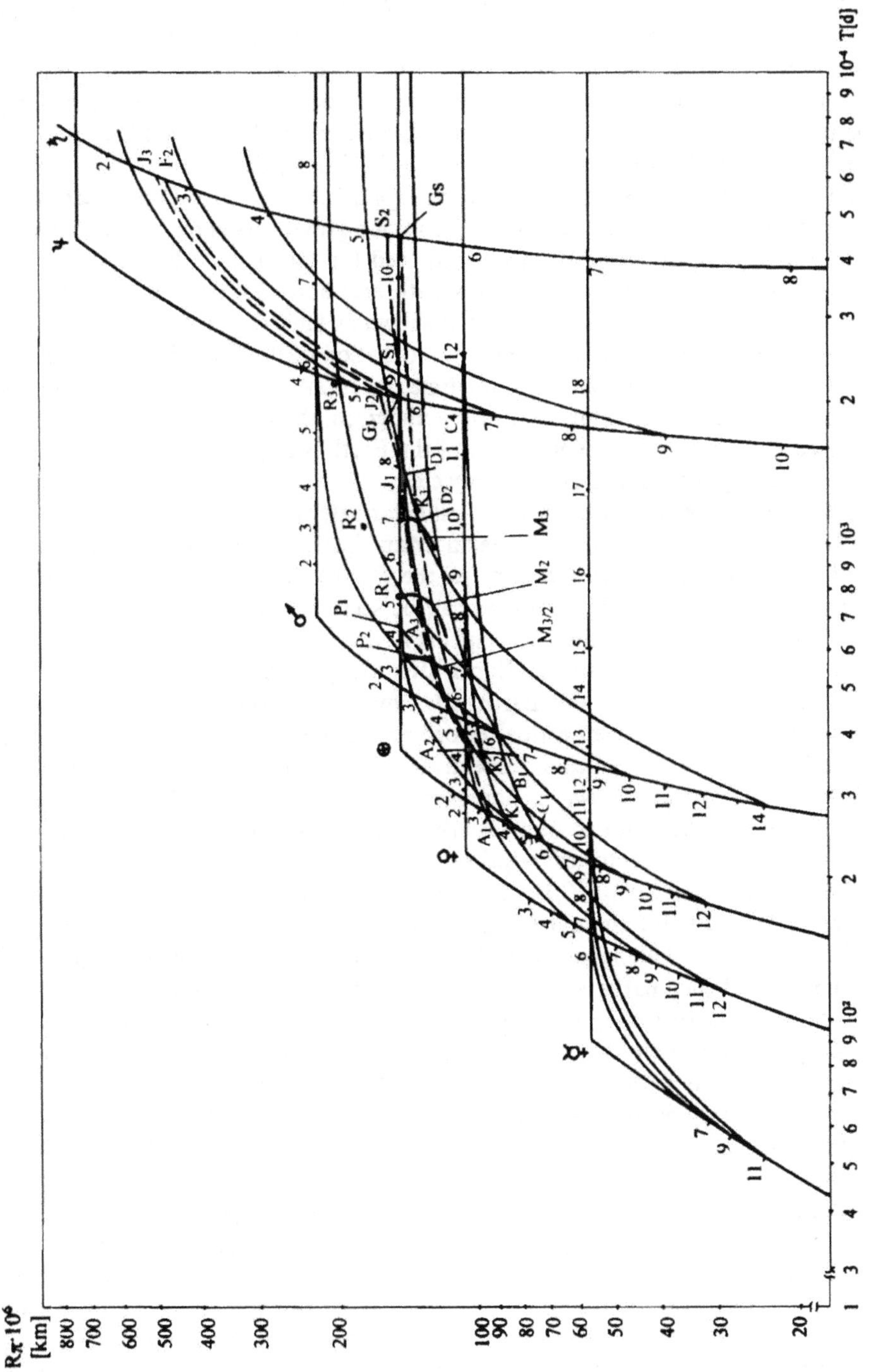

Figure 4.13 Formation of paths to Jupiter and Saturn.

velocity can be provided by a Venus flyby at a one-year recurrent trajectory with the start velocity at the Earth $v_{\infty E}=3.4$ km/s. However, nomographic analysis of this path shows that the gravitational capabilities of Venus and Earth are not sufficient for the spacecraft to reach Jupiter's orbit. It is obvious that velocity impulses (or maneuvers) will be necessary in addition to gravity assist maneuvers at Venus and Earth.

Thus, one more flyby of the Earth (path E–V–E–E–J, denoted in Table 4.5 as Venus–Earth2) allows the spacecraft to reach Jupiter's orbit. A two-year return orbit is formed during the first flyby at Earth. The second gravity assist maneuver at Earth provides the spacecraft with the energy required for the flight. The minimum (in terms of energy) characteristics of such a path are: $v_{\infty E}=3.4$ km/s, $v_{\infty V}=5.8$ km/s, $v_{\infty E}=8.85$ km/s, $v_{\infty J}=5.61$ km/s. This path is shown as the transfer A_1–A_2–A_3–G_J in the nomogram Figure 4.13. Accordingly, the duration of the flight is 3 years larger than that of the direct flight. At the same time, solutions with a lesser duration may exist. As can be seen from analysis of the nomogram, the amount of energy gained by the spacecraft during the second flyby of the Earth is so great that the spacecraft can be sent to Jupiter along a quicker trajectory.

Another variant of the path with two planetary flybys can be formed using the gravitational field of Mars. The path Earth–Mars–Earth–Jupiter, as the path with a flyby of Venus, must include a return (to the Earth) trajectory segment, and the minimum flyby velocity at the Earth, as can be seen from the nomogram, must be $v^{fl}_{\infty E}=8.85$ km/s. It is only on these (or higher) isolines that the Jupiter's orbit can be reached. Return trajectories to the Earth after flyby of Mars can have a period $T_{sc}=1.5$ years (to encounter the Earth after two revolutions), $T_{sc}=2$ years, and $T_{sc}=3$ years (to encounter the Earth after one revolution), that is, the spacecraft can return to the Earth in 3 or 2 years, respectively. These orbits returning to the Earth after a flyby of Mars are shown in nomogram 4.13 as curves M3/2, M2, M3 with planetary phasing taken into account. As the nomogram shows, the second flyby of the Earth on such trajectories requires start velocities of $v_{\infty E}=4.2$–4.8 km/s, $v_{\infty E}=5.6$–6.2 km/s, and $v_{\infty E}=7.1$–7.5 km/s, respectively. In this case, the flyby velocities of Mars should amount to $v_{\infty M}=8.0$–9.0 km/s, $v_{\infty M}=11.0$–12.0 km/s, and $v_{\infty M}=12.5$–14.2 km/s. The behavior of the isolines in the nomogram shows that the implementation of such paths is quite possible (the points representing the composite fragments of the path for $T_{sc}=1.5$ year are shown in the nomogram Figure 4.13 P_1–P_2–G_J). It can be seen that the Earth–Mars–Earth–Jupiter path leads to energy savings as compared with the direct flight to Jupiter or with a flight using the gravitation capability of Mars alone. The optimal energy characteristics of the path are as follows: $v_{\infty E}=4.5$ km/s, $v_{\infty M}=8.1$ km/s, $v_{\infty E}=8.85$ km/s, $v_{\infty J}=5.61$ km/s.

An evident drawback is the necessity for the spacecraft to enter a two- or three-year recurrent orbit, which makes the entire mission longer.

However, this path can be regarded as an option for the flight to Jupiter.

As for paths utilizing the gravitational fields of Venus and Mars, whatever the order of their flybys on the way to Jupiter, no energy savings can be attained (see Table 4.5). From the nomogram analysis it can be seen that the characteristics can be improved either by increasing the number of flyby revolutions at a certain planet, or by introducing powered-gravitational maneuvers of the spacecraft at planets.

Summing up the above results, we can mention as promising the schemes using the gravitational fields of Venus and Earth and Mars and Earth. The first variant requires two flybys of the Earth (or additional velocity impulses instead), whereas the second can be implemented with a single flyby of the Earth. Both options require the mission time to be increased by about 3 years, but the path via Venus and Earth requires minimum start velocities close to those of flights between the Earth group planets. Trajectories with a two-year increment of the flight time are possible via Mars, but their energy characteristics are worse.

It is worth mentioning that the path Earth–Mars–Mars–Jupiter yields the minimum encounter velocity at Jupiter.

The above nomographic analysis of flights to Jupiter shows the advantage of gravity assist maneuvers at Earth, Venus, or Mars (in different combinations) and the potential of studies of perturbation-recurrent trajectories for flights to remote planets.

4.1.5 Flights to Saturn

Saturn is a giant planet of the Jupiter group, and in terms of energy, the flight to it is even more difficult than that to Jupiter. Multiple-flyby paths allow one to reduce the energy expenditure for these flights [158–167].

Consider the nomographic analysis of different options of the flight paths to Saturn (see Table 4.6).

Direct flights to Saturn require rather high energy expenditures: even Hohmann's orbits yield the minimum $v_{\infty E}=10.20$ km/s, $v_{\infty S}=5.41$ km/s.

The nomogram shows that in the single-flyby options of the path, energy savings can be provided by the planets located on the path of Earth–Saturn flight, i.e., Mars and Jupiter.

A potential reduction of the flight velocity to $v_{\infty E}=4.50$ km/s and $v_{\infty S}=4.59$ km/s (with the minimum flyby velocity at Mars $v_{\infty M}=7.50$ km/s) can be provided by Mars.

Table 4.6 Flights to Saturn.

1st flyby planet	2nd flyby planet	$v_{\infty E}$, km/s	$v_{\infty 1}$, km/s	$v_{\infty 2}$, km/s	$v_{\infty S}$, km/s	Transfer type
—	—	10.20	—	—	5.41	$\pi-\alpha$
Mars	—	9.51	16.8	—	5.22	$\pi-\Pi^+-\alpha$
	Jupiter	8.05	13.8	5.4	2.40	$\pi-\Pi^+-\Pi^+-\alpha$
	Earth	7.9	14.0	10.2	5.41	$\pi-\Pi^+-\Pi^+-\alpha$
	Mars	8.45	14.7	14.7	5.2	$\pi-\Pi^+-\Pi^+-\alpha$
Jupiter	—	8.85	5.61	—	2.51	$\pi-\Pi^+-\alpha$
	Mars	8.90	6.0	8.5	4.61	$\pi-\Pi^+-\Pi^+-\alpha$
Venus	—	19.5	13.0	—	6.0	$N-\Pi^+-\alpha$
	Earth (2)	4.2	9.0	10.2	5.41	$\alpha-\Pi^+-\Pi^+-\Pi^+-\alpha$
	Mars	18.0	12.15	21.0	5.96	$N-\Pi^+-\Pi^+-\alpha$
	Venus	15.5	13.0	13.0	6.0	$N-\Pi^--\Pi^--\alpha$
	Jupiter	16.5	11.5	6.6	2.60	$N-\Pi^+-\Pi^+-\alpha$

However, an analysis of the nomogram using perturbation isochrones shows that such flight characteristics can be attained only by a multiple-revolution flyby of Mars (at least five revolutions around Mars are necessary, which makes the entire expedition excessively long). On the other hand, a reduction in the number of revolutions to three, for example (this solution is shown in the table) yields a relatively small energy saving for the entire flight because of the low gravitational capabilities of Mars. A single flyby at Mars on the way to Saturn yields even lower energy savings (about 0.7 km/s as compared with the direct flight). This path is denoted in nomogram 4.13 as $S_1 \to S_2$.

Low energy savings together with an increase in the flight time (by about 2 times) make this path unacceptable for the flight to Saturn.

The use of the gravity field of Jupiter on the way to Saturn fails to yield significant savings in the start velocity (indeed, the flight to Jupiter itself require considerable energy); however, the immense perturbation capabilities of Jupiter allow the arrival velocity to be more than halved (see Table 4.6). In the nomogram 4.13, the path Earth–Jupiter–Saturn is denoted as $\mathrm{G_J} \to \mathrm{G_s}$.

The use of Venus's field in single-flyby flights is not justified. As with the path E–V–J, the conditions of tangentiality of the boundary impulse are not satisfied in this case (the transfer type is $N-\Pi^{+}-\alpha$), which leads to a huge start velocity being required for the accomplishment of the mission.

Consider flights to Saturn with a combination of flybys of two planets. The nomogram shows that successive flybys of Mars and Jupiter (E–M–J–S path) yield a more significant saving in the transfer velocity. As compared with the optimal direct flight, the start velocity can be reduced by more than 2 km/s, and the arrival velocity, can be halved ($J_1 \to J_2 \to J_3$ transfer, see Figure 4.13).

The use of the Earth's gravitational field in the second flyby is very promising. Generally speaking, this can be achieved on the spacecraft orbits via Venus or Mars (paths E–V–E–S and E–M–E–S).

The results obtained show that the path E–V–E–S cannot be accomplished because of the limited perturbation capabilities of Earth and Venus and the fact that Saturn's orbit is far away. The accomplishment of this path requires either powered-gravitational maneuvers in planetary flybys or additional gravity assist maneuver of the spacecraft. Thus, one more flyby of the Earth allows the spacecraft to reach Saturn's orbit (scheme E–V–E–E–S). Analysis of the nomogram shows that such a path requires almost 2.5 times less energy than the direct flight. The duration of this flight will increase by about four years, for after the first flyby of the Earth, a three-year orbit of the spacecraft should be formed for a repeat gravity assist maneuver at the Earth. The characteristics of the optimal flight in this case are as follows: $v_{\infty \mathrm{E}} = 4.2$ km/s, $v_{\infty \mathrm{V}} = 9.0$ km/s, $v_{\infty \mathrm{E}}^{\mathrm{fl1}} = 10.2 \text{ km/s} = v_{\infty \mathrm{E}}^{\mathrm{fl2}}$, $v_{\infty \mathrm{S}} = 5.41$ km/s. The path Earth–Venus–Earth–Earth–Saturn is represented in nomogram 4.13 by the transfer $K_1 \to K_2 \to K_3 \to \mathrm{G_S}$.

Flights via Mars to Earth and then to Saturn (E–M–E–S) provide lower energy savings (in the nomogram, the path is denoted as the transfer $D_1 \to D_2 \to \mathrm{G_S}$). However, the energy saving in this case is perceptible (about 2.3 km/s). In this case, the spacecraft must first enter the initial swingby orbit of Mars with a period of about three years, and a gravity assist maneuver at the Earth allows the spacecraft to reach Saturn's orbit. Note that this path can be correlated with a powered-gravitational Earth–Earth–Saturn scheme with apocentral application of impulse on a three-year recurrent orbit for a repeat flyby of the Earth. This scheme, which was considered in detail in [159], is comparable in terms of energy expenditure with the E–M–E–S path with the difference that in the latter, the intermediate impulse is replaced by gravity assist maneuvering in the force field of Mars.

As far as the mission time is concerned, the flight E–V–E–E–S, which is undoubtedly beneficial in terms of energy (the energy savings are almost twofold), is as little as one year longer than the E–M–E–S path. Moreover, analysis of the nomogram shows that the perturbation potential of the Earth in the last flyby (E–V–E–E–S path) is not fully utilized, which leaves the possibility that quick paths to Saturn do exist.

Thus, as was in the case with the flights to Jupiter, this path with successive flybys of Venus and Earth can be considered most advantageous for reaching Saturn.

The estimates for other paths in the table are presented just for comparison; they are not acceptable because of energy considerations.

Underlining the extreme expediency of the repeat flyby of the Earth, we note that in these schemes the field of Venus or Mars provides additional impulse, allowing the repeat flight to be accomplished with the most beneficial velocities and the perturbation potential of the Earth to be utilized to a maximum possible extent for reaching planets as distant as Saturn. This circumstance confirms the potential of studies of alternatives (perturbation-recurrent and powered-recurrent) for flights to distant planets.

In addition to these promising schemes for flights to Saturn; all the variants of flights via Jupiter which are optimal in terms of energy should be mentioned as well, because after the flyby of this planet, it is not a problem to reach the orbit of Saturn.

4.1.6 Flights to Uranus

The problems of reaching distant planets—Uranus, Neptune, Pluto—require considerable energy, which makes it necessary for the designer to consider multiple-flyby schemes [56, 168–170].

Uranus is located at a huge distance from the Earth—more than 19 AU. Flights across such a distance are very time- and energy-consuming. It is sufficient to say that optimal direct flights Earth–Uranus can be accomplished with $v_{\infty E}=11.27$ km/s, $v_{\infty U}=4.66$ km/s, $T=16.5$ years.

Let us perform a nomographic analysis of possible paths to Uranus. Consider first the schemes with flybys first of a single and then of two planets (possible variants of solutions are presented in Table 4.7).

The nomogram shows that the perturbation effect of one of the intermediate planets (Mars, Jupiter, or Saturn) can be used in the flight to Uranus (it is obvious that flyby of any other planet is useless).

Mars, located on the path to Uranus, can improve the energy characteristics of the flight up to the values: $v_{\infty E}=5.6$ km/s, $v_{\infty U}=4.2$ km/s, provided that the flyby velocity at Mars is minimum $v_{\infty M}=9.1$ km/s. However, as was the case with flights to other

Table 4.7 Flights to Uranus.

1st flyby planet	2nd flyby planet	$v_{\infty E}$, km/s	$v_{\infty 1}$, km/s	$v_{\infty 2}$, km/s	$v_{\infty U}$, km/s	Transfer type
—	—	11.27	—	—	4.66	$\pi-\alpha$
Mars	—	11.05	18.5	—	4.60	$\pi-\Pi^{+}-\alpha$
	Earth	8.8	14.8	11.3	4.66	$\pi-\Pi^{-}-\Pi^{+}-\alpha$
	Jupiter	8.05	13.8	5.40	2.58	$\pi-\Pi^{+}-\Pi^{+}-\alpha$
	Saturn	9.51	16.8	5.22	2.20	$\pi-\Pi^{+}-\Pi^{-}-\alpha$
Jupiter	—	8.85	5.61	—	2.60	$\pi-\Pi^{+}-\alpha$
	Saturn	8.85	5.61	2.51	1.4	$\pi-\Pi^{+}-\alpha$
Saturn	—	10.20	5.41	—	2.34	$\pi-\Pi^{+}-\alpha$
Venus	Earth (3)	5.0	10.5	11.3	4.66	$\pi-\Pi^{+}(4)-\alpha$

planets of the Jupiterian group (e.g. Jupiter or Saturn), these savings can be attained at the expense of multiple flybys of Mars, which makes the mission time even greater. A single-flyby flight has unfavorable conditions for maneuvering, and therefore yields insignificant benefits in the spacecraft start and arrival velocities (see Table 4.7).

Far greater savings can be obtained on the spacecraft flight via Jupiter or Saturn (E–J–S or E–S–U paths). These are transfers $F_1 \to F_3$ and $E_1 \to E_2$ in nomogram 4.1. Considering the immense gravitational potential of the giant planets, it is sufficient to reach one of them for the spacecraft to pass the rest of the trajectory using the energy of the giant planet's perturbation field.

In this case, flights via Jupiter, which is nearer to the Earth, are better in terms of energy expenditure. As compared with direct flights, the departure velocity in this case is reduced by almost 2.5 km/s, and the arrival velocity by 2 km/s. The flight time along the path E–J–U is reduced to almost 5 years.

The characteristics of the flight via Saturn are worse than those for the path E–J–U, though they are still better than those for the direct flight. In the optimal variant, this mission will last for 8 years. To complete the comparison of single-flyby options of the flight to Uranus, we note that the best start velocities are obtained on the trajectories via Jupiter, and the best arrival velocities are obtained in the flights via Saturn—the planet nearest to Uranus.

Consider the paths to Uranus with two flybys.

A flyby at Mars before passing Jupiter or Saturn results in a reduction in the start velocity $v_{\infty E}$ of 0.7–0.8 km/s (as compared with the paths E–J–U and E–S–U). The arrival velocity at the target planet does not change in this case.

The flight with flybys of Jupiter and Saturn (E–J–S–U path) yields an additional reduction (almost twofold) in the arrival velocity due to Saturn, whereas the start velocity is determined by the energy required to reach Jupiter.

It is also possible to use the planets of the Earth group in flights to Uranus (in particular, a combination of Venus and Earth). However, more than a single gravity assist maneuver at the planets has to be performed in this case, because after the first return to the Earth on the path E–V–E– ... –U, the spacecraft should perform two more revolutions around the Earth in order to reach Uranus' orbit.

The energy required to accomplish this mission is: $v_{\infty E} = 5.0$ km/s, $v_{\infty V} = 10.5$ km/s, $v_{\infty \oplus 1,2,3} = 11.3$ km/s, $v_{\infty U} = 4.66$ km/s. The mission time in this case will increase by 8 years.

The Earth–Mars–Earth–Uranus path can be accomplished only with the inclusion of a four-year return leg to the Earth. In this case, the starting hyperbolic excess of the spacecraft velocity can be reduced by 2.4–2.5 km/s, and the mission time will increase by 4 years (as compared with the direct flight time).

More advantageous are combinations of flybys of the Earth group planets with flybys of giant planets. In this case, the path must include a flyby of Jupiter or Saturn. The energy expenses for such paths will be determined by the conditions of reaching Jupiter or Saturn via Venus or Mars. Thus, implementation of the Earth–Venus–Earth–Earth–Jupiter–Uranus path requires the start velocity $v_{\infty E} = 3.4$ km/s, and a similar path via Saturn, $v_{\infty S} = 4.2$ km/s.

Similar estimates can be made for the path Earth–Mars–Earth and further to Uranus via Jupiter or Saturn (see Tables 4.5 and 4.6).

Summing up the paths of flights to Uranus, we note the advantage of a Jupiter flyby (which is preferable) or a Saturn flyby on the way to Uranus. Inclusion of other Earth group planets in the path is justified only as a means of reducing the energy expenditure required to reach these giant planets.

4.1.7 Flights to Neptune and Pluto

Neptune and Pluto are the most distant planets of the solar system. The orbital characteristics of Pluto match the accepted orbital model to a lesser degree than the parameters of other planets. The considerable ellipticity of Pluto's orbit and its inclination ($\approx 17°$) make the nomographic assessment of the relevant flight schemes quite approximate.

Nevertheless, the nomographic analysis of the flight options to Pluto appears to be worth performing, because it allows a comparative estimate (other characteristics assumed equal) of the lower boundary of energy expenditure for different variants of the flight schemes to the most distant planet.

The direct Hohmann's flight to Neptune and Pluto requires a great deal of energy and time ($v_{\infty E} = 11.65$ km/s and 11.81 km/s, $T = 31$ years and 46 years, respectively).

Nomographic analysis of the options for reaching these planets is presented in Tables 4.8 and 4.9.

As can be seen from the above results, possible single-flyby options of the paths to these planets also require high energy expenditure.

The inclusion of Mars flyby yields almost no energy savings, because the velocities of passage of this planet are too high.

A flyby of any of the giant planets is more beneficial, and the path via Jupiter E–J–N or E–J–P is preferential. In this case, both the start and arrival velocities of the spacecraft can be significantly reduced (by 3 km/s) due to the gravity field of Jupiter (transfers $F_1 \rightarrow F_4$ and $F_1 \rightarrow F_5$ in the nomogram 4.1).

Table 4.8 Flights to Neptune.

1st flyby planet	2nd flyby planet	$v_{\infty E}$, km/s	$v_{\infty 1}$, km/s	$v_{\infty 2}$, km/s	$v_{\infty N}$, km/s	Transfer type
—	—	11.65	—	—	4.05	$\pi-\alpha$
Mars	—	11.52	19.0	—	4.0	$\pi-\Pi^{+}-\alpha$
	Jupiter	8.05	13.8	5.40	2.6	$\pi-\Pi^{+}-\Pi^{+}-\alpha$
	Saturn	9.51	16.8	5.22	2.18	$\pi-\Pi^{+}-\Pi^{+}-\alpha$
	Uranus	11.05	18.5	4.60	2.12	$\pi-\Pi^{+}-\Pi^{+}-\alpha$
Jupiter	—	8.85	5.61	—	2.62	$\pi-\Pi^{+}-\alpha$
	Saturn	8.85	5.61	2.51	1.70	$\pi-\Pi^{+}-\Pi^{+}-\alpha$
	Uranus	8.85	5.61	2.60	1.20	$\pi-\Pi^{+}-\Pi^{+}-\alpha$
Saturn	—	10.20	5.42	—	2.21	$\pi-\Pi^{+}-\alpha$
	Uranus	10.20	5.42	2.34	1.0	$\pi-\Pi^{+}-\Pi^{+}-\alpha$
Uranus	—	11.27	4.66	—	2.20	$\pi-\Pi^{+}-\alpha$

Table 4.9 Flights to Pluto.

1st flyby planet	2nd flyby planet	$v_{\infty E}$, km/s	$v_{\infty 1}$, km/s	$v_{\infty 2}$, km/s	$v_{\infty Pl}$, km/s	Transfer type
—	—	11.81	—	—	3.68	$\pi-\alpha$
Mars	—	11.7	21.5	—	3.63	$\pi-\Pi^{+}-\alpha$
	Jupiter	8.05	13.8	5.40	2.54	$\pi-\Pi^{+}-\Pi^{+}-\alpha$
	Saturn	9.51	16.8	5.22	2.10	$\pi-\Pi^{+}-\Pi^{+}-\alpha$
	Uranus	11.05	18.5	4.60	1.81	$\pi-\Pi^{+}-\Pi^{+}-\alpha$
	Neptune	11.52	19.00	4.00	2.35	$\pi-\Pi^{+}-\Pi^{+}-\alpha$
Jupiter	—	8.85	5.61	—	2.59	$\pi-\Pi^{+}-\alpha$
	Saturn	8.85	5.61	2.62	1.79	$\pi-\Pi^{+}-\Pi^{+}-\alpha$
	Uranus	8.85	5.61	2.60	1.15	$\pi-\Pi^{+}-\Pi^{+}-\alpha$
	Neptune	8.85	5.61	2.62	1.35	$\pi-\Pi^{+}-\Pi^{+}-\alpha$
Saturn	—	10.20	5.42	—	2.20	$\pi-\Pi^{+}-\alpha$
	Uranus	10.20	5.42	2.34	1.25	$\pi-\Pi^{+}-\Pi^{+}-\alpha$
	Neptune	10.20	5.42	2.21	1.12	$\pi-\Pi^{+}-\Pi^{+}-\alpha$
Uranus	—	11.27	4.66	—	1.80	$\pi-\Pi^{+}-\alpha$
	Neptune	11.27	4.66	2.20	1.10	$\pi-\Pi^{+}-\Pi^{+}-\alpha$
Neptune	—	11.65	4.05	—	2.40	$\pi-\Pi^{+}-\alpha$

Flybys of other, more distant planets of the Jupiter group yield lesser energy savings in the flights to Neptune and Pluto (as can be seen from the tables presented).

The arrival velocities at the target planets under consideration differ significantly depending on the chosen path. Thus, it can be seen that according to this criterion, the most advantageous are the paths utilizing a flyby of Uranus. In this case, the arrival velocity can be almost halved as compared with the direct flight paths.

When considering the double-flyby paths, one can see that only a flyby of Mars at the beginning of the flight allows the start characteristics of the flight to be somewhat improved. The use of other double-flyby combinations can result only in a reduction in the arrival characteristics at the target planet (as compared with the best single-flyby scheme via Jupiter). Thus, it is most beneficial to include a flyby of Uranus in flights to Neptune (paths Earth–Jupiter–Uranus–Neptune and Earth–Saturn–Uranus–Neptune). A flyby of Neptune or Uranus is preferable in the flight to Pluto (paths E–J–U–P, E–J–N–P, E–S–U–P, E–U–N–P). It should be noted that the results discussed here for flights via Jupiter (and for other planets of the Jupiter group as well) refer to the paths which are optimal in terms of energy expenditure (the condition of tangency between the transfer orbit and that of the target planet is satisfied); however, these paths have very long flight times (greater than that for direct Hohmann's paths), which obviously hampers the accomplishment of such flights. Jupiter is the main planet involved in the formation of flights to distant planets; its gravitational field is powerful enough to allow the spacecraft to be directed to these planets along faster high-energy trajectories (flyby of Jupiter with large v_∞). In this case, as can be seen from the nomogram, the conditions of arrival at the target planets are worse; however, the mission time is reduced considerably. Thus, flights to Neptune or Pluto via Jupiter can be accomplished in 7 and 8 years, respectively. At the same time, the arrival velocities at Neptune and Pluto will more than double (the spacecraft start velocity remaining the same).

As can be seen from the presented results (Tables 4.8 and 4.9), the energy expenditure for reaching the most distant planets of the solar system—Neptune and Pluto—along the optimal paths are determined by the energy required to reach Jupiter. Therefore, the problem of flying to these planets with minimum energy expenditure is immediately associated with the problem of selecting the best path to reach Jupiter.

Paths involving a flyby of a planet from the Earth group at the initial stage of the flight can be of interest in this respect. This can be one of the flights to Jupiter via Venus and Earth considered above, as well as via Mars and Earth. This kind of three-planet scheme can lead to considerable savings in flight energy. Thus, in the optimal case, the scheme E–V–E–E–J–N(P) requires as little as $v_{\infty E}=3.2$ km/s, whereas the path via Mars E–M–E–J–N(P) can be accomplished with $v_{\infty E}=4.5$ km/s. The time required for this mission will increase by about three years (as compared with the schemes of direct flight to Jupiter).

Maybe even greater are the benefits associated with the scheme including three flybys of Venus (see 4.1.4) and following on to Neptune and Pluto via Jupiter.

Energy savings (with a corresponding increase in the mission time) can be obtained using combined schemes, including flybys of other planets of the Jupiter group.

4.1.8 Nomographic analysis of the perturbation potential of the Earth in interplanetary flight schemes

Analysis of the above results shows the high efficiency of the paths where the spacecraft performs a gravity assist maneuver at the Earth.

Indeed, the gravitational field of the Earth has sufficiently high potential for changing the orbital parameters of the spacecraft and for sending it to regions of the solar system distant from the Earth.

In these beneficial flight options including passage of the Earth, the preliminary passage of one of the planets from the Earth group (Venus or Mars) is obligatory. The idea of forming this kind of scheme is that the spacecraft, starting from the Earth with a low high velocity $v_{\infty E}^{st}$ will approach the Earth in the next encounter with a higher velocity $v_{\infty E}^{fl}$. This makes the gravity assist maneuver at the Earth efficient and justifies the trajectory as a whole in terms of energy expenditure.

Thus, the planet of the intermediate flyby (Venus or Mars) is, as it were, a corrective mass, providing the return to the Earth with the most beneficial velocities.

Indeed, nomographic analysis of the parameters of such paths (e.g. Earth–Venus–Earth–Jupiter or Earth–Mars–Earth–Jupiter, etc.) shows that the effect of intermediate gravity assist maneuvers at Venus or Mars is quite low (in terms of the degree of variation in the spacecraft orbital parameters). However, it proves to be sufficient for the spacecraft to approach the Earth once more with the required velocity, to perform the gravity assist maneuver, and to be sent to the target planet along a trajectory with a higher energy. It is worth noting that the intermediate gravity assist maneuver can not only be accelerating (as for the flyby of Venus), but can also be decelerating as well (in the case of Mars flyby). Nevertheless, it does not negate the efficiency of the path in terms of energy, because the main benefits are due to the flyby of the Earth. These examples can be supplemented by other options of interplanetary trajectories considered earlier.

Unfortunately, when forming such schemes, one encounters the problem of selecting the "windows" for the spacecraft start and the dates of the first and second gravity assist maneuvers, because a certain mutual arrangement of planets is required. This is an obvious limitation in the use of such flight schemes.

In this connection, replacing the first flyby of a planet with segments of powered maneuvers of the spacecraft aimed at an optimal second passing of the Earth and accomplishment of the path is a sacrifice that allows us to circumvent the above problem [138].

Indeed, turning on the spacecraft engine on the way to the second Earth encounter allows us to avoid the problems of coordinating the planetary positions, because the second encounter with the Earth can be accomplished by selecting an additional intermediate impulse (or several impulses).

Let us use the nomograms to study the spacecraft trajectories, including a second encounter of the Earth, and to analyze variants of their application in different interplanetary flight schemes. Nomogram 4.14 shows curves determining the parameters of return segments of spacecraft trajectories to the Earth for two- and three-year start orbits (they are constructed based on the condition of the spacecraft encounter with the Earth after a decelerating impulse has been applied in the orbit apocenter for different central angles of the return segment).

Gravity assist maneuvers at the Earth (either acceleration or deceleration) result in the representing curves corresponding to the parameters of the new post-perturbation orbits shifting in the nomogram along the isolines $v_{\infty E} = const$. They will

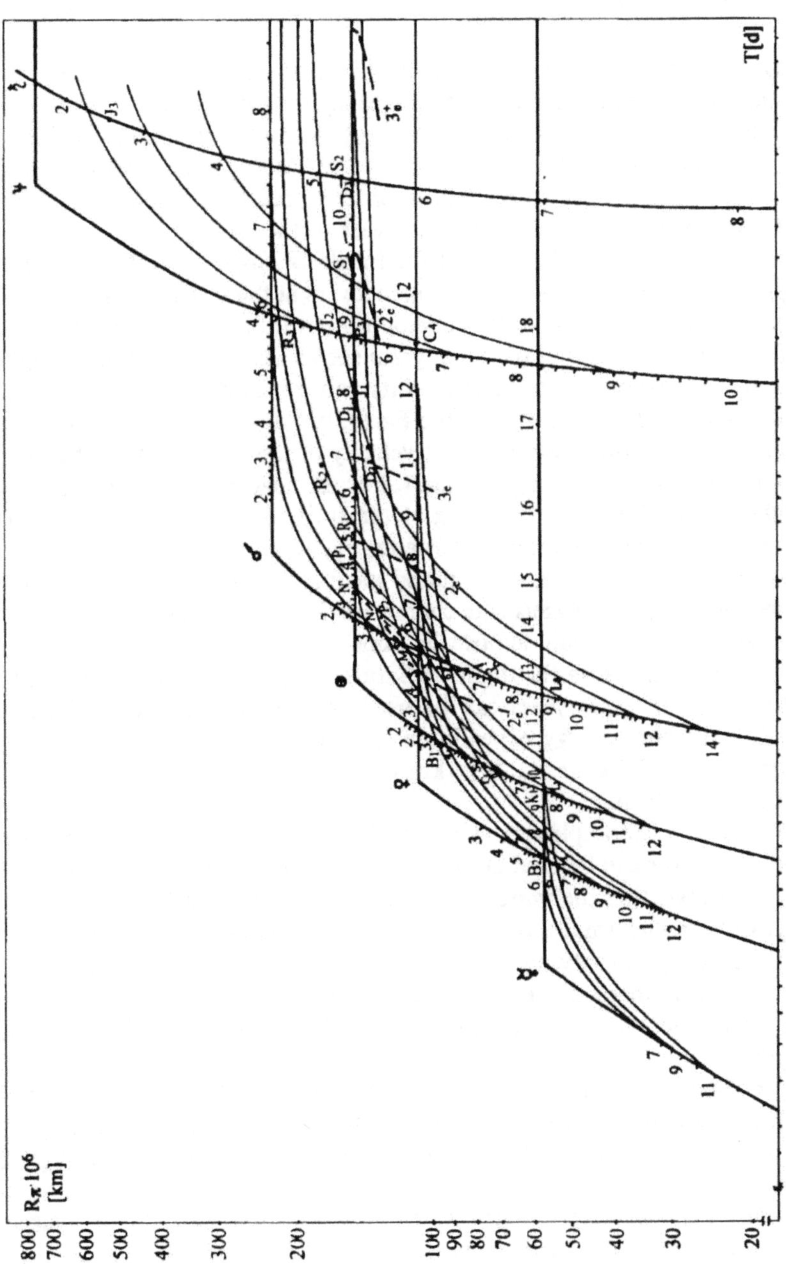

Figure 4.14 Formation of paths with multiple flybys of Earth.

shift to the right in the case of an acceleration gravity assist maneuver and to the left in the case of a deceleration gravity assist maneuver. In the nomogram these are constructed and denoted 2_e^+ and 2_e^- for two-year and 3_e^+ and 3_e^- for three-year orbits.

Analyzing the position of curves for deceleration ($2_e^-, 3_e^-$) and acceleration ($2_e^+, 3_e^+$) maneuvers in the nomogram, one can easily see that the use of deceleration maneuvers yields no notable benefits for reaching near planets (Mars and Venus), and paths with better characteristics can be found. Multiple deceleration gravity assist maneuver can be used at the Earth in order to reach near-solar regions, but this will be associated with a considerable increase in the mission time.

It is more beneficial to use the flyby of the Earth for flights to distant planets of the solar system. In this case, gravity assist maneuvers at the Earth can yield notable energy savings, and the increase in the mission time will be relatively small.

The behavior of the isolines in the nomogram shows that in the flights to Jupiter, both two-year and three-year recurrent orbits can be used; these orbits include acceleration gravity assist maneuvers at the Earth, but do not require an additional velocity impulse. In the first case, the start velocity of the spacecraft can be almost halved (relative to direct flights) $v_{\infty E} \geqslant 5$ km/s. For the spacecraft to reach Jupiter, it must pass the Earth along the return leg of the trajectory with $v_{\infty E}^{fl} \geqslant 9$ km/s. The spacecraft velocity at the encounter with Jupiter $v_{\infty J} \geqslant 5.6$ km/s.

In the case where $T_{ret} = 3$ years, the required starting velocity of the spacecraft will somewhat increase—$v_{\infty E} \geqslant 6.8$ km/s (the characteristics of the flyby and the encounter with the target planet remain almost the same).

As the nomogram shows, in the flights to Saturn, only spacecraft trajectories that return to the Earth in three years can be used (with a subsequent gravity assist maneuver). In this case, energy expenditure for the accomplishment of the flight ($v_{\infty E} \geqslant 6.8$ km/s, $v_{\infty E}^{fl} \geqslant 10.3$ km/s, $v_{\infty S} \geqslant 5.4$ km/s) is well below the expenditure for the direct flight ($v_{\infty E} \geqslant 10.2$ km/s).

In the flights to distant planets (Uranus, Neptune, Pluto) either a powered-gravitational maneuver can be used at the Earth on a two- or three-year return orbit (the nomogram shows that the impulse required for the spacecraft to reach Uranus at $T_{ret} \approx 3$ years is rather small), or orbits with return period $T_{ret} \approx 4$ years. In this case, with the starting velocity equal to 8 km/s and the mission time increased by 4 year, the orbits of these planets can be reached (for Uranus, $v_{\infty E}^{fl} \geqslant 11.3$ km/s, $v_{\infty U} \geqslant$ 4.6 km/s; for Neptune, $v_{\infty E}^{fl} \geqslant 11.7$ km/s, $v_{\infty N} \geqslant 4.1$ km/s; for Pluto, $v_{\infty E}^{fl} \geqslant 11.3$ km/s, $v_{\infty P} \geqslant 3.7$ km/s).

Thus, the nomographic analysis of the interplanetary flight schemes with a flyby of the Earth shows that they are most beneficial for flights to distant planets of the solar system, where they reduce the hyperbolic velocity required for the flight by 1.5–2 times.

It is obvious that the use of such schemes is not restricted to the variants considered above including gravity assist maneuver at the Earth. If necessary, their characteristics can be improved by the inclusion of a powered-gravitational maneuver or a flyby of one more planet (if possible). The principal advantage of such schemes is that they provide notable energy savings and extend the range of possible

Table 4.10 Spacecraft paths to distant planets using a flyby of the Earth (v_∞[km/s]).

Target planet		Direct flight	Two-year	Three-year	Four-year
Jupiter	$v_{\infty E}$	8.85	5.0	6.8	8.0
	$v^{fl}_{\infty E}$	—	9.0	9.0	9.0
	$v_{\infty J}$	5.6	5.6	5.6	5.6
Saturn	$v_{\infty E}$	10.2	—	6.8	8.0
	$v^{fl}_{\infty E}$	—	—	10.3	10.3
	$v_{\infty S}$	5.4	—	5.4	5.4
Uranus	$v_{\infty E}$	11.3	—	—	8.0
	$v^{fl}_{\infty E}$	—	—	—	11.3
	$v_{\infty U}$	4.6	—	—	4.6
Neptune	$v_{\infty E}$	11.6	—	—	8.0
	$v^{fl}_{\infty E}$	—	—	—	11.7
	$v_{\infty N}$	4.1	—	—	4.1
Pluto	$v_{\infty E}$	11.8	—	—	8.0
	$v^{fl}_{\infty E}$	—	—	—	11.8
	$v_{\infty P}$	3.7	—	—	3.7

spacecraft start dates. It should also be underlined that the use of these schemes makes it possible to standardize the requirements of the spacecraft acceleration block for flights to distant planets.

Below (in section 4.2) numerical studies are presented for a series of multi-purpose schemes incorporating gravity assist maneuvers of the spacecraft at the Earth.

4.1.9 Flights to the sun

Flights to the sun require considerable energy expenditure, which means that researchers use scheme options which include gravity assist maneuvers [56, 171, 174–180].

The suggested nomogram of flights in the solar system (Figure 4.1) makes it possible to analyze different variants of a space flight to the sun.

The nomogram shows that there are two possible ways of forming spacecraft trajectories in the near vicinity of the sun: by increasing the spacecraft start velocity or by using one or several gravity assist maneuvers. The first option implies an increase in the mission energy expenditure, whereas the second, implies an increase in the mission time.

Figure 4.15 shows numerical results of nomographic analysis of different flight paths to the sun. The analysis shows that an increase in the start velocity is advantageous in direct flights, flights via Venus and, especially, via Jupiter. It should be noted that a direct flight to the sun is very difficult to accomplish. Indeed, to reach the near-solar zone directly (~0.2–0.3 AU) requires considerable energy ($v_{\infty E}$= 10–14 km/s).

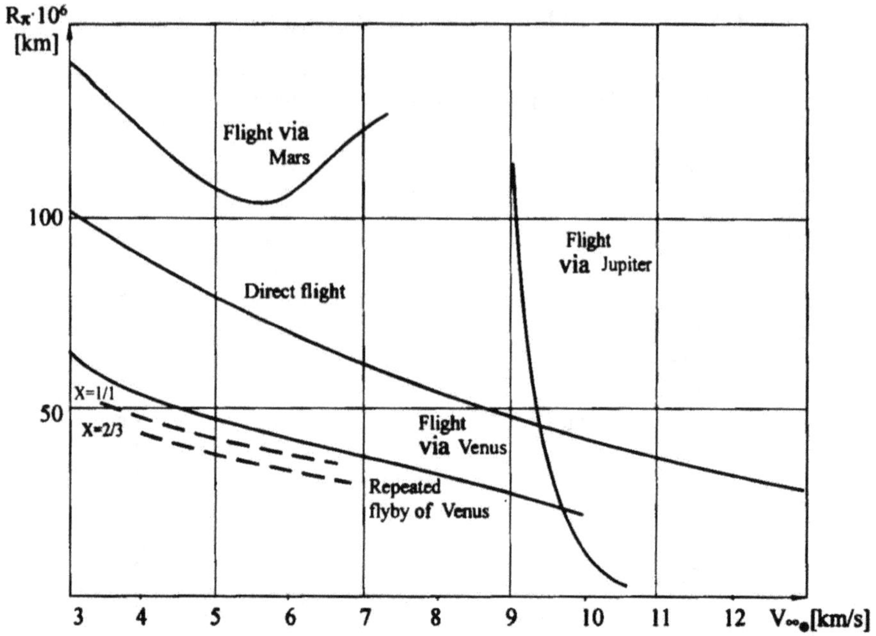

Figure 4.15 Optional spacecraft paths to reach near-solar regions.

Using planetary flybys allows an improvement in the energy characteristics of such a flight.

The results of analysis show that the greatest potential is provided by Jupiter, which has a powerful gravitational field. After a flyby of Jupiter, the minimum distance between the spacecraft and sun is 0.2–0.3 AU in the orbit pericenter. A small increase in the spacecraft start velocity ($v_{\infty E} \geqslant 9.5$ km/s) allows the spacecraft to travel even closer to the sun (down to 0.1 AU and even less). In the nomogram 4.1, this case corresponds to the transfer $Q_1 \rightarrow Q_2$.

If the velocity of approach to Jupiter is greater than 13.0 km/s, it is possible to send the spacecraft along the trajectory of falling on the sun. Similar possibilities are provided by other planets of the Jupiter group (Saturn, Uranus, Neptune). As can be seen from the nomogram, a flyby of any of these planets should be performed at lesser spacecraft velocities. However, in view of the vast distances to these planets, the duration of these missions to the sun increases substantially.

The role of the gravitational field of Venus in the flights to the sun is very promising. In this path, after the first passage (which yields a minimum distance of as little as 0.35–0.4 AU) it is advantageous to perform a second gravity assist maneuver at Venus. The second flyby of Venus requires the formation of a resonance orbit of the spacecraft with an associated increase in the mission time. The nomogram shows that the formation of the spacecraft orbit with the resonant ratio of the periods $X = 1:1$ ($T_{sc} = 224.7$ days) allows the spacecraft to travel even closer to the sun (approximately to 0.3 AU).

From the results presented (Figure 4.15) it follows that in the case under consideration, the efficiency of the second flyby of Venus is not too high. The condition of resonance of the orbital periods causes an increase in the flyby altitude at Venus, thus reducing the efficiency of the second gravity assist maneuver. It is more advantageous to form the resonant orbit with the resonant ratio of the periods $X = 2:3$ ($T_{sc} = 150$ days). In this case, the second gravity assist maneuver will make it possible to pass the sun at a distance less than 0.2 AU (transfer $Z_1 \rightarrow Z_2 \rightarrow Z_3$ in Figure 4.1). To do this, it is necessary to form a spacecraft orbit with a period $T = 150$ days after the first passage of Venus ($R_\pi = 40 \times 10^6$ km). After two revolutions, the spacecraft will encounter Venus, and after that will be able to approach the sun as close as 0.2 AU. The characteristics of this trajectory can be determined from the nomogram: $v_{\infty E} = 6.7$ km/s, $v_{\infty V} = 13.3$ km/s, $T_\Sigma = 550$ days. Generally speaking, the use of the planets of the Earth group in flights to the sun requires several flybys to be performed (it goes without saying that the periods of the spacecraft and planet must be divisible) in order to reduce the orbital perihelion.

Trajectories via Mercury do not yield significant benefits, because Mercury has a relatively weak gravitational field and cannot notably change the parameters of the flight orbit. However, nine flybys of Mercury allow the spacecraft to approach as near to the sun as 0.2 AU (transfer $A_1 \rightarrow A_9$ in the nomogram 4.1).

It appears fruitless to use the trajectories via Mars, because the approach in this case is even less than that in the direct flight with the same start velocity (Figure 4.15). Multiple flybys of Mars are also inefficient, because in order to reach the near-solar zone ($R_\pi \leqslant 0.2$ AU) the spacecraft start velocity must be increased ($v_{\infty E} \geqslant 8.5$ km/s) as well as the flyby velocity ($v_{\infty M} \geqslant 14$ km/s), and the mission time will increase significantly.

Consider variants of flights to the sun with flybys of two planets. The nominal potential of different combinations of some of these variants are shown in Table 4.11. The table presents the basic characteristics of different trajectories in the near-solar zones ($R_\pi = 0.2$ AU): hyperbolic excess of the spacecraft start velocity, flyby of the first and second planets, the number of gravity assist flybys that should be accomplished at the planets if several revolutions are required by the scheme. Analysis of the nomogram, in addition to these results, allows some qualitative conclusions to be made concerning the formation of these trajectories.

Table 4.11 Flights to the Sun.

1st flyby planet	2nd flyby planet	$v_{\infty E}$, km/s	$v_{\infty 1}$, km/s	$v_{\infty 2}$, km/s	Transfer type
—	—	13.8	—	—	$\alpha - N_\pi$
Mercury	Mercury (≥8)	7.65	9.55	9.55	$\alpha - \Pi^- - \Pi^- - N_\pi$
	Venus (≥2)	7.65	9.55	12.0	$\alpha - \Pi^- - \Pi^- - N_\pi$
	Earth (≥2)	8.25	11.0	13.8	$\alpha - \Pi^+ - \Pi^- - N_\pi$
Venus	Mercury (≥6)	3.40	6.05	7.50	$\alpha - \Pi^- - \Pi^- - N_\pi$
	Venus	6.70	13.30	13.30	$\alpha - \Pi^- - \Pi^- - N_\pi$
	Earth (≥2)	6.85	13.60	13.30	$\alpha - \Pi^+ - \Pi^- - N_\pi$

Consider some of these conclusions. The position and behavior of isolines of Mercury in the nomogram show that its use as a second flyby planet is not promising. Indeed, using a flyby of Mercury is inefficient in the trajectories Earth–Venus–Mercury–Sun and Earth–Mars–Mercury–Sun. Because of Mercury's rather weak gravity field, the flyby will not significantly reduce the perihelion altitude (by as little as 0.05–0.06 AU). A closer approach to the sun ($R_\pi \leqslant 0.2$ AU) requires multiple flybys of Mercury. Thus, in the flight via Venus, not less than 6 flybys at Mercury are necessary to attain the required approach to the sun.

Also of interest can be trajectories that use the gravity assist of Earth at the second passage (E–V–E–Sun and E–M–E–Sun). Thus, in recurrent trajectories to the Earth via Venus, the near-solar zones can be reached ($R_\pi = 0.30$–0.45 AU) at the expense of an increased spacecraft start velocity (up to $v_{\infty E} = 4$–7 km/s). In this case, a flyby of the Earth should be performed with greater velocity values ($v^{fl}_{\infty E} = 10$–14 km/s, transfer $L_1 \rightarrow L_2 \rightarrow L_3$ in Figure 4.1). The inclusion of one more flybys of the Earth in this scheme allows the spacecraft to approach the sun as close as 0.2–0.25 AU.

Flights via Mars and Earth somewhat impair the characteristics of such a trajectory—in terms of both mission time (two-year or three-year trajectories Earth–Mars–Earth have to be used) and energy requirements ($v_{\infty E} \geqslant 5.2$ km/s, $R_\pi \approx 0.5$ AU, for two-year orbits and $v_{\infty E} \geqslant 3.5$ km/s, $R_\pi \approx 0.4$ AU, for three-year orbits).

As can be seen from the nomogram, the inclusion of one more flybys of the Earth in such schemes will make it possible for the spacecraft to get as close to the sun as $R_\pi = 0.25$–0.30 AU.

It is interesting to compare the options for combining flybys of Venus and Mercury—the planets that are nearest to the sun. Nominal solutions presented in Table 4.11 (E–Me–V–Sun and E–V–Me–Sun) show that the second option is preferable in terms of energy requirements. Note that the first option makes it possible to approach even closer to the sun at the expense of an increase in the start velocity, whereas in the second option, such an increase in $v_{\infty E}$ is inefficient, because it results in a transfer into the domain of higher flyby velocities and a reduction in the perturbation effect of such flybys.

As for Jupiter and other trans-Jupiter planets, these can be used directly to reach the near-solar regions (primarily, Jupiter, as was shown earlier). Flying by any planet of the Earth group at the return leg appears inefficient, because in trajectories with energy as high as this, the perturbation effect of these planets is very low (it can be seen from the nomogram). On the initial leg of the trajectory (up to the giant planet), different combinations of flybys of Venus, Earth, and Mars can be used. This will allow almost halved energy requirements for the trajectory with the mission time increasing by 1–3 years.

It is worth noting that in these schemes, the problem of approaching the sun is to a great extent reduced to the problem of selecting a rational scheme for flying to Jupiter. In this case, if the arrival velocity at Jupiter is 13 km/s or more, the spacecraft can approach the sun as close as is required.

Thus, the trajectory Earth–Mars–Earth–Jupiter–Sun is possible with the use of two- and three-year orbits returning to the Earth (Figure 4.13). An acceleration gravity assist maneuver at the Earth allows the spacecraft to be sent to Jupiter and, after a gravity assist maneuver at Jupiter, to reach the near-solar zone ($R_\pi \approx 0.1$ AU). Energy requirements for such a trajectory are determined by the conditions of accomplishment of two- or three-year trajectory Earth–Mars–Earth ($v_{\infty E} \geqslant 4.5$ km/s). Similarly, recurrent orbits with an apocentral impulse and a flyby of the Earth can also be used (they were considered earlier).

It is impossible to reach Jupiter through the trajectory Earth–Venus–Earth–Jupiter–Sun with gravity assist maneuvers at Venus and Earth (as was shown earlier in nomogram 4.13). However, this trajectory can be implemented using a combination of powered-gravitational maneuvers near the flyby planets.

By and large, assessing trajectories of flights to the sun, we can select as most promising the trajectories via Jupiter and some trajectories including flybys of the planets of the Earth group. Among these, the trajectories via Venus, Mercury, or Earth allow the spacecraft to approach the sun as close as 0.2 AU.

For flights to zones closer to the sun (0.1 AU and less), one should consider the trajectories via Jupiter combined with Earth group planetary flybys used to reduce the energy requirements.

Flights via Jupiter are the most promising, though they impose more severe demands on the functioning of spacecraft systems during a long exploration mission.

4.2 Numerical studies of some promising schemes of planetary flights

Below we consider some promising multi-purpose interplanetary trajectories, in particular, flights to the largest outer planets (Jupiter and Saturn) using a powered-gravity assist maneuver at Mars and gravity assist maneuvers during the spacecraft's second encounter with the Earth.

Interest in such flight schemes is due to the possibility of enhancing the scientific efficiency of the mission by increasing the number of celestial bodies explored by a single spacecraft and reducing energy requirements as compared with direct flights to outer planets. Approaches to the selection and formation of multi-purpose planetary schemes are to a great extent determined by preliminary examination of the options based on the graphic-analytical method and the methods presented in 4.1.

Numerical studies of multi-purpose trajectories with powered-gravity assist flybys imply multiple determination of the parameters of planetocentric maneuvers. Therefore, we will first consider the methods for solving this problem, which is internal with respect to the problem of calculating the heliocentric segments of the flight.

4.2.1 Determination of the common pericenter for the trajectory of a planetary flyby

It is well-known [63, 64] that a nearly optimal value of the impulse of transfer velocity, given the values of incoming and outgoing velocities at infinity, can be attained if the velocity impulse is given in the point that is the common pericenter r_π of incoming and outgoing hyperbolas.

The value of r_π can be found from the following equation:

$$F(r_\pi)=v_{\infty in}+v_{\infty o}-\pi-\psi=0$$

where

$$\psi=\arccos(V^0_{\infty in}\cdot V^0_{\infty o}),\quad v_{\infty in}=\arccos\left(-\frac{\mu}{\mu+r_\pi V^2_{\infty in}}\right)$$

$$v_{\infty o}=\arccos\left(-\frac{\mu}{\mu+r_\pi V^2_{\infty o}}\right)$$

or in an expanded form:

$$F(r_\pi)=\pi+\psi-\arccos\left(-\frac{a_1}{a_1+r_\pi}\right)-\arccos\left(-\frac{a_2}{a_2+r_\pi}\right) \tag{4.1}$$

where $a_1=\mu/V^2_{\infty in}$, $a_2=\mu/V^2_{\infty o}$, and μ is the gravitational parameter of the planet. To develop an efficient algorithm to solve equation (4.1) let us examine the function $F(r_\pi)$ in the range of possible variations of the argument $0\leqslant r_\pi\leqslant\infty$.

Let us differentiate function $F(r_\pi)$ and write an expression for its first derivative:

$$\frac{dF}{dr_\pi}=\frac{\mu V_{\infty in}}{\sqrt{r_\pi(r_\pi V^2_{\infty in}+2\mu)}(\mu+r_\pi V^2_{\infty in})}+\frac{\mu V_{\infty o}}{\sqrt{r_\pi(r_\pi V^2_{\infty o}+2\mu)}(\mu+r_\pi V^2_{\infty o})}$$

$$=\frac{a_{in}V_{\infty in}}{r_\pi V_{\pi in}(a_{in}+r_\pi)}+\frac{a_o V_{\infty o}}{r_\pi V_{\pi o}(a_o+r_\pi)} \tag{4.2}$$

where V_π is the velocity in the pericenter of the hyperbola.

It can clearly be seen that $dF/dr_\pi>0$ whatever the values of the incoming velocity $V_{\infty in}$ and outgoing velocity $V_{\infty o}$ at infinity.

Differentiating expression (4.2) with respect to r_π yields the expression for the second derivative in the form:

$$\frac{d^2F}{dr_\pi^2}=-\mu\left[\frac{V^2_{\infty in}(a_{in}+r_\pi)^2+r_\pi^2V^2_{\infty in}}{r_\pi^3V^3_{\pi in}V_{\infty in}(a_{in}+r_\pi)^2}+\frac{V^2_{\infty o}(a_o+r_\pi)^2+r_\pi^2V^2_{\infty o}}{r_\pi^3V^3_{\pi o}V_{\infty o}(a_o+r_\pi)^2}\right]$$

Thus, the second derivative of the function $F(r_\pi)$ is negative whatever the values of r_π, $V_{\infty in}$, $V_{\infty o}$ and, therefore, the function $F(r_\pi)$ is convex upwards.

At $r_\pi=0$ we have $F(0)=\psi+\pi$ and $dF/dr_\pi=\infty$, i.e. the curve $F(r_\pi)$ is tangent to the ordinate axis in the point $r_\pi=0$, and takes a negative value. In the case when $r_\pi\to\infty$, the dependence $F(r_\pi)$ asymptotically tends to the value $F(\infty)=\psi>0$. Thus, equation $F(r_\pi)=0$ has a single solution in the set of admissible values $r_\pi\in(0,\infty)$, because $F(r_\pi)$ is a monotonically growing function.

Let us denote by $r_\varkappa^*$ the solution of equation $F(r_\varkappa)=0$, then $F(r_\varkappa)<0$ for $0\leqslant r_\varkappa\leqslant r_\varkappa^*$ and $F(r_\varkappa)>0$ for the interval $r_\varkappa^*<r_\varkappa<\infty$.

From here a rule for selecting the initial estimate $r_{\varkappa 0}$ can be derived: it is expedient to select $r_{\varkappa 0}<r_\varkappa^*$, then the equation $F(r_\varkappa)=0$ can be solved by Newton's iteration method.

It is reasonable to take as a initial estimate the minimum admissible radius of the flyby of the planet (safe radius) $r_{\varkappa\min}$, and if $F(r_{\varkappa\min})<0$, then the iterations continue according to the formulae:

$$r_{\varkappa i+1}=r_{\varkappa i}-F(r_{\varkappa i})\left(\frac{dF}{dr_\varkappa}\right)^{-1},$$

$$\frac{dF}{dr_\varkappa}=\frac{a_{\rm in}}{(a_{\rm in}+r_{\varkappa i})\sqrt{r_{\varkappa i}^2+2r_{\varkappa i}a_{\rm in}}}+\frac{a_{\rm o}}{(a_{\rm o}+r_{\varkappa i})\sqrt{r_{\varkappa i}^2+2r_{\varkappa i}a_{\rm o}}}.$$

If $F(r_{\varkappa\min})>0$, then the radius of the common pericenter we are searching for is less than the safe radius. In this case we will reduce the initial estimate $r_{\varkappa 0}$ until the condition $F(r_{\varkappa 0})<0$ is satisfied, and then we will use the above algorithm to find the common pericenter.

If the common pericentral radius $r_\varkappa^*$ obtained from equation (4.1) is less than the safe flyby radius ($r_{\varkappa\min}$) the maneuver cannot be performed in the common pericenter. In this case, the optimal value of the transfer velocity impulse ($V_{\rm M}$) can be found by a one-dimensional search for the minimum in the outgoing leg of incoming hyperbola, determined by the values of $V_{\infty\varkappa}$ and $r_{\varkappa\min}$. This scheme makes it possible to perform a guaranteed flyby of the planet, to determine the orbit based on measurements performed during the approach to the planet up to the orbital pericenter, and to combine the impulse of the powered maneuver with correction of the trajectory in order to reduce the total energy requirement. For the selected flyby scheme we have:

$$V_{\rm M}^*=\begin{cases}\sqrt{\dfrac{2\mu}{r_\varkappa^*}+V_{\infty\rm in}^2}-\sqrt{\dfrac{2\mu}{r_\varkappa^*}+V_{\infty\rm o}^2} & \text{for } r_\varkappa^*\geqslant r_{\varkappa\min}\\ \min V_{\rm M} & \text{for } r_\varkappa^*<r_{\varkappa\min}\\ r_{\varkappa\min}<r<\infty. & \end{cases}$$

Thus, the problem of optimizing a planetocentric maneuver is regarded as an internal problem with respect to heliocentric legs and can be solved whatever the variations in the parameters of heliocentric trajectories.

4.2.2 Flights to external planets with a Mars gravity assist

In [137] the efficiency of the gravity assist of Mars was shown for the flight to Venus under the assumption that the orbits of Earth, Mars, and Jupiter are circular and coplanar. However, the use of a simplified model of planetary motion fails to allow one to estimate the effect of periodicity of the start dates, eccentricity, and inclination of planetary orbits on the characteristics of interplanetary flight legs, nor does it allow one to study variations in the energy and other parameters of Earth–Mars–Jupiter (E–M–J) trajectories as functions of the starting year.

The use of a flight scheme to Jupiter with a gravity assist of Mars for the US spacecraft Galileo could save ≈ 300 m/s in the characteristic velocity [151].

Optimal trajectories were sought based on the criterion of minimum total expenditures:

$$\mathrm{opt}V_{\Sigma}=\min(V_{\mathrm{p}}+V_{\mathrm{M}}),\quad |V_{\infty\mathrm{in}}|=|V_{\infty\mathrm{o}}|$$

$$V_{\mathrm{M}}=\begin{cases}0 & \text{at } r_{\pi}\geqslant r_{\pi\min}\\ V_{\mathrm{M}}^{*} & \text{at } r_{\pi}<r_{\pi\min}\end{cases} \tag{4.3}$$

where

$$r_{\pi}=\frac{\mu}{V_{\infty\mathrm{in}}^{2}}\left(\operatorname{cosec}\frac{\psi}{2}-l\right),\quad \cos\psi=V_{\infty\mathrm{in}}^{0}\cdot V_{\infty\mathrm{o}}^{0}\qquad 0\leqslant\psi\leqslant\pi$$

r_{π} is the pericentral radius of flyby trajectory of Mars; $r_{\pi\min}$ is the radius of safe flight; μ is the gravitational parameter of Mars.

As can be seen from (4.3) the search for the optimal trajectories was conducted in the class of orbits satisfying the condition of equality of incoming and outgoing velocities at infinity for the planet.

Powered maneuver in the activity sphere of Mars was applied in the case where the search for optimum led to the domain of variables, where $r_{\pi}<r_{\pi\min}$. In these cases the value of the pericentral radius was set equal $r_{\pi\min}$, and the maneuver was conducted in the outgoing leg of the planetocentric trajectory.

For the flight scheme under consideration

$$V_{\mathrm{M}}^{*}=\min_{r\geqslant r_{\pi\min}}V_{\mathrm{M}}\quad \text{for } 0\leqslant v<\pi-\arccos\frac{\mu}{\mu+r_{\pi\min}V_{\infty}^{2}}$$

where v is the true anomaly of the point of maneuver accomplishment.

The results of the calculation show that the additional angle of rotation of the velocity vector at infinity is rather small. The derivative of the maneuver velocity with respect to the rotation angle for the family of constant-energy trajectories equals

$$\left.\frac{\partial V_{\mathrm{M}}}{\partial\psi}\right|_{V_{\infty}=\mathrm{const}}=\frac{\mu e^{2}V}{(\rho-r)V_{r}^{2}+(\rho-r)V_{m}^{2}-rV_{n}V_{\infty}} \tag{4.4}$$

where r, V, V_r, V_m are the distances from the center of the planet, velocity, and its radial and transverse components; e and ρ are the eccentricity and focal parameter of the orbit.

Function (4.4) has one extremum in the outgoing leg of hyperbola, and

$$\left.\frac{\partial V_{\mathrm{M}}}{\partial \psi}\right|_{r=r_{\pi}}=\sqrt{\frac{\mu}{p}}e \geqslant \left.\frac{\partial V_{\mathrm{M}}}{\partial \psi}\right|_{r\to\infty}=V_{\infty}.$$

Based on this property we get a simple expression for calculating the optimal powered maneuver:

$$V_{\mathrm{M}}^{*}=2V_{\infty\,\mathrm{in}}\cdot\sin\frac{\psi-\psi^{*}}{2}$$

where

$$\sin\frac{\psi}{2}=\frac{\mu}{\mu+V_{\infty\,\mathrm{in}}^{2}r_{\pi}^{*}}.$$

The suggested technique was used to calculate the trajectories of the Earth–Mars–Jupiter flight. The duration of the Earth–Mars flight varied from 60 to 160 days, and that of the Mars–Jupiter flight, from 500 to 1700 days. The safe flight radius with respect to Mars was assumed to be 3600 km, the altitude of the circular orbit of an artificial satellite of the Earth was 200 km.

For Earth–Mars–Jupiter trajectories, the interval between the optimal start dates is equal to the least common multiple of synodic periods of Earth, Mars, and Jupiter and amounts to 2.2 years, i.e. it is virtually equal to the periodicity of optimal Earth–Mars trajectories. Therefore, the possibility of accomplishing an Earth–Mars–Jupiter flight is twice as rare as that of carrying out direct Earth–Jupiter flights, for which the periodicity of optimal dates is 1.09 years.

Table 4.12 gives the principal characteristics of Earth–Jupiter trajectories in a 12-year interval which are optimal in terms of energy requirements. In individual start years, trajectories with a gravity assist of Mars allow a saving of 100 to 670 m/s (relative to direct flights) in the total characteristic velocity for reaching Jupiter. The velocities at infinity of approach to Jupiter are found to decrease as compared with direct flights. This is due to an increase in the pericentral radius of the heliocentric orbit after the powered-gravity assist maneuver at Mars.

For some start years, the flight time to Jupiter for trajectories E–M–J increases more than twice relative to direct flights.

The characteristics of interplanetary trajectories E–M–J were supposed to repeat in a time interval, denoted as the period of great oppositions and equal to the least common multiple of the astral periods and all mutual synodic periods of revolution of the planets. For E–M–J trajectories, the minimum value of the great opposition period is determined primarily by the period of Jupiter's revolution and amounts to 12 years. However, the synodic and sidereal periods of revolution of the planets are approximately divisible, and therefore, their configurations in 12 years are essentially different, which results in the failure of the characteristics of E–M–J trajectories to recur in the time interval equal to the period of great oppositions.

Table 4.12 Earth–Jupiter trajectories.

T_s	Sep 8, 1989	Oct 12, 1990	Nov 11, 1991	Dec 11, 1992	Jan 7, 1994	Feb 8, 1995	Mar 14, 1996	Apr 21, 1997	May 20, 1998	Jul 7, 1999	Aug 8, 2000	Sep 12, 2001
T_f	771	780	794	867	918	791	735	721	776	934	801	766
T_e	Oct 19, 1991	Nov 30, 1992	Jan 13, 1994	Apr 27, 1995	Jul 13, 1996	Apr 9, 1997	Mar 19, 1998	Apr 12, 1999	Jul 14, 2000	Jan 26, 2002	Oct 18, 2002	Oct 18, 2003
$V_{\infty 1}$, km/s	9.44	9.57	9.48	9.16	8.69	8.86	9.06	9.19	9.17	8.91	9.19	9.46
V_p, km/s	6.72	6.80	6.74	6.53	6.24	6.35	6.47	6.56	6.54	6.38	6.56	6.73
$V_{\infty 2}$, km/s	6.61	6.74	6.77	6.17	5.93	6.73	7.12	7.02	6.32	5.62	6.16	6.69
$\delta_{\infty 1}$, deg	35.2	39.5	34.5	21.4	−8.1	−25.6	−37.3	−37.3	−25.4	5.4	24.1	36

Studies of optimal Earth–Mars–Saturn trajectories show that in some start years, savings in the acceleration velocity from the Earth (relative to Earth–Saturn trajectories) may range from 300 m/s to 1200 m/s. In this case, the flyby of Mars is performed with no powered maneuver. In other start years, direct trajectories are most efficient in terms of energy. However, optimal Earth–Mars–Saturn trajectories are characterized by an increase in the total flight time by a factor of 2 to 3 as compared with direct flights.

4.2.3 Flights to external planets using repeated flybys of the Earth

Presented below are the results of numerical examination of the flight scheme for reaching outer planets with powered maneuver in the heliocentric segment of the flight and subsequent flyby of the Earth. Consider the method of optimization of such flight trajectories and examine their ballistic characteristics using flights to Jupiter and Saturn as an example.

A flight scheme to one of the external planets with a repeat flyby of the Earth is shown in Figure 4.16. At the moment T_1 the spacecraft enters a transition heliocentric orbit. Near the aphelion of the orbit, a powered maneuver is performed at the moment T_2 in order to transfer the spacecraft to the trajectory for an encounter with the Earth on the specified date T_3 or T'_3. Such a maneuver leads to an increase in the velocity at infinity of the encounter with the Earth as compared with the velocity at infinity of the departure from the Earth received by the spacecraft on the start date T_1. A gravity assist of the Earth on the date T_3 (T'_3) makes it possible to increase the heliocentric energy of the spacecraft and to transfer it to the orbit of encounter with the external planet at the moment T_4.

Specifying the heliocentric radius vector of the point where powered maneuver $\bar{R}_2$ is to be conducted and the time T_2 of arrival at this point, we can determine the transfer orbit $T_1 \to T_2$ as well as other legs of the interplanetary trajectory $T_2 \to T_3$, $T_3 \to T_4$ by solving Euler–Lambert equations. To join the trajectories in the flyby of the Earth it is convenient to introduce the heliocentric vector $\bar{R}_E$ of the point of intersection of incoming and outgoing hyperbolas, in which the powered maneuver is performed. In the case where a purely ballistic flyby is performed, $\bar{R}_E$ is the radius vector of the pericenter of flyby hyperbola.

For the flight scheme under consideration, the total velocity required to reach the external planet is

$$V_\Sigma = V_1 + V_2 + V_3$$

and if the spacecraft is to be placed in the orbit of the planetary satellite, this velocity is

$$\bar{V}_\Sigma = V_1 + V_2 + V_3 + V_4$$

where $V_1 = f(T_1, \bar{R}_2, T_2)$ is the velocity of acceleration from the orbit of the Earth's artificial satellite; $V_2 = f(T_1, \bar{R}_2, T_2, T_3)$ is the velocity of heliocentric powered

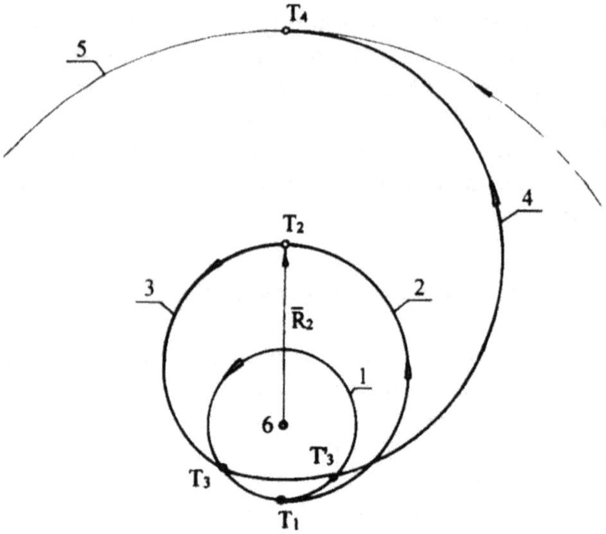

Figure 4.16 Diagram of Earth–Earth–Jupiter flight (1) orbit of Earth, (2) transfer orbit, (3) orbit of encounter with Earth, (4) orbit approaching Jupiter, (5) orbit of Jupiter, (6) the Sun.

maneuver; $V_3 = f(\bar{R}_2, T_2, T_3, T_4, \bar{R}_E)$ is the velocity of the powered maneuver during the flyby of the Earth; $V_4 = f(T_3, T_4)$ is the velocity required to transfer the spacecraft in to the specified orbit of a satellite of the external planet.

The problem of searching for the optimal trajectory in terms of energy reduces to minimizing the total velocity $\text{opt}V_\Sigma = \min V_\Sigma$ with respect to variables T_1, $\bar{R}_2$, T_2, T_3, T_4, $\bar{R}_E$.

Tentative calculations showed that it was possible to significantly reduce the problem dimensionality and to select the domains of local extremums of the cost function. This made it possible to develop a computational procedure allowing mass-scale optimization of trajectories without weakening the requirements to the accuracy of results.

Let us make some assumptions to simplify the examination of the scheme. In the general case, the flight trajectory from Earth to an outer planet lies beyond the ecliptic plane. An advantage of the flight scheme under consideration is that it allows the spacecraft to leave the ecliptic plane without expending energy. To do this, the transfer orbit is located in the ecliptic plane, and the necessary escape from it is performed with gravity assist of the Earth during a flyby at the moment T_3 (T_3'). The Earth's orbit is almost circular ($e = 0.017$), therefore, in terms of energy it is beneficial to direct the velocity vector (at infinity) of departure from the Earth at the moment T_1 collinear to the orbital velocity of the Earth. This provides a maximum energy value for the transfer orbit.

As a result, instead of five variables T_1, $\bar{R}_2$, T_2, three variables are introduced that determine the moment of departure from the Earth, the period of rotation of the transfer orbit, and the moment of powered maneuver—T_1, T, T_2.

Once the heliocentric legs of trajectories $T_2 \to T_3$ and $T_3 \to T_4$ have been determined, solutions of the Euler–Lambert equation are used to find the incoming $\bar{V}_{\infty in}$ and outgoing velocities $\bar{V}_{\infty o}$ at infinity at the Earth, which uniquely determine the plane of the heliocentric trajectory of the Earth's flyby:

$$\bar{C}^{o} = \frac{\bar{V}_{\infty in} \times \bar{V}_{\infty o}}{\sin\psi}, \quad \cos\psi = \bar{V}_{\infty in} \cdot \bar{V}_{\infty o},$$

where $\bar{C}^{o}$ is the unit vector of the integral of areas for heliocentric flyby orbit. Hence, $\bar{C}^{o} \cdot \bar{R}_E = 0$ and the problem of calculation of the geocentric maneuver is solved in the flyby plane.

Analysis shows that the presence of multiple extremums in the function $V_\Sigma = f(T_i)$ is due to the following local optimums: the minima of velocity V_Σ depend on the period T and lie near the values of transfer orbit periods that are approximately divisible by the Earth's period of rotation. Furthermore, the trajectories $T_2 \to T_3$ of encounter with the Earth have local minima, depending on which point, T_3 or T_3', is selected. These trajectories will be referred to as type 1 or 2, respectively. And finally, trajectories $T_3 \to T_4$ reaching the outer planet have two local minima lying in the first and second half-revolutions of this orbit.

Therefore, the calculation procedure was organized as follows. The total duration of the flight $T_\Sigma = T_4 - T_1$ was specified as a parameter of the problem and remained constant in the process of optimization. In each of the above mentioned domains, minimization followed the scheme

$$\text{opt}V_\Sigma = \min_{T_1}\left(\min_{T}\left(\min_{T_2}\left(\min V_\Sigma(T_3, T_4)\right)\right)\right) \tag{4.5}$$

$$T_4 = T_1 + T_\Sigma, \quad T_\Sigma = const$$

by the method of directed item-by-item examination using quadratic interpolation to find extremums.

In accordance with (4.5), the procedure regularly passed from the internal optimization cycles to external ones, and the optimal value of the internal parameter, obtained for the previous value of the external one, was used as an initial approximation for the new value of the external parameter. The global optimum was determined as minimum among all possible values. It was found convenient to use the following values as initial: $T = 2$ or 3 years, $T_2 = T_1 + T/2$, $T_3 = T_1 + T - 50$ days, $T_3' = T_1 + T + 50$ days, $T_1 = T_{1dir}$, where T_{1dir} is the optimal start date for direct flights from Earth to the outer planet, and T_Σ was considered in the range from 3.5 to 9.0 years with an interval of 0.5 year.

The suggested procedure was used to calculate Earth–Earth–Jupiter trajectories (E–E–J). In calculations, the altitudes of the circular orbit of artificial satellite of Earth and safe flyby at the Earth were set equal to 200 km. The value of deceleration impulse at Jupiter was determined for the orbit of an artificial satellite of Jupiter with the pericentral radius equal to $2R_J$ ($R_J = 71{,}500$ km is Jupiter's radius) and the revolution period of 60 days.

First of all, it is interesting to examine in more detail the trajectories E–E–J for the start window in 1990. This start year corresponds to the maximum energy requirements for Earth–Jupiter direct trajectories in the 12-year cycle (see Table 4.12).

Table 4.14 gives the dynamics of variations in the principal characteristics of trajectories E–E–J optimal in terms of V_Σ criterion depending on the total flight time in the case of starting from Earth in 1990. Comparison with direct Earth–Jupiter trajectories shows that the E–E–J flights provide a 1.85 km/s saving in the velocity necessary to reach the planet, and the total flight time increases by 1.5–2.3 times.

The above data show that there is an optimal value of V_Σ in terms of flight time, and a further increase in the flight time leads to an increase in the required velocity V_Σ. It should also be noted that the period of revolution along the transfer orbit is stable: it is near 740 days for the optimal trajectories of the 1st type, and near 765 days for the trajectories of the 2nd type. A $\pm$10-day variation in the period of revolution along the transfer orbit relative to its optimal value results in an increase in V_Σ by ~40 m/s.

If the flight time is 4–5.5 years, optimal trajectories E–E–J realize a purely ballistic flyby of the Earth, and $r_\pi > r_{\pi\min}$, i.e. $V_3 = 0$.

The limiting velocities at infinity for a flyby of the Earth along trajectories of the type 2 ($T_\Sigma = 4$ years) lie in the range 9.35–10.5 km/s, and are slightly greater than the velocities of departure from the Earth for direct trajectories.

In the case of "quick" flight trajectories ($T_\Sigma = 3.5$ years), the velocity impulse in the flyby amounts 1.12–2.09 km/s and is applied in the common pericenter of incoming and outgoing hyperbolae; the altitude of the common pericenter above the Earth's surface exceeds 200 km.

Note that the trajectories of the 1st type include flyby of the Earth from the side facing the sun, whereas the trajectories of the 2nd type, are from the shadowed side.

Analysis of the isoline fields for Earth–Jupiter trajectories showed that for the trajectories E–E–J of the 2nd type, the flight to Jupiter is accomplished by the E–J trajectories of the 1st half-revolution, and for the 2nd-type trajectories, by the 2nd half-revolution.

It can be seen from Table 4.14 that for the optimal E–E–J trajectories, the velocity impulse V_2 is applied within $\pm$50 days from the moment of the spacecraft passing the aphelion of transfer orbit, and the impulse magnitude varies within 0.5 to 1.0 km/s. Radius vector of aphelion of the transfer orbit for $T = 2$ years can be as long as 330 million km, therefore, when moving near the apocenter, studies of lower zones of the asteroid belt can be conducted.

It is important to note that optimal trajectories for flyby–descent flight schemes as well as for schemes with the spacecraft orbiting as an artificial satellite of Jupiter virtually coincide. This is due to a low sensitivity of the deceleration velocity impulse V_4 to variations in the velocity value at infinity of approaching Jupiter $V_{\infty J}$. Thus, for the incoming hyperbola with the pericentral radius $r_\pi = 2R_J$ and $V_{\infty J} = 7$ km/s.

$$\frac{\partial V_4}{\partial V_{\infty J}} = \frac{V_{\infty J}}{V_\pi} = 0.16,$$

where V_π is the velocity in the pericenter of incoming orbit.

Table 4.13 Earth–Saturn trajectories (first half-loop).

T_s	Dec 20, 1981	Jan 15, 1983	Jan 15, 1984	Jan 19, 1985	Jan 29, 1986	Feb 9, 1987	Feb 21, 1988	Mar 15, 1989	Mar 17, 1990	Mar 30, 1991
T_f	1713	2782	2416	2063	1837	1692	1599	1540	1479	1443
T_e	Aug 29, 1986	Aug 28, 1990	Aug 27, 1990	Aug 13, 1990	Feb 9, 1991	Sep 28, 1991	Aug 8, 1992	May 23, 1993	Apr 4, 1994	Mar 12, 1995
$V_{\infty 1}$, km/s	11.15	10.37	10.25	10.28	10.38	10.52	10.68	10.84	11.00	11.14
V_p, km/s	7.89	7.34	7.26	7.28	7.34	7.44	7.55	7.67	7.78	7.88
$V_{\infty 2}$, km/s	6.16	5.38	5.20	5.42	5.96	6.53	7.01	7.35	7.72	7.91
V_T, km/s	1.15	0.95	0.91	0.96	1.10	1.26	1.40	1.50	1.63	1.69
$\delta_{\infty 1}$, deg	25	−6	−9	−12	−18	−25	−31	−35	−39	−42

Table 4.13 (continued).

T_s	Apr 11, 1992	Apr 25, 1993	May 7, 1994	May 21, 1995	Jun 3, 1996	Jun 19, 1997	Jul 9, 1998	Jul 18, 1999	Jul 26 2000
T_f	1414	1407	1381	1398	1429	1549	1880	1991	1684
T_e	Feb 24, 1996	Mar 2, 1997	Feb 16, 1998	Mar 19, 1999	Mar 2, 2000	Sep 15, 2001	Aug 20, 2003	Dec 29, 2004	Mar 6, 2005
$V_{\infty 1}$, km/s	11.26	11.35	11.38	11.37	11.29	11.13	10.82	10.31	10.38
V_p, km/s	7.96	8.02	8.05	8.04	7.99	7.87	7.65	7.30	7.35
$V_{\infty 2}$, km/s	8.04	7.98	8.06	7.79	7.45	6.68	5.79	5.71	6.07
V_T, km/s	1.73	1.71	1.74	1.65	1.54	1.30	1.05	1.03	1.13
$\delta_{\infty 1}$, deg	−43	−42	−40	−36	−31	−25	−13	8	15

Table 4.14 Characteristics of Earth–Earth–Jupiter trajectories starting in 1990–1991.

Type	T_Σ, years	T_1	T days	T_2-T_1 days	T_3	V_Σ km/s	V_1 km/s	V_2 km/s	V_3 km/s	V_4 km/s	$V_{\infty J}$ km/s
1	3.5	13.01.91	740	340	23.11.92	5.99	4.34	0.53	1.12	1.73	10.90
	4.0	13.01.91	750	325	10.11.92	5.34	4.37	0.97	0	0.95	7.17
	4.5	15.01.91	740	350	24.11.92	4.99	4.34	0.65	0	0.74	5.79
	5.0	22.01.91	735	368	6.12.92	4.89	4.33	0.55	0	0.73	5.69
	5.5	30.01.91	735	398	15.12.92	4.89	4.33	0.55	0	0.81	6.25
2	3.5	24.11.90	765	323	4.01.93	7.06	4.43	0.56	2.09	3.07	15.40
	4.0	31.10.90	765	373	19.12.92	5.13	4.44	0.69	0	1.18	8.44
	4.5	2.11.90	760	360	16.12.92	4.99	4.42	0.57	0	0.79	6.14
	5.0	4.11.90	755	428	21.12.92	5.0	4.41	0.59	0	0.71	5.60
	5.5	16.11.90	765	313	25.12.92	4.99	4.43	0.55	0	0.77	5.99
	6.0	21.11.90	755	366	6.01.93	5.07	4.39	0.50	0.16	0.87	6.70

Calculations show (see Table 4.15) that the transfer orbit with a period of rotation $T \approx 3$ years is not useful in E–E–J trajectories, because in this case the acceleration velocity from the orbit of the Earth's artificial satellite is ~5.2 km/s, and considering the velocity impulse in aphelion $V_2 = 0.2$ km/s, the loss in terms of V_Σ amounts to 0.4 to 0.5 km/s relative to trajectories with $T=2$ years. In this case, the total flight time increases by ~1 year.

To study the pattern of variations in the optimal characteristics of E–E–J trajectories for different start years and to compare them with the characteristics of E–J direct flights, calculations were conducted for a 12-year period from 1982–1994. The optimization was performed for both types of trajectories with the flight duration of 4 years and $T \approx 2$ years. These parameters provide significant savings in V_Σ and increase the flight time by 1.3–2 years relative to direct trajectories. The results of calculations are presented in Table 4.16.

It should be noted that the characteristics of optimal E–E–J trajectories recur with a period of 12 years, equal to the period of the Earth–Jupiter great oppositions. Therefore, the results presented in Table 4.16 can be used to plan the flights along the path Earth–Earth–Jupiter from 1990 to 2000. It can be seen from the table that within the starting time interval under consideration, the value of V_Σ varies from 4.98 to 5.24 km/s for the trajectories of type 2 and from 5.11 to 5.50 km/s for the trajectories of the type 1. Note that throughout the interval of possible start times under consideration, the type 2 trajectories are more advantageous for flyby–descent schemes of flight to the planet than type 1 trajectories, and the saving in V_Σ in different years varies from 70 to 340 m/s. This is due to a lesser velocity impulse that is applied for type 2 trajectories in the vicinity of the aphelion of the transfer orbit. However, as can be seen from Table 4.16, the type 2 trajectories feature higher velocities of approach to Jupiter. Therefore, for missions aimed at placing the spacecraft in the orbit of an artificial satellite of the planet, the selection of the type of E–E–J transfer trajectory depends on the year of departure from Earth.

Table 4.15 Earth–Earth–Jupiter trajectories with a transfer orbit with 3-year period.

Type	T_Σ years	T_1	T days	T_2-T_1 days	T_3-T_1 days	V_Σ km/s	V_1 km/s	V_2 km/s	V_3 km/s	V_4 km/s	$V_{\infty J}$ km/s
	5.0	31.01.91	1100	530	1068	5.41	5.19	0.20	0.02	0.97	8.9
1	5.5	4.02.91	1100	490	1069	5.38	5.19	0.18	0.01	0.75	8.7
	6.0	6.02.91	1100	510	1068	5.39	5.19	0.19	0	0.75	8.8
	5.0	7.12.90	1110	655	1125	5.44	5.21	0.22	0.02	1.12	9.1
2	5.5	9.12.90	1110	555	1122	5.38	5.21	0.17	0	0.78	8.7
	6.0	12.12.90	1110	475	1123	5.42	5.2	0.20	0.02	0.74	8.9

For the trajectories with $T_\Sigma=4$ years, the dates of start from Earth along the optimal trajectories in terms of energy requirements each year are later than the start dates to Jupiter along direct 1st-half-revolution trajectories (the heliocentric angular flight distance is less than 180°). For the type 2 trajectories, the lag is 15 to 20 days, for the type 1 trajectories, it can be as large as 90 days.

There are two windows for start dates to Jupiter each year; they are separated by ~75 days and the corresponding total flight times are equal.

The energy characteristics of optimal trajectories are stable within a 12-year interval. Thus, the total velocity impulses V_Σ for E–E–J trajectories never differ by more than 260 m/s, whereas for direct trajectories, this difference amounts to 560 m/s. In addition to that, these trajectories impose no rigid restrictions on the moments of start, maneuver, and flyby of Earth.

For the E–E–J flight scheme, the boost velocity impulse from the orbit of the Earth's artificial satellite is ~4.5 km/s, which differs insignificantly from the velocity impulses typical of the flights to Venus and Mars. Taking into account the duration of the powered segment in the case of boost from the orbit of the Earth's artificial satellite, the savings in characteristic velocity relative to direct E–E–J trajectories, for which the boost velocities V_1 vary from 6.23 to 6.80 km/s, are even greater.

To assess the efficiency of Earth–Earth–Saturn flights, consider first the principal characteristics of direct Earth–Saturn flights. The period of Saturn's rotation around the sun is known to be 30 years, the eccentricity of its orbit equals 0.056, and the inclination of the orbital plane to the ecliptic is 2.5°. In flights to such orbits, the energy required to fly away from Earth is determined mainly by the inclination of the transfer trajectory, whereas the effect of eccentricity on the boost velocity is insignificant. The boost velocities from Earth are minimal in the start years such that the encounter with the planet takes place in the vicinity of ascending or descending nodes, and are maximum when Saturn is encountered in the points of its orbit at a maximum distance from the ecliptic plane. Therefore, when studying the energy characteristics of trajectories of flight to Saturn, it is sufficient to consider start dates within a 15-year interval.

Table 4.13 gives principal parameters of optimal Earth–Saturn trajectories in 1981–2000 requiring minimum boost velocity at Earth: $V_{\infty \eta}$ is the velocity at ininity

Table 4.16 Earth–Earth–Jupiter trajectories starting in 1982–1994.

Type	T_1, days	T days	T_2-T_1 days	T_3, days	T_V, days	V_Σ km/s	V_1 km/s	V_2 km/s	V_4 km/s	$V_{\infty J}$ km/s
1	8.04.82	740	400	11.02.84	7.04.86	5.28	4.38	0.90	0.90	6.86
2	26.01.82	760	370	16.03.84	25.01.86	4.98	4.40	0.58	1.07	7.86
1	17.05.83	745	326	17.03.85	16.05.87	5.33	4.41	0.95	0.87	66.8
2	2.03.83	760	360	23.04.85	1.03.87	5.02	4.42	0.60	1.01	7.56
1	24.06.84	740	390	25.04.86	23.01.88	5.32	4.40	0.92	0.84	6.52
2	5.04.84	760	390	30.05.86	5.04.88	5.04	4.44	0.60	0.98	7.36
1	6.08.85	740	320	7.06.87	5.08.89	5.21	4.40	0.81	0.82	6.37
2	17.05.85	765	333	8.07.87	16.05.89	5.08	4.47	0.61	0.97	7.28
1	11.11.86	735	368	12.07.88	10.09.90	5.21	4.37	0.84	0.82	6.38
2	24.06.86	765	343	14.08.88	23.06.90	5.12	4.48	0.64	0.99	7.44
1	15.10.87	735	350	7.08.89	14.10.91	5.41	4.38	1.03	0.85	6.58
2	29.07.87	765	353	19.09.89	28.06.91	5.18	4.47	0.71	1.05	7.76
1	14.11.88	745	353	5.09.90	13.11.92	5.50	4.47	1.13	0.89	6.83
2	29.08.88	765	413	23.10.90	28.08.92	5.24	4.47	0.77	1.12	8.13
1	13.12.89	745	353	6.10.91	12.12.93	5.47	4.36	1.11	0.93	7.06
2	30.09.89	765	403	21.11.91	29.09.93	5.20	4.46	0.74	1.17	8.37
1	13.01.91	750	325	10.11.92	12.01.95	5.34	4.37	0.97	0.95	7.17
2	31.10.90	765	373	19.12.92	30.10.94	5.13	4.44	0.69	1.18	8.44
1	9.02.92	740	340	16.12.93	8.02.96	5.11	4.35	0.76	0.95	7.16
2	29.11.91	760	350	15.01.94	28.11.95	5.04	4.41	0.63	1.16	8.36
1	9.03.93	740	360	15.01.95	8.03.97	5.15	4.37	0.78	0.92	7.01
2	28.12.92	760	400	15.02.95	27.12.96	5.88	4.40	0.60	1.12	8.14
1	13.04.94	740	370	16.02.96	12.04.98	5.29	4.39	0.90	0.89	6.84
2	27.01.94	755	428	21.03.96	26.01.98	5.02	4.39	0.63	1.07	7.88

of arrival at Saturn; V_T is the impulse velocity of transfer to orbit of Saturn's artificial satellite with pericentral and apocentral radii: $r_\pi = 160{,}000$ km, $r_\alpha = 5{,}750{,}000$ km.

From the above data it follows that the characteristics of optimal direct Earth–Saturn trajectories are widely scattered. Thus, the flight time varies between 3.8 and 5.65 years, the boost velocity from the orbit of Earth's artificial satellite varies from 7.28 to 8.05 km/s, and the velocity at infinity of arrival at Saturn is 5.42 to 8.06 km/s. These variations in the parameters of trajectories are due to the different conditions of approach to Saturn. For example, in the case of starting in 1985 or 2000, the planet is encountered in the descending and ascending nodes of the orbit, and in the case of starting in 1993–1995, it is encountered in the zone of Saturn's maximum deviation from the ecliptic plane.

The search for optimal Earth–Earth–Saturn (E–E–S) trajectories was based on the procedure given earlier for the Earth–Earth–Jupiter trajectories. Let us analyze variations in the energy characteristics of E–E–S trajectories versus the total flight time and the flight time along individual legs of the trajectory.

The boost impulse at Earth remains virtually constant, because of the constancy of the transfer orbit period, which amounts to ~750 days and ~780 days for orbits of the 1st and 2nd type, respectively.

For trajectories with $T \approx 2$ years, a powered maneuver is required during flyby of Earth whatever the value of T_Σ; in this case, the value of velocity for optimal trajectories amounts to 0.7 to 1.0 km/s, and the velocity impulse of maneuver in the aphelion of transfer orbit lies between 0.7 and 1.18 km/s.

Similar trajectory characteristics were studied for the transfer orbit with a period of 3 years. The use of such a transfer orbit yields somewhat greater energy savings in V_Σ (~300 m/s) in optimal trajectories with $T_\Sigma = 8$ to 9 years. However, in faster trajectories with $T_\Sigma = 6$ to 7 years, the velocity at infinity of arrival at Saturn increases significantly, which results in increasing V_Σ and $\tilde{V}_\Sigma$.

It is worth mentioning that for $T \approx 3$ years and total flight time of 7 years, the Earth flyby can be accomplished without powered maneuver. For $T_\Sigma = 6$ years, the maneuver impulse V_2 amounts to 0.61–0.83 km/s. A comparison of the data presented shows that for quick trajectories with $T_\Sigma = 6$ to 7 years, it is beneficial to use schemes E–E–S with $T \approx 2$ years, and for trajectories optimal in terms of energy with $T_\Sigma = 8$–9 years, the transfer orbit period should be ~3 years.

Radii of aphelions of transfer orbits with periods of 2 and 3 years amount to 340 and 470 million km, respectively. In these cases, the duration of the spacecraft's movement within the asteroid belt more than 330 million km from the sun equals 170 and 680 days, respectively.

To construct the calendar of flights to Saturn following the scheme E–E–S, we selected the class of quick trajectories with the total flight time of 6 and 7 years. The optimization was made to yield the minimum V_Σ for transfer orbits with $T \approx 2$ years, among the trajectories of the 2nd type. These parameters provide considerable energy savings with acceptable increase in the mission time.

The results of calculations for start dates ranging from 1985 to 2000 are shown in Figure 4.17. These data show that the total velocity required to reach Saturn varies within the 15-year interval under consideration from 6.17 to 6.61 and from 5.96 to 6.47 for T_Σ equal to 6 and 7 years, respectively. Transferring from a trajectory with $T_\Sigma = 6$ years to one where $T_\Sigma = 7$ years yields savings in V_Σ varying from 50 to 310 m/s, depending on the start year, together with a decrease in the velocity at infinity of approach to Saturn of 1.5–2.0 km/s. This results in a 470–600 m/s reduction in the deceleration impulse to enter the orbit of an artificial satellite of Saturn, or in a 300–390 m/s reduction in the velocity of entering the atmosphere.

Figure 4.17 shows that the application of the Earth–Earth–Saturn scheme allows a notable smoothing of scatter in the main characteristics of flight trajectories to Saturn.

Comparative analysis of the above data allows the following conclusions for different flight schemes to Saturn.

Direct schemes of Earth–Saturn flights require a boost impulse at Earth of 7.28–8.05 km/s, the flight time being 3.8–5.65 years, the deceleration velocity, 0.96– 1.74 km/s, and the velocity of entering the atmosphere, 35.68–36.35 km/s.

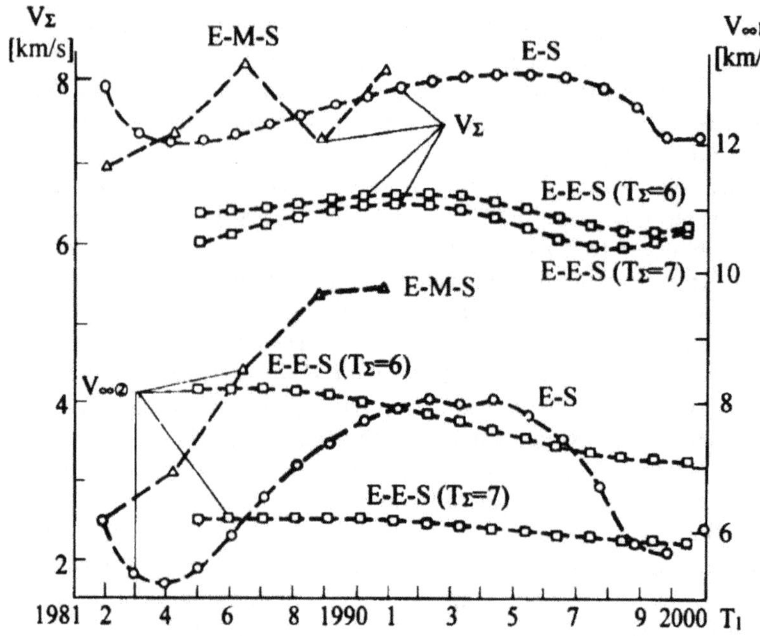

Figure 4.17 Main characteristics of Earth–Saturn, Earth–Mars–Saturn and Earth–Earth–Saturn trajectories starting in 1981–2000.

There is no point in applying the scheme of flight to Saturn with a gravity assist at Mars, because though in some years it allows energy savings of 0.3–1.2 km/s, the flight time in this case is several times greater.

The advantages of using the E–E–S scheme are as follows:

(i) for trajectories with $T_\Sigma = 6$ and 7 years, the scheme makes it possible to reduce the energy required to reach Saturn by 0.9–1.67 and 1.18–1.93 km/s, respectively, relative to direct flights;

(ii) it provides stability of energy and time characteristics of trajectories (thus, the difference between the total energy requirements for a 15-year period does not exceed 500 m/s);

(iii) it provides stability of the boost impulse at Earth, which amounts to ~4.5 km/s, i.e. its value is close to the boost impulses required to reach Mars or Venus. Taking into account the length of the powered boost segment from the orbit of Earth's artificial satellite, the amount of energy saved as compared with direct flight is even greater, and the scheme of starting from the orbit of Earth's artificial satellite is simpler. Finally, it should be noted that the energy required for the E–E–S flight is equal to or somewhat less than the energy required to reach Jupiter along direct trajectories. Therefore, the application of such a flight scheme allows similar boost to be used to deliver the spacecraft to Jupiter and Saturn.

4.3 Examination of periodical recurrent trajectories with multiple flybys of Mars and Earth

Recurrent trajectories of a space flight along a closed Earth–Mars–Earth route attract attention in the context of a possible manned flight to Mars.

The preparation of such a flight can include a long-term mission of an automated spacecraft with multiple flybys of Mars and Earth. An automated interplanetary "shuttle" spacecraft placed at an orbit, which alternatively comes close to Mars and Earth, can provide long-term opportunity to study the zones near these planets and the physical processes taking place in the Solar System. Again, the orbits like these (they are also called *cyclic orbits*) can be used for transport operations in manned missions. In these cases, a spacecraft, placed in such an orbit and equipped with a device for docking onto lighter spacecrafts and a life-support system, can be used to convey people and equipment from low near-earth orbits to the target planet in future interplanetary missions [115–120].

The periodicity of recurrence of the optimal dates for launching a spacecraft in to the Earth–Mars–Earth path is almost a multiple of the orbital period of Earth, and amounts to slightly more than two years ($\Delta T_S \simeq 2.13$ year); however, even a small change in the relative positions of the planets, as well as the perturbation effect of a planetary flyby, can notably affect the characteristics of the spacecraft orbit, thus hampering the implementation of such a mission. It is only numerical studies of these schemes based on a rigorous mathematical model that allow one to reveal the optimal characteristics of the path: launch windows, flight calendar, dates of planetary flybys, orbital characteristics of trajectories, and, most importantly, the energy expenditure required for the accomplishment of this kind of multipurpose spacecraft orbit.

The numerical solutions discussed below refer to the trajectories along the selected path, which are optimal in terms of energy and meet the restriction on the total mission time. The numerical studies were based on the universal algorithm for designing multi-purpose spacecraft orbits in the Solar System discussed in Chapter 2. Given the path M_j: 10003.1–10004.2–10003.2–10004.2 . . ., and the launch date $T_S = T_1$, the algorithm allows us to successively form a complex interplanetary trajectory L_j satisfying the criterion of minimum total characteristic velocity $V_{\Sigma\min}$.

A numerical examination of a multi-revolution complex trajectory, passing successively near Mars and Earth (Earth–Mars–Earth–Mars–Earth–Mars–Earth path) was conducted for a wide interval of launch dates ($T_S = 1992–2012$). The energy resources and the time of the mission were supposed to be limited $T_\Sigma \leqslant 10$ km/s, $T_\Sigma \leqslant 7$ years.

The examination revealed 10 launch windows for the spacecraft to follow the given path. The corresponding trajectories are not equivalent in terms of energy expenditure and provide a minimum total characteristic velocity for the relevant dates T_S, the periodicity of the launch-window recurrence being $\Delta T_S = 2.06–2.25$ years. The period of recurrence of the main characteristics of the complex multi-revolution trajectory is about 15 years and is equal to the period of the great opposition of Mars and Earth.

The total duration of the mission along the selected path (three complete revolutions of the "shuttle" vehicle between Earth and Mars) is quite stable ($T_\Sigma=6.15$–6.32 years) and shows virtually no dependence on the spacecraft launch date. At the same time, variations in the flight time for individual legs of the path are more pronounced: the time of the Earth–Mars leg varies from 0.23 to 0.50 years, and that for the Mars–Earth leg from 1.54 to 1.76 years (with the following combination of the segments of elliptic orbits: the flight along the first half-revolution at the Earth–Mars leg and along the second half-revolution at the Mars–Earth leg).

Table 4.17 gives the characteristics of the mission under consideration. The data show that the best conditions for the accomplishment of the mission (in terms of $V_{\Sigma\min}$) are in windows launching in November 1994, August 2005, and October 2009 (the corresponding values of V_Σ are 5.02, 4.96, and 4.89 km/s). The least favorable launch date for this mission is the year 2001. In this case the total energy expenditures are to high $V_{\Sigma\min}=8.61$ km/s (other parameters of the scheme are not optimal either).

The fuel reserve required for the spacecraft to implement the mission will be minimum if the spacecraft is launched in 2005 or 2009 (the ΔV_Σ is 0.39 and 0.42 km/s, respectively). The impulse required to place the spacecraft on the given path is rather stable and amounts to $\Delta V_1=4.41$–4.81 km/s throughout the date interval under consideration. The exception is the year 2001, when $\Delta V_1=5.20$ km/s and $\Delta V_\Sigma=3.41$ km/s.

Distribution of the values of the characteristic velocity in the case where powered and gravitational maneuvers are accomplished in the mission are represented by the values of ΔV_i. Here, subscripts 1, 3, and 5 denote the values of ΔV_i applied at Earth, and subscripts 2, 4, and 6, the values of ΔV_i applied in the flyby of Mars. They are small in the optimal variants of the mission and are sufficient to ensure that the shuttle spacecraft functions on the trajectory between Earth and Mars during more than six years with the reserve of the characteristic velocity on board being less than 0.5 km/s. It is interesting to assess the conditions of performing the powered and gravity assist maneuvers of the spacecraft at Mars and Earth throughout the multi-revolution trajectory. Table 4.18 gives the values of hyperbolic velocity excess of orbiting the spacecraft and approaching the planets $v_{\infty i}^-$ for optimal trajectories in the mission under study as well as the radii of application of velocity impulses from the planetary centers $R_{\pi i}$ ($i=3, 5, 7$; $i=2, 4, 6$ for Mars and Earth, respectively).

The incoming velocities at Mars at the first flyby $v_{\infty 2}^-$ range from 6.99 to 11.80 km/s, at the second flyby $v_{\infty 4}^-$, from 7.02 to 11.71 km/s, and at the third flyby $v_{\infty 6}^-$, from 5.37 to 11.99 km/s. The interval of incoming velocities at Earth is narrower: $v_{\infty 3}^-$ ranges from 5.37 to 6.41 km/s; $v_{\infty 5}^-$ from 5.39 to 6.30 km/s; $v_{\infty 7}^-$ from 5.24 to 9.75 km/s. The flyby altitudes at Earth are: $R_{\pi 3}=6573$–15592 km, $R_{\pi 5}=6536$–$13{,}734$ km. These elevations at flybys of Mars are: $R_{\pi 2}=10714$–158420 km, $R_{\pi 4}=6693$–45048 km, $R_{\pi 6}=4552$–159776 km. It is worth noting that greater flyby altitudes correspond as a rule to the paths which are optimal in terms of energy.

Table 4.17 Optimal project ballistic characteristics for flights along "shuttle" Earth–Mars–Earth path (three revolutions) $h_l = 200$ km.

Launch date, T_s	V_Σ	ΔV_1	ΔV_2	ΔV_3	ΔV_4	ΔV_5	ΔV_6	Δt_1	Δt_2	Δt_3	Δt_4	Δt_5	Δt_6	T_Σ
	km/s							Year						
1992.75	5.111	4.481	0.090	0.070	0.390	0.060	0	0.440	1.620	0.470	1.620	0.470	1.580	6.170
1994.78	5.022	4.421	−0.020	0.480	0.010	0.010	−0.060	0.410	1.600	0.440	1.680	0.380	1.760	6.270
1996.94	5.461	4.410	0.021	0.390	0.050	0.560	0.020	0.410	1.640	0.410	1.700	0.380	1.760	6.300
1999.07	6.620	4.390	−0.030	0.120	0.510	0.660	0.910	0.350	1.700	0.380	1.740	0.410	1.740	6.320
2001.32	8.610	5.200	0.320	0.220	0.800	0.720	1.340	0.230	1.740	0.380	1.760	0.500	1.680	6.290
2003.51	5.890	4.810	0.290	0.390	0.240	0.120	0.030	0.230	1.760	0.410	1.740	0.470	1.600	6.210
2005.66	4.960	4.570	0.100	−0.090	−0.090	0.010	0.010	0.320	1.700	0.440	1.660	0.440	1.700	6.260
2007.74	5.080	4.430	0.050	0.170	0.280	0.050	0.100	0.410	1.620	0.470	1.640	0.410	1.740	6.290
2009.81	4.890	4.470	0.060	0.030	−0.050	−0.250	0.030	0.440	1.640	0.440	1.640	0.410	1.540	6.150
2011.88	5.210	4.550	0.090	0.250	−0.230	0.060	0.020	0.440	1.640	0.410	1.700	0.380	1.740	6.310

Table 4.18 Optimal characteristics of planetocentric segments of flight along Earth–Mars–Earth path (three revolutions).

T_S	$v_{\infty 1}^{+}$	$v_{\infty 2}^{-}$	$v_{\infty 3}^{-}$	$v_{\infty 4}$	$v_{\infty 5}^{-}$	$v_{\infty 6}^{-}$	$v_{\infty 7}^{-}$	$R_{\pi 2}$	$R_{\pi 3}$	$R_{\pi 4}$	$R_{\pi 5}$	$R_{\pi 6}$
	km/s							year				
1992.75	5.405	7.759	5.760	8.816	5.661	10.449	5.408	31559	15317	17438	13734	71456
1994.88	5.271	9.203	5.373	10.791	6.301	10.899	5.989	158420	15592	20002	9899	9596
1996.94	5.234	10.758	5.424	11.645	5.602	11.645	5.784	57932	11972	45048	7058	11004
1999.07	5.205	11.806	5.781	11.127	5.452	8.446	5.364	44877	7692	9579	6741	4552
2001.32	6.881	11.079	5.767	8.405	5.396	5.347	5.237	10714	6780	6693	7638	6427
2003.51	6.112	8.432	5.875	7.021	6.081	7.188	5.296	26553	6373	29555	7317	19802
2005.66	5.604	6.995	6.168	7.596	6.195	9.366	8.424	125786	9896	21119	9876	9688
2007.74	5.294	7.330	6.632	8.500	6.236	11.344	9.755	72296	15429	14255	9457	6390
2009.81	5.377	8.639	5.810	10.276	6.154	11.333	5.914	29784	13205	19889	9619	159766
2011.88	5.551	10.015	5.691	11.714	6.102	11.995	5.848	26787	10788	36462	6556	22892

The above results confirm the possibility of ballistic realization of multi-revolution shuttle flights of a spacecraft between Earth and Mars and allow these flights to be considered as a means for long-term studies performed by automated spacecrafts on Earth–Mars–Earth missions, which can be a stage in the preparation and implementation of a manned mission to Mars. In the future, spacecrafts used as shuttle orbits will fulfill various functions of transport and navigation support in the implementation of such missions.

5. MULTI-PURPOSE FLIGHTS IN PLANETARY-SATELLITE SYSTEMS

5.1 Natural planetary satellites and their perturbation potential

Multi-purpose flights near planets that have a system of natural satellites are of great interest. Such schemes were used in the analysis and development of the well-known projects "Fobos" [113], "Galilei" [148–157], "Kassini" [163–167]. Studies on the perturbation effect of the Moon were also published [122–130]. Flybys of Martian satellites, in view of their minor mass, exerted no noticeable effect on the spacecraft orbits; flybys of massive satellites of a large planet allow a purposeful alteration of the spacecraft near-planetary orbit, thus extending the range of research.

When considering planetary missions, we should first mention the potential of the four satellites of Jupiter (Galilean satellites), whose masses are comparable with those of planets such as Mercury. Large natural satellites are known also at Saturn, Uranus, and Neptune. The gravitational characteristics of these satellites allow them to be considered objects for gravity assist maneuvering in the near-planetary zone.

Table 5.1 presents some characteristics of 12 known satellites of Jupiter (at the time of writing at least 17 satellites have been discovered). Jupiter with its satellites is, as it were, a miniature solar system. This system occupies a region with a radius of 30 million km, and has a rather compact inner system and vast outer system of satellites. The four inner satellites, called the Galilean satellites (I–IV) and satellite V orbit at relatively small distances from the planet ($R=1.81\cdot10^5$–$1.88\cdot10^6$ km) in almost circular orbits with a very low inclination to the equatorial plane of the planet.

All the Galilean satellites have the same periods of rotation both around the planet and around their axes; therefore, they always face Jupiter with the same side.

The outer system can be divided into two groups: satellites VI, VII, and X, which, like the five inner satellites, rotate counterclockwise when viewed from the north pole of the ecliptic plane, and satellites VIII, IX, XI, and XII, which move clockwise (reverse movement). Orbits of satellites VIII and IX are subject to rather strong perturbations, so their order with respect to Jupiter can vary over several years.

When flying to Jupiter and moving within its sphere of activity, the spacecraft should pass through the outer system of satellites, whereas the inner system of more massive satellites can play a more significant role in gravity assist maneuvers.

The satellites of Saturn are less studied because of their greater distance. A peculiarity of the system of Saturn satellites is that its six satellites are in resonance in pairs. Table 5.2 presents some characteristics of the 10 largest satellites of Saturn (at the time of writing at least 19 satellites are known).

Table 5.1 Physical and gravitational characteristics of the satellites of Jupiter [29].

Satellite	Average distance from Jupiter [km]	Period of satellite [days]	Eccentricity of the orbit	Inclination to the orbit of Jupiter [degrees]	Inclination to the equator of Jupiter [degrees]	Gravitational constant [km^3/s^2]	Radius of the sphere of activity [km]	Diameter [km]	Acceleration of gravity at the surface [cm/s^2]
V Amaltea	$181 \cdot 10^3$	0.498	0.0030	3.57	0.455	—	—	160	—
I Io	$421.6 \cdot 10^3$	1.769	0.0000	3.138	0.027	4826.62 ± 507	7945	3470	131.11
II Europa	$670.9 \cdot 10^3$	3.551	0.0003	3.565	0.468	3141.737 ± 88	9786	3160	126.11
III Ganymede	$1070 \cdot 10^3$	7.155	0.0015	3.222	0.183	10350 ± 165	2442	5000	155
IV Callisto	$1883 \cdot 10^3$	16.689	0.0075	3.965	0.253	6448.16 ± 672	37563	4700	53
VI Gimalia	$11480 \cdot 10^3$	250.57	0.1580	28.750	—	—	—	120	96.20
VII Elara	$11737 \cdot 10^3$	259.65	0.2070	27.966	—	—	—	40	—
X Lisitea	$11720 \cdot 10^3$	259.22	0.1070	28.466	—	—	—	12	—
XII Ananke	$21200 \cdot 10^3$	631.0	0.1690	147.3	—	—	—	11	—
XI Karme	$22600 \cdot 10^3$	692.0	0.2070	163.616	—	—	—	15	—
VII Pacife	$23500 \cdot 10^3$	735.0	0.738	148.066	—	—	—	11	—
IX Sinone	$23700 \cdot 10^3$	758.0	0.275	156	—	—	—	14	—

Table 5.2 Physical and gravitational characteristics of the satellites of Saturn [29].

Satellite	Average distance from Saturn [km]	Period of satellite [days]	Eccentricity of the orbit	Inclination to the orbit of Saturn [degrees]	Inclination to the equator of Saturn [degrees]	Gravitational constant [km^3/s^2]	Radius of the sphere of activity [km]	Diameter [km]	Acceleration of gravity at the surface [cm/s^2]
X Janus	$151.4 \cdot 10^3$	0.695	0.007	—		—	—	—	—
I Mimas	$185.5 \cdot 10^3$	0.942	0.0202	26.74		—	—	500	3.92
II Enceladus	$238.0 \cdot 10^3$	1.370	0.0045	26.74		—	—	570	6.86
III Tethys	$294.7 \cdot 10^3$	1.888	0.0000	26.74		43.035	1226.0	800	20.58
IV Diona	$377.4 \cdot 10^3$	2.737	0.0022	26.74		68.46	1956.0	850	41.05
V Rhea	$527.0 \cdot 10^3$	4.518	0.0009	26.74		146.67	1461.0	1400	—
VI Titan	$1222.0 \cdot 10^3$	15.945	0.0292	26.74		9390.0	43819.0	4850	156.96
VII Hyperion	$1481.1 \cdot 10^3$	21.277	0.1042	26.00		—	—	350	—
VIII Japetus	$3561.3 \cdot 10^3$	79.331	0.0283	16.18		92.17	22580.0	1330	27.44
IX Phoebe	$12952 \cdot 10^3$	550.48	0.1630	174.42		—	—	300	—

It is interesting to assess the changes in the orbital characteristics of the spacecraft resulting from the flyby of natural satellites of planets.

As mentioned earlier (see Chapter 1), the efficiency of a gravity assist maneuver can be characterized by the maximum vector change in velocity Δv_{max} and the maximum change in the spacecraft orbital energy.

Table 5.3 shows the values of these characteristics for the Galilean satellites of Jupiter and the five largest satellites of Saturn. Here v_c is the circular orbital velocity of the satellite of Jupiter.

The results presented illustrate the maximum perturbation potential of the Jupiter and Saturn satellites, when flybys are made in the orbits of the corresponding optimal energy level h_{opt}.

As can be seen, the greatest energy at Jupiter can be obtained by a flyby of Io; flybys of Ganymede, Europa, and Callisto produce a lesser effect. In the sphere of activity of Saturn, the greatest effect is the perturbation effect of Titan, whereas the potential of Tethys, Diona, and Rhea are approximately equal.

The above results allow us to conclude that the most efficient flyby of any of the natural satellites of Jupiter is that performed along the orbit whose energy constant is close to that of the orbit of the flyby body.

It is interesting to assess the efficiency of gravity assist maneuvers for a wide range of parameters of the incoming orbit, taking into account that the efficiency of a gravity assist maneuver is affected by the initial energy level of the incoming orbit, the incoming trajectory angle, and the flyby altitude.

Figures 5.1 and 5.3 present the curves of maximum increase in the spacecraft orbital energy resulting from a gravity assist maneuver at the satellites of Jupiter and Saturn for a wide range of energy of the initial incoming orbits. The curves were constructed using formula (1.2.2) at the values of the corresponding relative flyby height $m=1$. The corresponding optimal arrival angles at the Galilean satellites of Jupiter are shown in Figure 5.2, and at the satellites of Saturn in Figure 5.4.

Table 5.3

Satellites	v_c, km/s	Δv_{max}, km/s	h_{opt}, km^2/s^2	Δh_{max}, km^2/s^2
Satellites of Jupiter:				
Io	17.584	1.831	−274.22	64.39
Europa	13.741	1.448	−166.50	39.80
Ganymede	10.880	1.885	−99.10	41.01
Callisto	8.200	1.676	−58.21	27.50
Satellites of Saturn:				
Tethys	11.340	0.268	−125.90	6.08
Diona	10.610	0.306	−109.70	6.50
Rhea	8.460	0.397	−68.50	6.72
Titan	5.560	1.810	−37.60	20.20
Japetus	3.270	0.338	−9.40	2.20

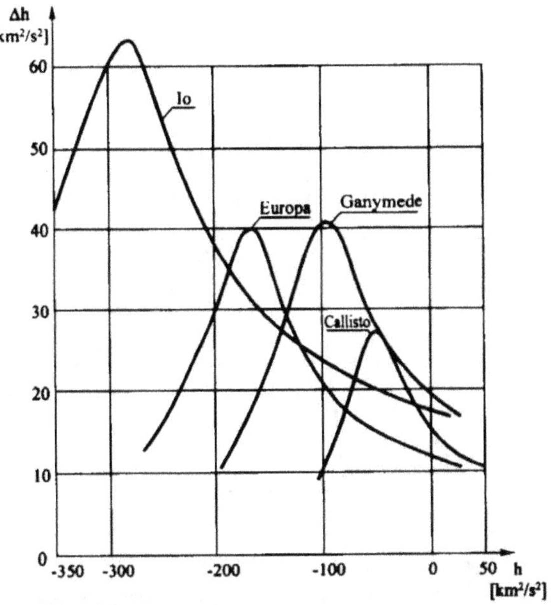

Figure 5.1 Maximum change in the spacecraft orbital energy due to gravity assist flybys of satellites of Jupiter.

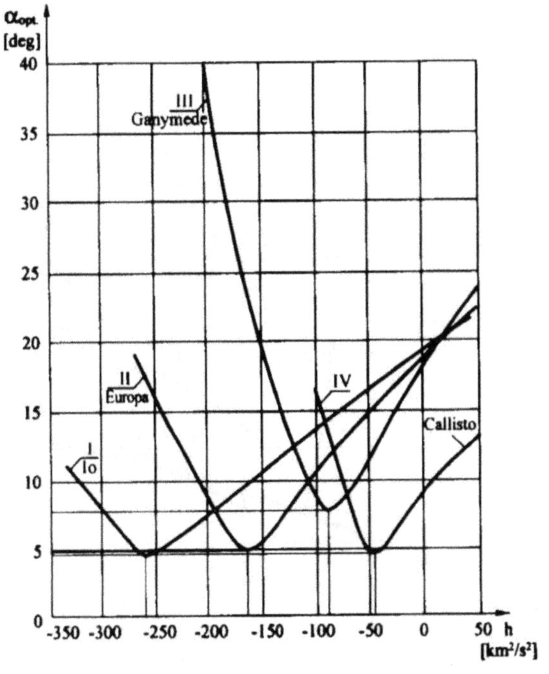

Figure 5.2 Optimal angles of approaching Jupiter satellites.

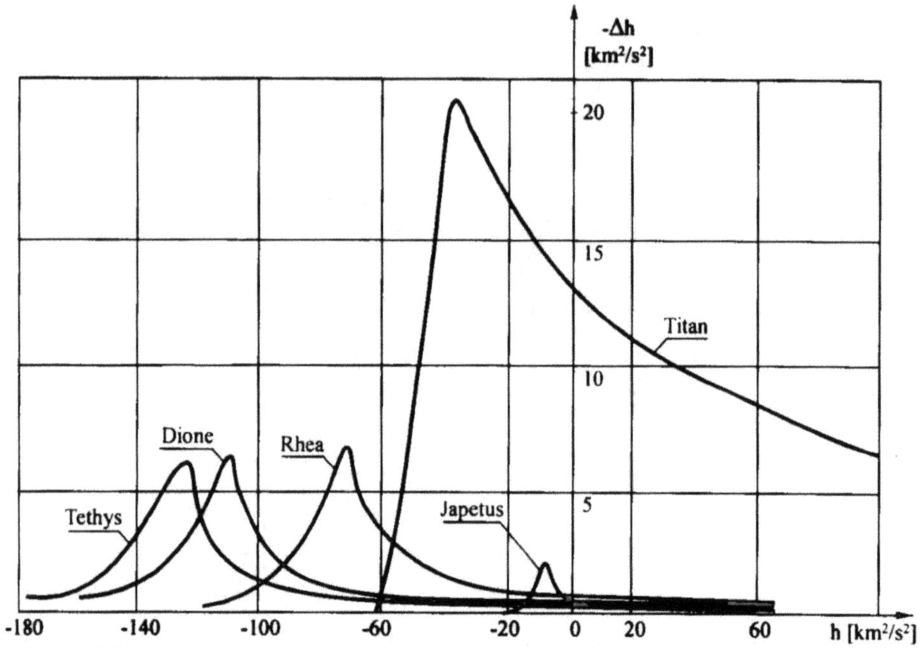

Figure 5.3 Maximum change in the spacecraft orbital energy due to a gravity assist flybys of satellites of Saturn.

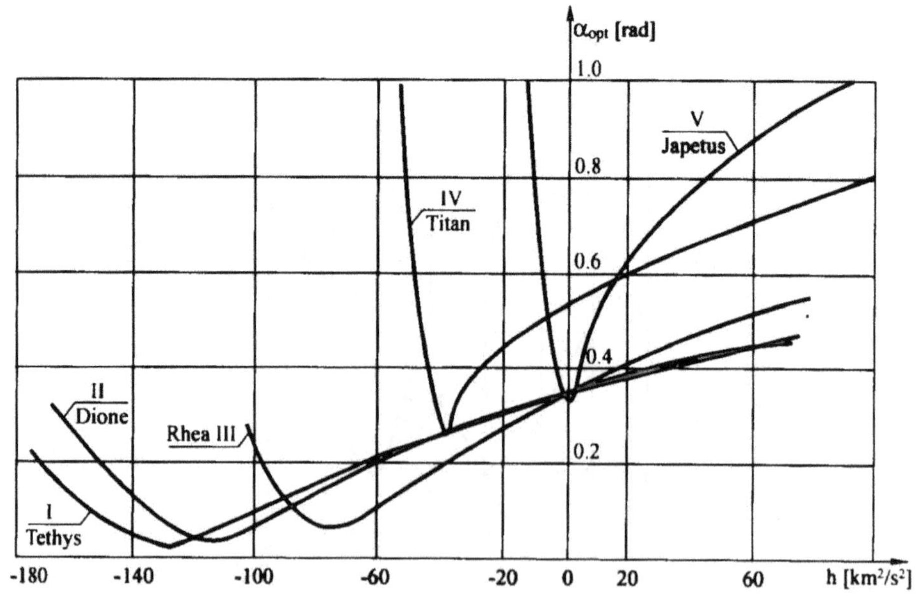

Figure 5.4 Optimal angles of approaching satellites of Saturn.

The results show that the optimums of the curves of the spacecraft orbital energy increment due to gravity assist maneuvers are clearly defined (they are located near the values of orbital energy of the corresponding flyby satellite). A significant reduction in the efficiency of the maneuver is found to result from deviation of the energy constant of the spacecraft arrival orbit from the optimal values (in particular, if they fall into the domain of hyperbolic flyby orbits). Thus, the most efficient gravity assist maneuver for deceleration in hyperbolic and high-energy elliptic arrival trajectories (semi-major axis $a > 10^6$ km) is a Ganymede flyby (in particular for incoming hyperbolas corresponding to the initial energy of 30–50 km^2/s^2 or interplanetary trajectories of the spacecraft approach to Jupiter $v_{\infty J} = 5.5$–7 km/s). For elliptic planetocentric orbits ranging within $0.7 \cdot 10^6$ km $< a < 10^6$ km, the most efficient is the flyby of Europa. Finally, in the case of elliptic orbits with semi-major axis $a < 0.7 \cdot 10^6$ km, the most expedient is the flyby of Io.

Similar zones can be selected for the satellites of Saturn. The graphs presented show that the most efficient is the flyby of Titan (it can be performed both in hyperbolic and elliptic arrival orbits). Considerably less efficient are the effects of Tethys, Diona, and Rhea, though their flyby can be useful in some ranges of the initial arrival energy. The potential of Japetus is also rather low; its preferential flyby domains lie in the range of high-energy elliptic orbits.

The relationships presented in Figure 5.2 show the optimal spacecraft arrival angles at the satellites of Jupiter vary within 5°–20°. Those angles for Saturn vary within a wider range (Figure 5.4). The values of optimal incoming angles α_{opt} are significantly affected by deviations of the spacecraft arrival energy h from its optimal values h_{opt}.

The curves in Figure 5.5 (in the case of a flyby of Ganymede, a satellite of Jupiter) illustrate the behavior of the efficiency of a gravity assist maneuver Δh in a wide range of parameters of spacecraft incoming orbits: arrival angles α, orbital energy h, relative flyby altitude m. The graph shows that the efficiency of a gravity assist maneuver of the spacecraft decreases significantly when the incoming orbit parameters deviate from the optimal values. The most sensitive to such deviations are the trajectories with optimal incoming energy; this fact imposes some navigational requirements on the use of this kind of orbit.

Such relationships can be useful in the analysis of spacecraft orbits utilizing gravity assist maneuvers at planetary satellites.

5.2 Formation of spacecraft orbits in planetary satellite systems

As was shown above, using flybys of several natural planetary satellites results in significant changes in the spacecraft orbital parameters with no velocity impulses applied (and, therefore, no fuel spent). This circumstance allows the spacecraft gravity assist maneuvers to be used for impulseless control of the orbits of exploration spacecraft trajectories in planetary satellite systems.

The objectives of such flybys, in addition to the possibility of gaining additional data during the flyby of the natural satellite, can be as follows:

- a change (either decrease or increase) in the spacecraft orbital period;
- a change in the pericenter (or apocenter) of the orbit;

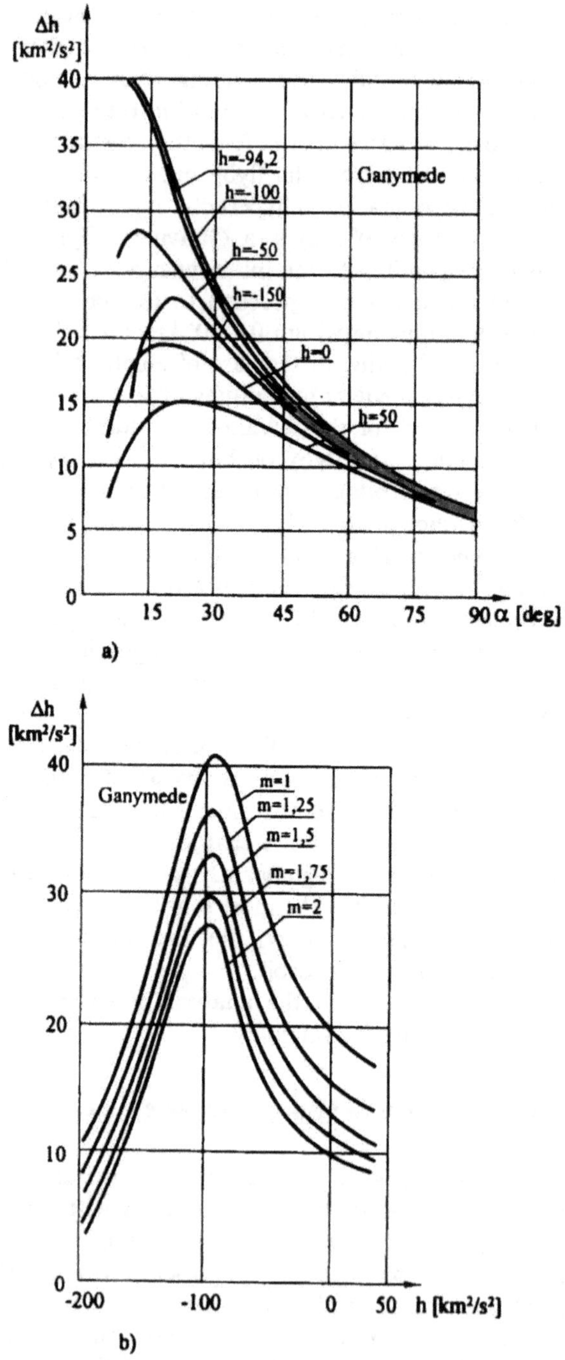

Figure 5.5 Changes in the spacecraft orbital energy as a function of (a) angle of approaching and (b) relative flyby altitude *m* at Ganymede.

- rotation of the apsides;
- a change in the angle of the spacecraft orbital plane inclination.

The schemes of a spacecraft flight in planetary satellite systems including gravity assist maneuvers at natural planetary satellites can be constructed and analyzed based on the approaches and algorithms discussed in Chapter 2.

5.2.1 Perturbation changes in the periods of orbit

As was shown in 2.2, a gravity assist maneuver at an attracting natural satellite of a planet can notably affect the spacecraft orbital period. A series of maneuvers at one or more natural satellites allows one to perform the required transformation of the spacecraft orbit. Analysis of such transfers conducted for the Galilean satellites of Jupiter (Figures 2.11–2.13) implies the combined use of the isoline fields of relative velocities $\Delta v = const$ and isochrones $T^* = const$, determining the post-perturbation periods of the orbits.

Similar isoline fields can be constructed for any planetary satellite system, thus allowing graphical analysis of post-flyby orbits across a wide range of spacecraft orbital characteristics.

Figure 5.6 shows the combined field of isolines $\Delta v = const$ for the five largest satellites of Saturn: Tethys, Diona, Rhea, Titan, and Japetus. These isolines,

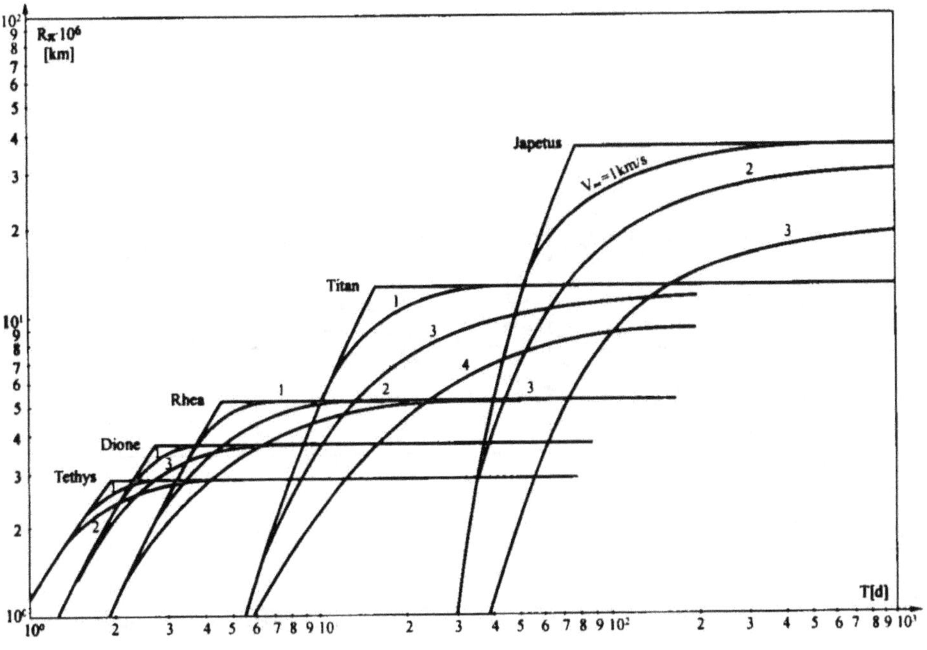

Figure 5.6 Isolines of relative velocities $\Delta v = const$ for satellites of Saturn.

constructed in coordinates of orbital elements R_π and T, reflect changes in the pericentral distances and periods of the spacecraft orbit in the planetary satellite system of Saturn. Each of the points in the field is associated with an orbit of the spacecraft with a certain value $\Delta v = v_\infty$, and changes in the orbital elements in the course of gravity assist maneuver correspond to the behavior of the isolines on the presented plot (in accordance with the invariance condition).

Analysis of isochrones $T^* = const$ for the satellites of Saturn shows that Titan, the largest of the satellites of Saturn, has the highest potential for changing the spacecraft orbital parameters (see Figure 5.7).

The use of isochrone for each of the natural satellites of the planet together with the combined fields of isolines makes it possible to analyze individual schemes of spacecraft orbits and to make some qualitative estimates about flight schemes in planetary satellite systems.

Thus it is evident, for example, that during the flight to a natural satellite of the planet (typical for the satellites of both Jupiter and Saturn) it is beneficial to transform the spacecraft orbits in the domain with a large pericentral distance R_π and small rotation period T (top left part of the nomogram). This provides the spacecraft encounter with the corresponding natural satellite of the planet at low velocity Δv. It should also be mentioned that a single flyby of a natural planetary satellite brings about an abrupt reduction in the period. This change is especially distinct when the period T of the initial orbit is large. It appears that the above fact counts in favor of schemes with large initial orbital periods. Transferring the spacecraft from a hyperbolic incoming orbit to such an orbit requires relatively low

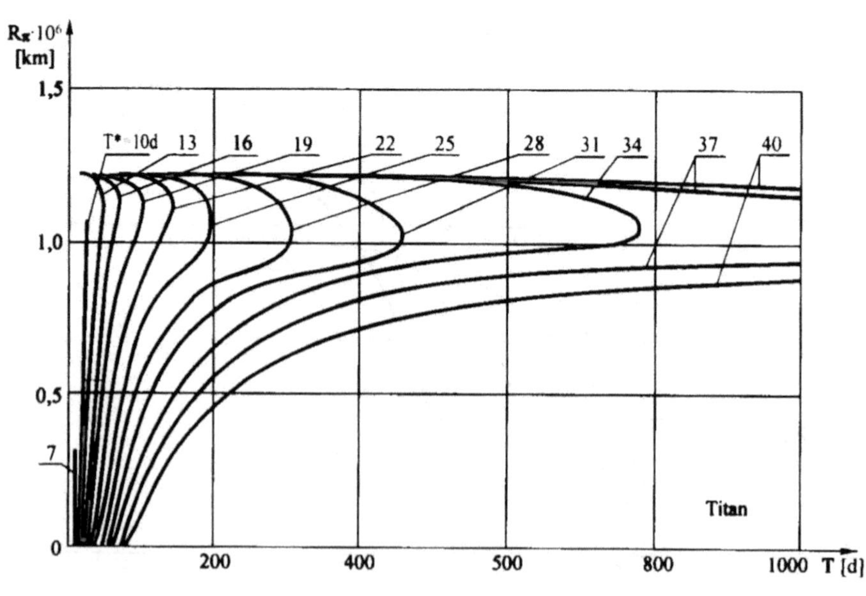

Figure 5.7 Isolines of isoperiodic orbits $T^* = const$ in the case of flyby of Titan.

fuel expenditure. Further on, using a flyby of a natural planetary satellite at the first revolution will allow a significant reduction in the period of revolution of the spacecraft.

5.2.2 Gravity assist rotation of the spacecraft orbit

A gravity assist maneuver near an attracting body makes it possible to rotate the spacecraft orbit within the plane of its movement and to change the inclination of the orbit.

Such maneuvers are considered in detail in Chapter 2 by using as an example a flyby of Ganymede, a natural satellite of Jupiter. A flyby of Ganymede (Figure 2.9) makes it possible to rotate the apsides of the spacecraft orbit by 25°–30° and even more. Similar or somewhat less is the potential of other Galilean satellites of Jupiter.

Figure 5.8 shows the isolines of the rotation angles of apsides of spacecraft orbits brought about by flybys of Titan, the largest satellite of Saturn.

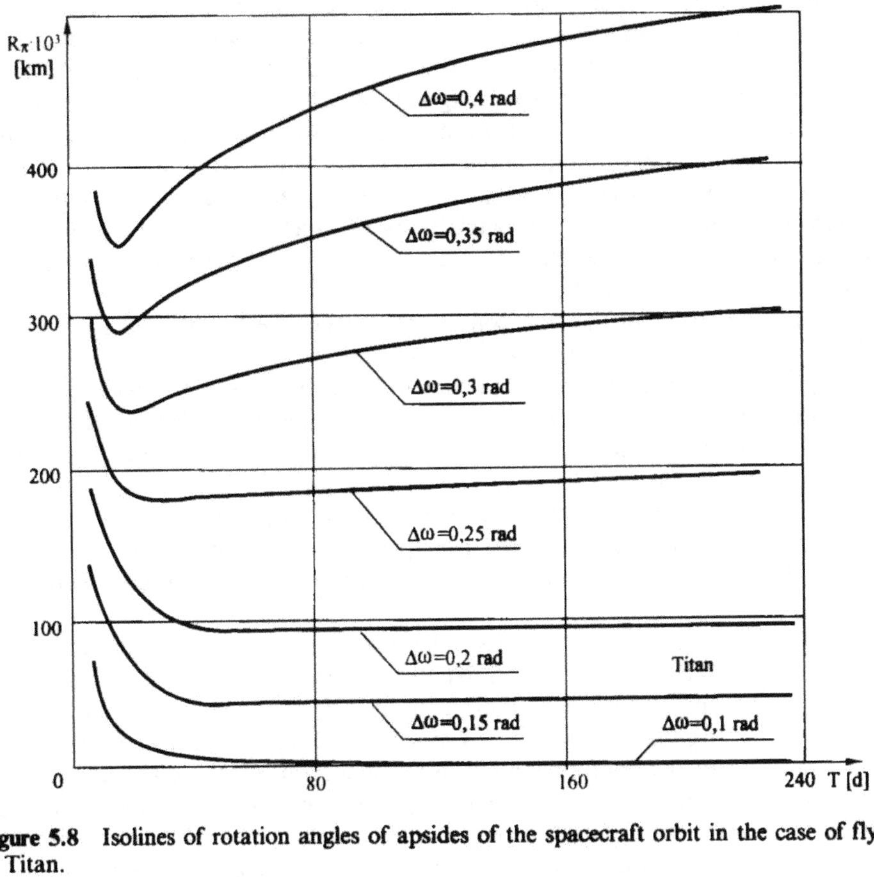

Figure 5.8 Isolines of rotation angles of apsides of the spacecraft orbit in the case of flyby of Titan.

When analyzing gravity assist rotations of the spacecraft orbit apsides, it is worth mentioning that the most significant rotations can be attained at highly elongated orbits with pericenters close to the orbital radius of the natural satellite that is passed by.

Note that the rotation potential of the apsides during a flyby of an attracting body is somewhat greater due to the rotation of the body itself on its orbit during the gravity assist maneuver (during the time in which the spacecraft passes the sphere of activity of the attracting body). However, this effect will be perceptible only in the case of flybys of bodies with strong gravity fields (and, therefore, large spheres of activity). In the case of flybys of the satellites of Jupiter and Saturn, whose spheres of activity are small, the effect of additional rotation of the spacecraft velocity is small, and the angular changes in the apsides can be taken from the plots presented.

An important feature of the use of gravity assist maneuvers at a natural planetary satellite is the possibility to change the inclination of the spacecraft orbital plane. Such a problem can appear, for example, when it is necessary to eliminate the initial inclination of the spacecraft orbital plane to the orbital plane of a natural satellite or when one wants to increase the spacecraft orbital inclination.

Figure 5.9a,b,c presents the curves of maximum changes in the inclination of the spacecraft orbit as a result of a single flyby of the natural satellites of Jupiter and Saturn. These curves were constructed using relationships 1.2.10–1.2.11 for different values of velocity v_∞ at the corresponding natural satellite of the planet. As can be seen from the results presented, the highest potential of changing Δi are typical of the satellites of Jupiter–Callisto and Ganymede (up to 9°–10°). Among the satellites of Saturn, Titan features the highest potentialities in changing the inclination of the spacecraft orbit (up to 16°). A single flyby of Japetus allows a rotation of the spacecraft orbital plane by angles of up to 5°. The rest of the natural satellites of Saturn yield Δi of the order of 1°–2°.

It should be noted that the best conditions for the spacecraft's approach to the satellite are characterized by the spacecraft velocity of $v_\infty = 1\text{–}2$ km/s.

Taking into account that the orbital planes of the satellites of Jupiter and Saturn are almost coplanar, a single flyby of one of the satellites is sufficient to transfer the spacecraft from the orbital plane of the flyby satellite into the orbital plane of another natural satellite.

In addition, the effect of rotating the spacecraft orbital plane can be used to direct the spacecraft into the regions of the near-planet space of interest to researchers (e.g. with the aim of exploring high-latitude or equatorial zones of planets, Saturnian rings, etc.).

Figures 5.9d,e also present graphical relationships $i_{max} = f(v_\infty)$ for natural satellites of distant planets—Uranus and Neptune. These relationships illustrate the potential of gravitational rotations of orbits in the sphere of activity of these planets.

It is worth noting that multiple flybys of a natural planetary satellite allow us to obtain even greater angles of inclination in the spacecraft orbital plane; this can be utilized when designing the exploration trajectories in the gravity fields of large planets. The construction of such orbits requires a combined analysis of the isolines $v_\infty = const$ and $T^* = const$ in the main coordinates R_π, T being transformed.

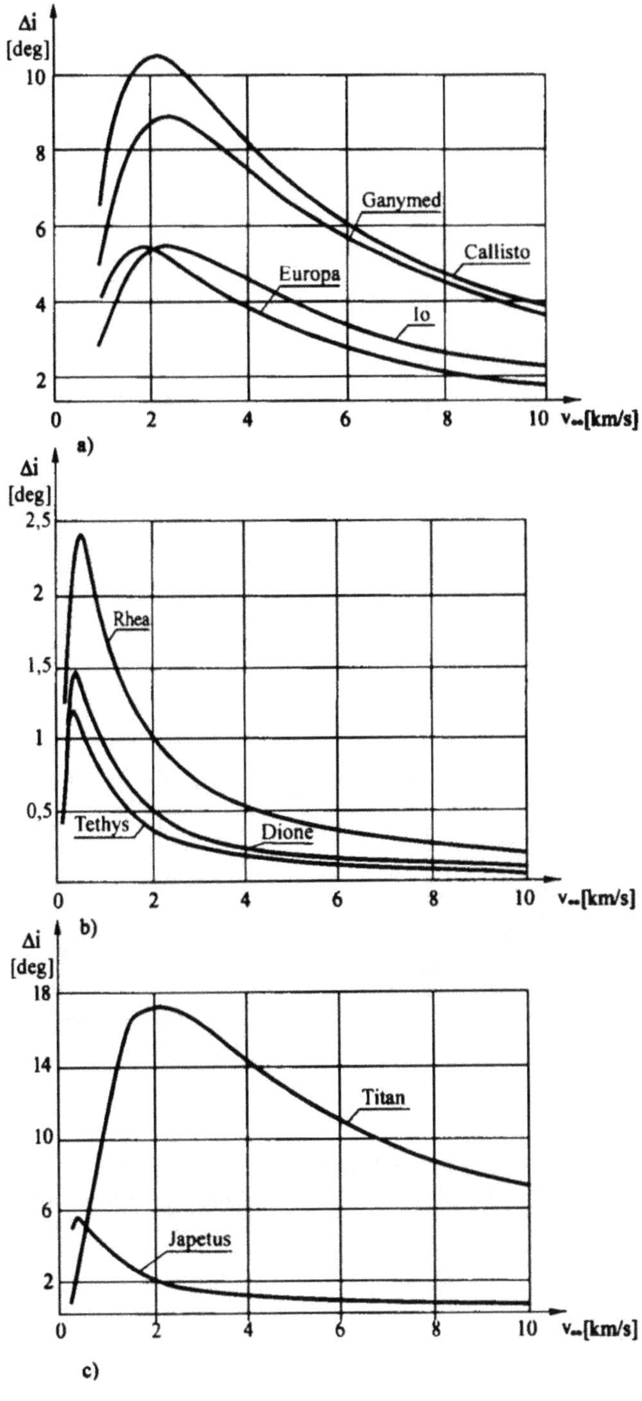

Δi
[deg]
10
8
6
4
2
0
2
4
6
8
10
v∞[km/s]
Ganymed
Callisto
Europa
Io
a)
Δi
[deg]
2,5
2
1,5
1
0,5
0
2
4
6
8
10
v∞[km/s]
Rhea
Tethys
Dione
b)
Δi
[deg]
18
14
10
6
2
0
2
4
6
8
10
v∞[km/s]
Titan
Japetus
c)

Figure 5.9

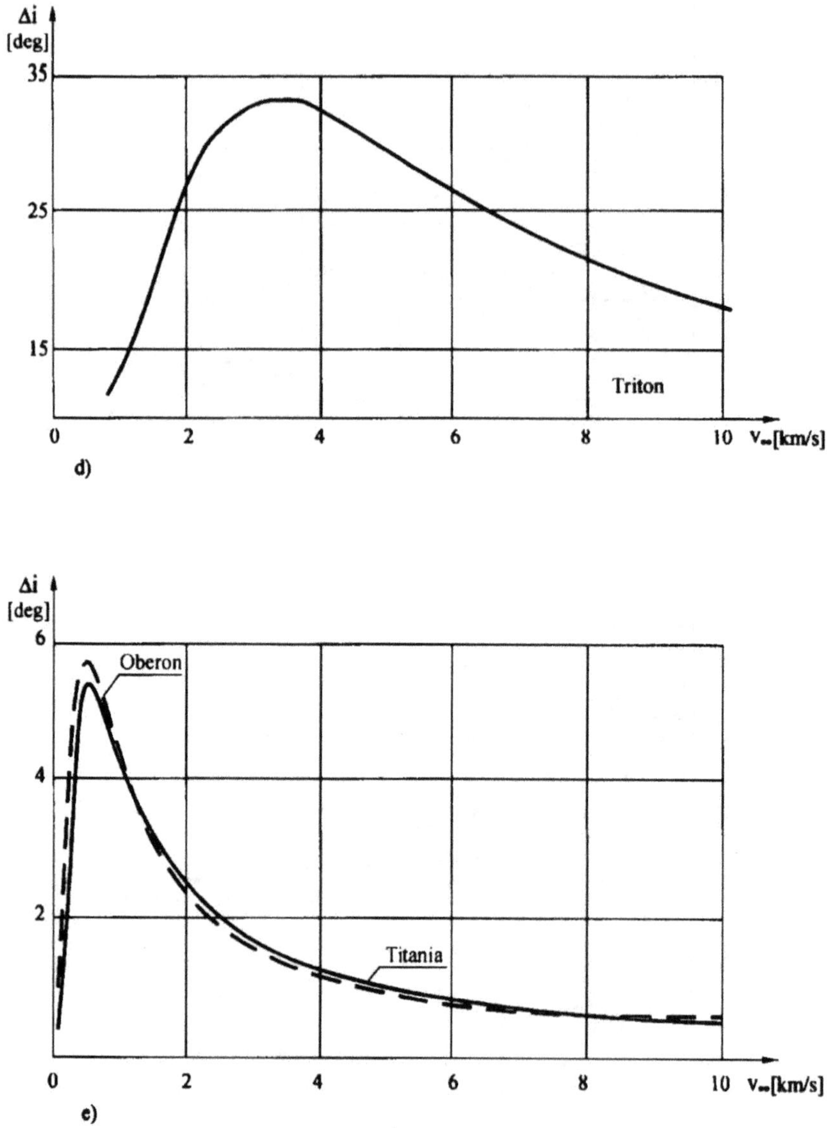

Figure 5.9 Maximum change in the angle of spacecraft orbit inclination in the case of flyby of satellites of large planets: (a) Galilean satellites of Jupiter; (b) and (c) satellites of Saturn; (d) Triton (satellite of Neptune); (e) Oberon and Titania (satellites of Uranus).

5.3 The use of gravity assist maneuvers to orbit the spacecraft as an artificial planetary satellite

The above numerical results show that it is possible in principle to use the gravity fields of the satellites of planets to reduce the orbital energy of the spacecraft by gravity assist maneuvers in the near-planetary region.

A flyby of a natural satellite can be performed in either a hyperbolic or an elliptic planetocentric orbit. As can be seen from the behavior of the curves in Figures 5.1 and 5.3, the relative effect of the spacecraft deceleration in the second case is more perceptible. However, the gravity assist maneuver in a hyperbolic incoming orbit can be useful as well, because it allows a reduction in the spacecraft energy and an increase in the pericentral distance of the orbit to prepare for the next deceleration maneuver (powered or atmospheric).

Consider different variants of the spacecraft entering the orbit of an artificial satellite of a planet using deceleration gravity assist maneuvers.

5.3.1 On the possibility of capturing spacecraft in an artificial satellite orbit without velocity impulses

The condition of such a capture as a result of gravity assist flybys of one or several natural planetary satellites is the requirement that the post-perturbation velocity of the spacecraft be less than the escape velocity for the given planet. The flyby of a satellite or satellites in the planetary satellite system can be performed in either the descending or ascending leg of the spacecraft hyperbolic trajectory.

The feasibility of orbiting the spacecraft as an artificial satellite of the planet with a flyby of a natural planetary satellite (or their combination) can be assessed by Formula 1.2.2 describing the perturbation change of the spacecraft orbital energy due to the flyby of the natural satellite. The condition of capture in this case is

$$v_\infty^2 - \sum_i \Delta h_i < 0,$$

where v_∞ is the hyperbolic excess of the spacecraft velocity when approaching the planet; Δh_i is the perturbation change of the orbital energy of the spacecraft due to flyby of the ith natural satellite of the planet.

This estimate can be also obtained from the plot (Figures 5.1, 5.5) for the case of maximum decrease in the spacecraft orbital energy due to flybys of satellites of Jupiter or Saturn.

Figure 5.10 shows numerical results of examining the perturbation effect of a single flyby of Galilean satellites in a hyperbolic descending trajectory and that of successive flybys of two satellites in different combinations. Each case, corresponding to the flyby of a single satellite, is denoted by a single letter (I—Io, G—Ganymede, and so on), and the case of successive flybys of two satellites is denoted by two letters (e.g. GI denotes successive flybys of Ganymede and Io).

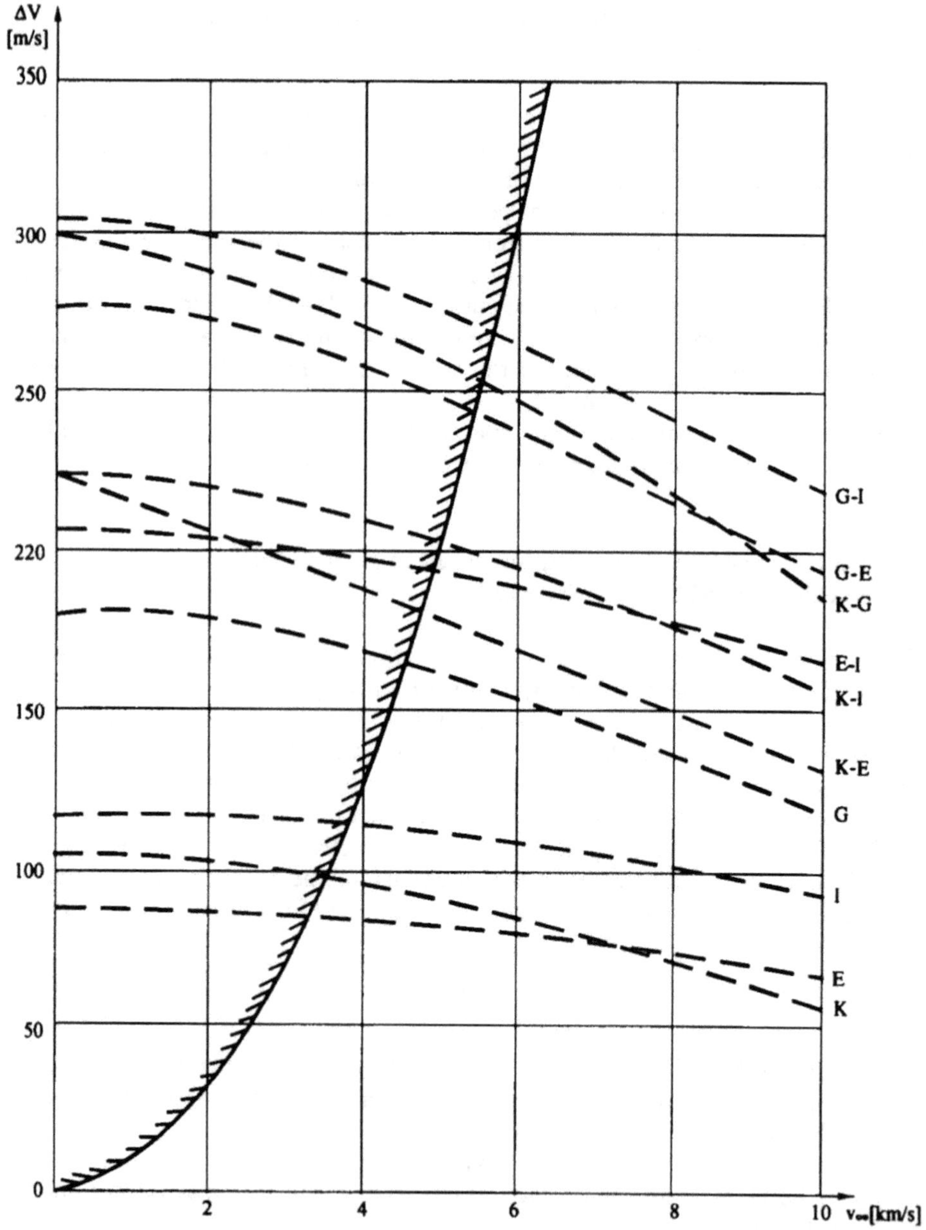

Figure 5.10 Maximum deceleration effect of flyby of natural satellites of Jupiter along the descending leg of hyperbola (equivalent to the velocity of pericentral deceleration).

The plot shows the maximum possible efficiency of the spacecraft maneuvers with the gravity assist of natural satellites of Jupiter reduced to the equivalent velocity of the pericentral powered deceleration:

$$\Delta v=\left(\frac{2k}{R_\pi}+h\right)^{1/2}-\left(\frac{2k}{R_\pi}+h-\sum_i \Delta h_i\right)^{1/2} \tag{5.1}$$

The pericentral distance of the orbit of the natural satellite of Jupiter was assumed to be $R_\pi = R_J + H_{atm}$. Also shown is the boundary of the domain of impulseless capture into a strongly extended ($e \approx 1$) orbit of a natural satellite of Jupiter.

As can be seen from Figure 5.10, the effect of perturbation deceleration is most pronounced in low-energy hyperbolic orbits. As the energy constant of the trajectory grows, the efficiency of the perturbation maneuvers of the spacecraft declines (from 90–175 m/s to 50–125 m/s for a single flyby and from 210–300 m/s to 125–225 m/s for a double flyby variant). For a nominal trajectory to Jupiter ($v_\infty = 6$ km/s) the saving of the characteristic velocity of pericentral deceleration amounts to 75–100 m/s for flyby of a single satellite and 170–270 m/s for flybys of two satellites. The most beneficial here is the flyby of Ganymede or the combination Ganymede–Io. As can be seen from Figure 5.10, an impulseless capture into a near-Jupiterian orbit requires a hyperbolic velocity excess at Jupiter $v_\infty < 4.5$ km/s (the option of flyby along the incoming hyperbola of Ganymede) or $v_\infty < 5.6$ km/s (the option of a combined flyby of Ganymede and Io).

Taking into account that the direct-flight interplanetary trajectories of a spacecraft with a high-thrust engine yield $v_\infty \geqslant 5.6$–5.8 km/s, we can conclude that neither of the above options of perturbation maneuvers at natural satellites of Jupiter can provide impulseless capture of the spacecraft into an orbit of a natural satellite of Jupiter.

At the same time, one should take into consideration the possibility of reducing v_∞ at Jupiter by using trajectories to Jupiter with a flyby of Mars (this can be seen from the behavior of the isolines in Figure 4.1) or on flights to Jupiter, by using a spacecraft with a low-thrust engine.

In the first case, the arrival velocity of the spacecraft can be reduced by 0.5–0.7 km/s, and as follows from Figure 5.10, there are options for the impulseless capture of the spacecraft into an orbit of an artificial satellite of Jupiter by performing perturbation maneuvers during flybys of two natural satellites (in the combinations Ganymede–Io or Ganymede–Europa).

In the second case, the scheme of impulseless capture with a flyby of one of Galilean satellites of Jupiter can be accomplished.

As for the feasibility of an impulseless capture of the spacecraft into an orbit of an artificial satellite of Saturn, the most promising in this case appears to be the largest Saturnian natural satellite—Titan. Indeed, as can be seen from Figure 5.3, only Titan is capable of reducing the spacecraft velocity in a hyperbolic incoming orbit to the values below the parabolic velocity (for $v_\infty < v_{\infty\,max}$, $v_{\infty\,max} = 3.45$ km/s). Other satellites do not possess sufficient gravity to capture the spacecraft into an orbit of an artificial satellite of Saturn ($v_{\infty\,max} \leqslant 0.8$ km/s). Their gravity fields can be useful only in elliptic orbits with relatively low energy levels ($h \leqslant -60$ km^2/s^2).

The direct flights Earth–Saturn are known to yield a hyperbolic velocity excess at Saturn $v_\infty \geqslant 5.44$ km/s. Using paths with a flyby of Mars allows a reduction of this limit by 0.4–0.5 km/s, and only a flyby of Jupiter yields acceptable velocity values $v_\infty \geqslant 2.6$ km/s at Saturn. This means that in the path Earth–Jupiter–Saturn, a flyby of Titan can provide an impulseless capture of the spacecraft into an orbit of artificial satellite of Saturn.

Thus, in direct flights to Jupiter and Saturn, it is not possible to facilitate an impulseless capture into an orbit around the planet using of flybys at its natural satellites. This can be accomplished in paths using a flyby of Mars in the flight to Jupiter and a flyby of Jupiter in the flight to Saturn. In the first case, the capture into an orbit of an artificial satellite of Jupiter becomes possible after successive flybys of Ganymede and Io or Ganymede and Europa, in the second case, a single flyby of Titan allows an impulseless capture of the spacecraft into an orbit of artificial satellite of Saturn.

One may also mention the very favorable possibilities presented by flybys of Jupiter and Saturn for reducing the energy of spacecraft approaching the other trans-Jupiterian planets—Uranus and Neptune, which have large natural satellites. Direct flights to these planets give lesser values of v_∞ than those at Jupiter and Saturn; and the introduction of a flyby allows an even greater reduction in the incoming velocity and makes it feasible to enter an artificial satellite orbit of these planets without power. In the cases where purely impulseless capture cannot be implemented, it is expedient to use gravity assist together with other means of deceleration to orbit the spacecraft as an artificial satellite of the planet, thus improving the energy parameters of such orbiting.

5.3.2 Combined schemes of orbiting the spacecraft as an artificial satellite of a planet

Gravity assist in combination with other means of spacecraft deceleration (e.g. powered or atmospheric deceleration) allows a flyby of a natural satellite of the planet to be also performed along an elliptic planetocentric orbit with the aim of putting the spacecraft into a specified orbit of a natural satellite of the planet.

Such a flyby can be performed either within the first half-revolution of the ellipse (from the pericenter to apocenter) or within the second half-revolution (from apocenter to pericenter); it can be either single or multiple.

The efficiency of such gravity assist maneuvers in orbiting the spacecraft as a satellite of Jupiter is illustrated by Table 5.5. The saving in the characteristic velocity V_c was estimated based on the double-impulse scheme of the spacecraft entering the orbit of a natural satellite of Jupiter (with the deceleration impulses applied in the apsidal points of the transfer ellipse $R_\pi = R_p + H_{atm}$, $R_\alpha = R_s$). The saving of ΔV in V_c is related to the pericentral point of the spacecraft orbit in accordance with Formula 5.1.

Comparing these estimates with those for the perturbation effect in a hyperbolic trajectory (Figure 5.10) one can see that the ΔV saving in the elliptic orbit is greater than that due to a maneuver in a hyperbolic incoming orbit. This can be explained

Table 5.5

Orbit of natural satellite of Jupiter	Via Io, ΔV, m/s	Via Europa, ΔV, m/s	Via Ganymede, ΔV, m/s
Europa	492.5	—	—
Ganymede	270	325	—
Callisto	173	189	380

by a shift in the spacecraft energy into the domain of flyby orbits with large perturbation change. An even greater effect can be achieved by using multiple flybys of the Galilean satellites.

In this case, each gravity assist flyby or power–gravity maneuver in the vicinity of the flyby satellite should result in an orbit with a period divisible by the period of rotation of the given natural satellite of Jupiter (a so-called resonant orbit) for the subsequent encounter with it so as to perform the next gravity assist maneuver. The obvious advantages of such a scheme are that it allows multiple use of the benefits of gravity assist maneuver, and the expenditure for powered maneuvers will be determined only by putting the spacecraft into the initial orbit of the artificial planetary satellite.

To study such paths we will use the nomogram of planetocentric flights (Figure 2.13) for four Galilean satellites of Jupiter.

The approach to the nomographic examination of such schemes is discussed in detail in Chapter 2.

Recall briefly its peculiarities. If we have the orbit of an artificial planetary satellite which the spacecraft is to enter, it is represented in the nomogram by a point in the coordinates of parameters R_π and T. An analysis of the nomogram (behavior of isolines) shows what number of flybys and at which planet it is necessary to leave with minimum V_c. Motion along the isolines of the nomogram is implemented in the inverse direction—from the final orbit to the initial one—using of isochrones impulseless gravity assist maneuvers of the spacecraft.

Consider the spacecraft entering a low near-Jupiter orbit with a low rotation period ($R_\pi = 0.4 \cdot 10^6$ km, $T = 5$ days). Such an orbit can be of considerable interest in the examination of the zones near Jupiter, its magnetosphere, and the radiation of the planet. Putting the spacecraft into such an orbit immediately from the incoming hyperbolic trajectory would require very high energy expenditure ($\Delta V = 4$ km/s for $v_\infty = 6$ km/s).

At the same time, the use of gravity assist flybys to put the spacecraft into the given orbit of an artificial planetary satellite may provide trajectories which are more acceptable in terms of energy expenditure.

The position of the point A representing the final orbit in the nomogram (Figure 2.13) shows that, in principle, a flyby of any of the Galilean satellites (except for Callisto) or a combination of such flybys can be used. We will restrict our study to the resonant orbits with a flyby of one of the natural satellites of Jupiter.

As can be seen from the nomogram, isolines of possible gravity assist maneuvers pass through this point: $v_\infty = 5.2$ km/s for Io, $v_\infty = 7.3$ km/s for Europa, and

$v_\infty = 4.6$ km/s for Ganymede. Parameters of the initial orbits from which the successive reduction will be performed are determined by the choice of points in the relevant isolines of the nomogram. Moving along each isoline and satisfying the conditions of divisibility of the periods of the spacecraft orbit to be formed and the orbit of the natural satellite of the planet, we can reach the point A (representing the final orbit).

Thus, the designer can choose the initial orbit, taking into account the different requirements for the flight in the sphere of activity of the planet (concerning energy, navigation, and investigations). However, the initial point must in any case lie on the corresponding isoline $v_\infty = const$.

In the example in question, the requirement to minimize energy expenditure for placing the spacecraft in an orbit of an artificial planetary satellite implies that highly-extended initial orbits should be selected, for example: $R_\pi = 421\,200$ km, $T = 10$ days for subsequent flybys of Io; $R_\pi = 620\,000$ km, $T = 150$ days for flybys of Europa; $R_\pi = 1\,020\,000$ km, $T = 120$ days for flybys of Ganymede.

The characteristic velocities of the spacecraft entering these initial orbits can be determined accordingly: $\Delta V = 2.9$ km/s for the first scheme, $\Delta V = 0.85$ km/s for the second scheme, $\Delta V = 1.1$ km/s for the third scheme (for the hyperbola incoming to Jupiter $V_\infty = 6$ km/s). It is evident that a spacecraft entering a given near-planetary orbit using multiple gravity assist flybys requires significantly less energy than a direct powered craft entering the same orbit. In this case, the most efficient are the second and third options, which allow the spacecraft to enter highly-extended elliptic orbits (with subsequent reduction to the required final orbit).

The chain of subsequent resonant orbits can be constructed using the method suggested above (see Chapter 2), considering the conditions of divisibility of the orbital periods of the spacecraft and the natural satellite of the planet.

For example, for the above-considered variant of the flight path via Ganymede, the following orbit chain can be formed:

$$\begin{matrix} R_\pi\ [\text{km}] & 1.02\cdot10^6 \\ T\ [\text{day}] & 120 \end{matrix}\Bigg\}\rightarrow\begin{matrix} 0.99\cdot10^6 \\ 43.0 \end{matrix}\Bigg\}\rightarrow\begin{matrix} 0.92\cdot10^6 \\ 21.45 \end{matrix}\Bigg\}\rightarrow\begin{matrix} 0.8\cdot10^6 \\ 10.7 \end{matrix}\Bigg\}\rightarrow\begin{matrix} 0.681\cdot10^6 \\ 7.5 \end{matrix}\Bigg\}\rightarrow\begin{matrix} 0.4\cdot10^6 \\ 5.0 \end{matrix}$$

that is, four successive flybys of Ganymede allow the spacecraft to enter the given low orbit of an artificial satellite of Jupiter. The energy expenditure in this case will be far less than that for the direct placement of the spacecraft in the given orbit.

Other orbits that can be formed with gravity assist are of no less interest. In these cases, the sites of gravity assist maneuvering can be used in combination with other means of spacecraft deceleration (e.g. powered or atmospheric), or can be based purely on gravity assist flybys (as in the above case).

5.4 Multi-purpose flight schemes to natural planetary satellites

One of the promising problems of astronautics is that of a spacecraft reaching a natural satellite of the planet with the aim of orbiting as its satellite or landing at its surface. This problem, concerning the formation of a trajectory in the gravity fields of various bodies, is one of the most complicated in interplanetary flight. Its solution is known to be associated with examining the external and internal problems using piecewise-conic approximation of the interplanetary trajectory legs within the spheres of activity of attracting bodies.

The nomogram developed allows the examination of the spacecraft path options, including gravity assist flybys in a planetary satellite system (i.e. the internal problem of the flight), and the selection of most beneficial paths for reaching the required satellite.

5.4.1 Nomographic analysis of spacecraft flights to a natural planetary satellite with gravity assist maneuvers

One of the possible flight schemes to a natural satellite of a planet in a planetary satellite system uses multiple flybys of one or several other natural satellites of the planet. Such paths, which were considered in the problems of orbiting the spacecraft as an artificial satellite of a planet, are attractive due to the possibility of decelerating the spacecraft with no fuel spent and, also because of the higher scientific value of the path due to the possibility of under taking additional exploration of the satellite during the flyby.

From the above nomograms it can be seen that an intersection of the isolines $v_{\infty j}=const$ can be chosen such that the spacecraft approaches the target natural planetary satellite with low v_{∞} having accomplished a flyby with high v_{∞}. As can be seen from Figure 5.11, two variants of the schemes using resonant orbits are possible:

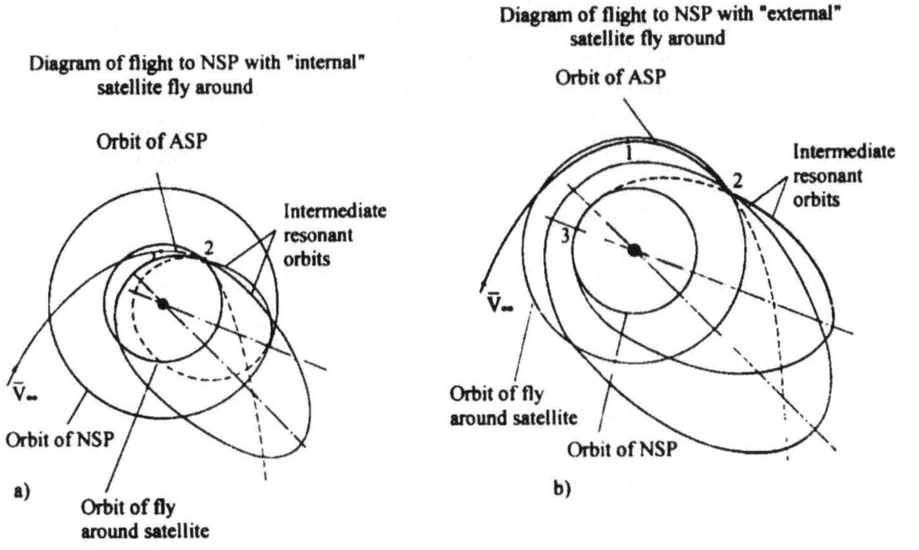

Figure 5.11 Schematic diagram of multiple maneuvering to fly to a natural satellite of a planet: (a) flyby of an inner satellite; (b) flyby of an outer satellite.

the scheme with a flyby of an "outer" satellite and the scheme with a flyby of an "inner" satellite (with respect to the target satellite). In the first case, the spacecraft approaches the target satellite at the pericenter of the spacecraft orbit (or near it), whereas in the second case, this takes place at the apocenter (or near it).

In general terms, the scheme of flight to a natural planetary satellite with multiple gravity assist flybys of other natural satellites of the planet can be represented as follows: (i) a velocity impulse at the pericenter of the incoming hyperbola places the spacecraft in a highly-extended orbit of an artificial satellite of the planet; (ii) multiple gravity assist maneuvers at one of the natural satellites of the planet are used to reduce the spacecraft orbital period; (iii) the spacecraft approaches the target satellite and by using a powered maneuver is captured in an orbit of an artificial satellite of the natural planetary satellite.

We take the total characteristic velocity ΔV_{Σ} at the site of powered deceleration as the criterion for assessing the schemes to be formed. Evaluation of this criterion will require us to estimate the velocity impulses for placing the spacecraft in the initial orbit of an artificial satellite of the planet and for its capture in the given orbit of an artificial satellite of the natural satellite of the planet.

The graphic-analytical method and nomogram of transfer orbits in a planetary satellite system is a tool for examining the trajectory options.

In conducting such an analysis we assume, in accordance with the prerequisites of the method, that all maneuvers of the spacecraft are coplanar to the orbital plane of the natural planetary satellite, the incoming hyperbola is specified by the value of v_{∞}, the pericentral altitude of the incoming hyperbola is considered not fixed (it can be selected based on the conditions of forming the next trajectory leg, e.g. the best approach to the natural planetary satellite for flyby). It is also assumed that the first gravity assist maneuver of the spacecraft is made at the point where the spacecraft orbit intersects with the orbit of the flyby natural planetary satellite within the first half-revolution of the spacecraft orbit. No coordination in time is made between the initial positions of the natural satellite of the planet and the spacecraft in the trajectory. It can be performed in an acceptable time interval taking into account that the periods of rotation of the satellites of Jupiter and Saturn considered here are small (the corresponding synodic periods of the natural satellites of the planets are presented in Tables 5.6 and 5.7).

The nomogram of combined isolines (Figure 2.13) and isochrones of perturbation transformation for the natural satellite of the planet being flown by (Figures 2.11, 2.12)

Table 5.6

Satellite	T_{sid} [day]	T_{syn} [day]			
		Io	Europa	Ganymede	Callisto
Io	1.769	—	3.53	2.35	1.97
Europa	3.551	3.53	—	7.07	4.51
Ganymede	7.155	2.35	7.07	—	12.55
Callisto	16.689	1.97	4.51	12.55	—

Table 5.7

Satellite	T_{sid} [day]	T_{syn} [day]				
		Tethys	Diona	Rhea	Titan	Japetus
Tethys	1.887	—	6.25	3.34	2.20	1.99
Diona	2.737	6.25	—	6.99	3.32	2.83
Rhea	4.517	3.34	6.99	—	6.34	4.81
Titan	15.945	2.20	3.32	6.34	—	19.88
Japetus	79.331	1.99	6.34	4.81	19.88	—

allow an assessment of the option of the flight to a natural planetary satellite. In the class of schemes we consider here (including multiple gravity assist maneuvers) the expenditure will be determined by the energy required for the spacecraft transfer from the incoming hyperbola to the initial orbit of an artificial planetary satellite (in the first part of the scheme) and for its placing in the orbit of an artificial satellite of the natural planetary satellite (in the last part of the multiple-revolution path). These two powered legs are separated by a multi-revolution trajectory of the spacecraft, some of the revolutions including flybys of the natural planetary satellite (the sequence of these revolutions should be selected by moving the mapping point in the nomogram between its initial and final positions). It should be mentioned that the point representing the orbit of the spacecraft approaching the target natural planetary satellite in accordance with the schemes under consideration should lie on the line $K_j\pi_j$ (for the schemes of the first kind) or on the line $K_j\alpha_j$ (for the schemes of the second kind).

Examination of the schemes can utilize the procedures of "direct motion" along isolines—from the point of leaving the hyperbolic incoming orbit to the spacecraft's arrival at the target natural satellite of the planet. In this approach, the value of the first deceleration impulse (transfer from hyperbolic orbit to the selected orbit of the artificial satellite of the planet) determines only the conditions of the spacecraft's approach to the natural satellite of the planet for a flyby. This approach fails to account for the effect of this factor on the formation of the flight trajectory as a whole and total energy expenditures for the flight. The principle of "reverse" motion appears more suitable, i.e. we select the best suitable orbit approaching the target satellite (based on the requirement of providing low values of $v_{\infty NSP}$), and calculate the maneuvers of the path under study in the inverse order using the presented plots and nomograms of gravity assist and powered transformations of the spacecraft orbit (Figure 2.13). The result is the initial elliptic orbit and the initial impulse required at the point of transfer between the hyperbola and the orbit of the artificial planetary satellite.

As an example of analysis and formation of such a path, consider the flight of a spacecraft to Io, one of the Galilean satellites of Jupiter. Let the spacecraft be in a hyperbolic incoming orbit to Jupiter $v_\infty = 6$ km/s. A direct flight to Io (a transfer immediately from the incoming hyperbola) requires very large energy expenditure ($\Delta v = 5.7$ km/s).

Flights using flybys of the natural satellites of Jupiter require lesser values of the incoming velocity to the target satellite: for the path via Europa $v_{\infty \min}=1.75$ km/s, for the path via Ganymede $v_{\infty \min}=2.3$ km/s, for the path via Callisto $v_{\infty \min}=4.9$ km/s. It is evident that in all these cases the first kind of scheme should be implemented (see Figure 5.11).

Parameters of the initial spacecraft orbit for the accomplishment of this scheme should be selected so as to make the orbit highly extended with a large spacecraft rotation period, and to make the pericentral distance as small as possible (from the point of view of minimizing the energy expenditure for placing the spacecraft in this orbit). A sufficiently large period of initial orbit can be selected (in accordance with the designer's requirements). The pericentral distance of the orbit should be selected from the condition of the spacecraft encounter with the intermediate satellite: for the path via Europa, $R_\pi \leqslant 0.67 \cdot 10^6$ km; for the path via Ganymede, $R_\pi \leqslant 1.068 \cdot 10^6$ km; for the path via Callisto, $R_\pi \leqslant 1.882 \cdot 10^6$ km. At the same time, it can be seen from the nomogram that an increase in the initial period, as well as lowering the pericenter of the spacecraft initial orbit, results in a transfer to the isolines of a high-velocity flyby of the intermediate satellite. This in turn leads to an increase in the velocity of approach to the target satellite (this can be seen from the behavior of the nomogram isolines).

Thus, the problem of selecting the optimal trajectory in terms of energy expenditure can be formulated as the problem of selecting an initial point in the nomographic plane (to represent the initial orbit of the spacecraft as an artificial satellite of the planet) that will provide a minimum to ΔV_Σ. Here ΔV_Σ denotes the total velocity expenditure for placing the spacecraft in the orbit of an artificial planetary satellite and then in the orbit of artificial satellite of the natural planetary satellite:

$$\Delta V_\Sigma = \Delta V_1(v_{\infty p}, R_{\pi 1}, T_1) + \Delta V_2(v_{\infty NSP}, R_{\pi 2}, T_2)$$

Analysis of the nomogram shows that the most efficient in terms of energy is the path via Europa, because of the least energy consumed at the sites of powered maneuvering.

Thus, for the velocities of approaching Io $v_{\infty NSP}=1.75$–4 km/s ($\Delta V_2=0.5$–2.15 km/s) the flyby of Europa should be made at the velocities v_∞^{fl} from 2 to 7 km/s, the period of the initial orbit can vary from 6.2 to 90 days, and the pericentral distance, from $0.67 \cdot 10^6$ to $0.565 \cdot 10^6$, respectively ($\Delta V_1=4.5$–1.2 km/s).

The selection of the specific trajectory—"hyperbolic" orbit of the spacecraft, "resonant" orbits of Europa flyby; orbit of the artificial satellite of Io—should be made in the isoline field of the flyby of Europa, taking into account the divisibility of the orbital periods of this satellite and the spacecraft.

One of the possible implementations of the flight path to Io can be presented as the following chain:

$$\begin{matrix} T\ \text{[days]} \\ R_\pi \cdot 10^6\text{[km]} \end{matrix} \quad \left.\begin{matrix} 9.46 \\ 0.525 \end{matrix}\right\} \rightarrow \left.\begin{matrix} 7.1 \\ 0.475 \end{matrix}\right\} \rightarrow \left.\begin{matrix} 5.9 \\ 0.450 \end{matrix}\right\} \rightarrow \begin{matrix} 5.1 \\ 0.426 \end{matrix}$$

This means that as the result of a pericentral deceleration impulse $\Delta V_1 = 3.0$ km/s, the spacecraft will transfer from the incoming hyperbola to the orbit with the period $T = 9.46$ days and pericentral altitude $R_\pi = 0.525 \cdot 10^6$ km. Next, after three gravity assist flybys at Europa (with $v_\infty^{fl} = 5$ km/s) the spacecraft will enter the orbit $T = 5.1$ days, $R_\pi = 0.426 \cdot 10^6$ km. Then, by a deceleration impulse $\Delta V_2 = 2.15$ km/s at the apocenter of the orbit, the spacecraft will enter a highly-extended orbit of an artificial satellite of Io. The total energy requirements for these maneuvers will amount to $\Delta V_\Sigma = 5.15$ km/s.

The solution presented is evidently not optimal either in terms of energy or in terms of the number of flybys of natural planetary satellites (in this case, Europa), because the initial orbit and flyby isoline are selected arbitrarily within the range of admissible solutions. Analysis of the nomogram shows that minimization of the total impulse is associated with two contradictory requirements: on the one hand, in order to reduce the impulse ΔV_2 for placing the spacecraft in the orbit of an artificial satellite of the natural planetary satellite one should make the velocity of the flyby of the natural satellite of the planet as small as possible; however, this will result in an increase in the impulse ΔV_1 required to place the spacecraft in the initial orbit. On the other hand, the need to reduce ΔV_1 makes one select a highly-extended initial orbit with a low pericenter, which leads to a flyby of the intermediate natural satellite of the planet at a higher velocity and to an according increase in ΔV_2.

Thus, the problem of forming such a path is an optimization problem: it is necessary to select the initial spacecraft orbit and the number of flybys and "idle" revolutions of the spacecraft in the "resonance" trajectory so as to provide min ΔV_Σ.

The nomogram allows one to perform an approximate analysis of optimal options of the trajectory. By varying the coordinates of the initial mapping point on the nomographic plane, one can find its position so as to provide minimum value to the total characteristic velocity. Thus, the optimal flight isoline will be selected to be used as the basis for the formation of the trajectory.

Thus, in the case of the variant considered above, the best path energy parameters can be obtained using the following chain of orbits for $v_\infty^{fl} = 5$ km/s.

$$\begin{matrix} T\,[\text{days}] \\ R_\pi \cdot 10^6\,[\text{km}] \end{matrix} \quad \left.\begin{matrix} 9.46 \\ 0.6 \end{matrix}\right\} \rightarrow \left.\begin{matrix} 7.1 \\ 0.575 \end{matrix}\right\} \rightarrow \left.\begin{matrix} 5.9 \\ 0.545 \end{matrix}\right\} \rightarrow \left.\begin{matrix} 4.8 \\ 0.514 \end{matrix}\right\} \rightarrow \left.\begin{matrix} 4.1 \\ 0.465 \end{matrix}\right\} \rightarrow \begin{matrix} 3.55 \\ 0.421 \end{matrix}$$

$$\Delta V_1 = 3.5\ \text{km/s},\ \Delta V_2 = 1.26\ \text{km/s},\ \Delta V_\Sigma = 4.76\ \text{km/s}.$$

Similar studies can be made for any other options for reaching the natural satellite of the planet using the multiple-flyby scheme. For example, for the second kind of scheme (using a flyby of an inner satellite with respect to the target natural planetary satellite of the planet), optimization of the flight trajectory to Callisto via Europa yields the following chain of orbits:

$$\begin{matrix} T\,[\text{days}] \\ R_\pi \cdot 10^6\,[\text{km}] \end{matrix} \quad \left.\begin{matrix} 21.1 \\ 0.647 \end{matrix}\right\} \rightarrow \left.\begin{matrix} 17.6 \\ 0.641 \end{matrix}\right\} \rightarrow \left.\begin{matrix} 14.05 \\ 0.631 \end{matrix}\right\} \rightarrow \left.\begin{matrix} 10.5 \\ 0.616 \end{matrix}\right\} \rightarrow \begin{matrix} 9.1 \\ 0.597 \end{matrix}$$

$$v_\infty^{fl} = 5.1\ \text{km/s},\ \Delta V_1 = 2.25\ \text{km/s},\ \Delta V_2 = 1.1\ \text{km/s},\ \Delta V_\Sigma = 2.62\ \text{km/s}.$$

Such an analysis of the solutions of multiple-flyby trajectories based on the nomogram is approximate and can be used to quickly assess a scheme in hand.

The accurate determination of the parameters of the trajectory being formed requires the solution of a numerical optimization problem.

The solution of such a problem requires the numerical determination of the parameters of such a trajectory (the required number of revolutions, moments at which the spacecraft encounters the natural satellite of the planet, altitudes of the spacecraft flyby over the surface of the satellite) so as to minimize the energy expenditure ΔV_Σ.

5.4.2 Numerical modeling of multi-revolution flight schemes to natural planetary satellites

The suggested universal algorithm for forming multi-purpose missions can be applied to the examination of any trajectory option in planetary satellite systems. Given the conditions under which the spacecraft reaches the target planet's sphere of activity (date, velocity, and location of the spacecraft) one can calculate an option for the multi-purpose mission. To do this, it is sufficient to specify the code chain and restrictions on the selected mission (see Section 2.3).

However, in the comparative analysis of different options of multi-purpose missions in planetary satellite systems, it is reasonable to avoid relating such numerical studies to specific dates of interplanetary flight schemes and to use universal models of studies.

Presented below is an algorithm for examining multi-revolution flight schemes in planetary satellite systems; the algorithm is implemented as one of the regimes of the universal algorithm.

The orbits of the natural planetary satellites are supposed to be circular and coplanar and are specified in the central field of the planet (with a gravitational constant K_p) by their radii R_i. The radii of the corresponding natural satellites of the planet are r_i, their gravitational constants are $K_i (i=1,\ldots,l$ is the number of the satellite in the planetary satellite system in question). The spacecraft orbit is also assumed to be coplanar to the orbital planes of the natural planetary satellites and is specified by the pericentral distance r_π and the period of rotation T. Then the angle of the spacecraft's is encounter with the natural satellite of the planet is determined as

$$\alpha = \frac{\pi}{2} - \beta \tag{5.2}$$

where

$$tg^2\beta = \frac{a(1-e^2)}{a^2e^2-(a-R_i)^2};$$

$$a = \sqrt[3]{\frac{K_p T^2}{4\pi^2}}, \quad e = 1 - \frac{r_\pi}{a}.$$

Flyby of the ith natural satellite of the planet at a relative distance of $m=r_p/r_i \geqslant 1$ from the satellite surface determines a new orbit of the spacecraft, which has energy constant h^* and period T^* (see 1.2.2). For the second encounter between the spacecraft and the natural planetary satellite the following condition must be met

$$T^* = T_i \frac{k_1}{k_2} \tag{5.3}$$

where k_1 is the number of complete revolutions of the natural planetary satellite until the second encounter with the spacecraft; k_2 is the number of complete revolutions of the spacecraft until the encounter with the natural satellite of the planet; k_1 and k_2 are relatively prime integers.

It is evident that varying the relative flyby altitude m ($m \geqslant 1$) of the spacecraft over the surface of the natural satellite of the planet, one can meet condition (5.3) and use the flyby to form the spacecraft orbit so as to provide its second encounter with the natural planetary satellite, facilitating further modification of the spacecraft orbit.

Parameters of the post-flyby orbit are determined by the relationships

$$\alpha^* = -\frac{K_p}{h^*}, \quad T^* = 2\pi\sqrt{\frac{a^{*3}}{K_p}},$$

$$e^* = \sqrt{1 - \frac{h^*}{K_p^2} R_i^2 \left(V_1 \cos\alpha + \frac{\Delta h}{2V}\right)^2}, \tag{5.4}$$

$$r_\pi^* = a^*(1 - e^*)$$

and can be used to form the next flyby revolution with respect to the natural satellite of the planet.

Thus, a chain of orbits can be formed

$$\{r_{\pi_0}, T_0\} \rightarrow \{r_{\pi_1}, T_1\} \rightarrow \ldots \{r_\pi^*, T^*\} \rightarrow \ldots \{r_{\pi_k}, T_k\} \tag{5.5}$$

with a subsequent increase or decrease in the energy constant of the spacecraft.

It is evident that the feature of multiple gravity assist maneuvers described above can be used for both acceleration and deceleration of the spacecraft to the specified energy level in the vicinity of the planet with a system of natural satellites.

Below we consider trajectories with multiple decelerations at natural satellites in the vicinity of the planet (e.g. Jupiter or Saturn) with the aim of reaching one of the satellites and entering the given orbit.

In the general form, the deceleration scheme in this case is as follows.

1. The spacecraft flies along a hyperbolic trajectory from the sphere of activity of the planet to the pericenter;
2. Deceleration velocity impulse ΔV_1 is applied in the pericenter of the orbit to place the spacecraft in an elliptic orbit;

3. Successive flybys of natural planetary satellites results in a chain of elliptic orbits;
4. Apsidal approach (pericentral or apocentral) to the target natural satellite of the planet in the last revolution of the elliptic orbit and the spacecraft entering the orbit of artificial satellite of the natural satellite of the planet by applying a deceleration impulse (ΔV_2) at the pericenter of the near-satellite hyperbolic orbit.

Recall that there can be two types of such schemes: (i) involving a flyby of a natural inner satellite with respect to the target natural satellite of the planet (apocentral approach) and (ii) with a flyby of an outer natural satellite with respect to the target natural satellite of the planet (pericentral approach).

Thus, the criterial function for such a trajectory can be written as

$$\Delta V_\Sigma = \Delta V_1 + \Delta V_2 \Rightarrow \min. \tag{5.6}$$

The components of the criteria function are determined as

$$\Delta V_1 = \left(\frac{2K_p}{r_\pi} + V_\infty^2\right)^{1/2} - \left(\frac{2K_p}{r_{\pi 0}} - \frac{K_p}{(K_p T_0^2/4\pi^2)^{1/3}}\right)^{1/2} \tag{5.7}$$

where V_∞ is the hyperbolic excess of the spacecraft velocity when approaching the planet (its gravitational constant is K_p); $r_{\pi 0}$, T_0 are the parameters of the initial elliptic orbit, which the spacecraft enters after the application of the decelerating velocity impulse provided that the pericentral altitudes of the hyperbolic and elliptic orbits coincide $r_\pi = r_{\pi 0}$. Another component of the criteria function ΔV_2 is determined as the deceleration impulse required to place the spacecraft in an orbit of an artificial satellite of the natural planetary satellite (circular or elliptic)

$$\Delta V_2 = \left[\frac{2K_l}{r_l} + \left(\sqrt{\frac{2K_p}{R_l} - \frac{K_p}{a}} - \sqrt{\frac{K_p}{R_l}}\right)^2\right]^{1/2} - \left(\frac{2K_l}{r_l}\right). \tag{5.8}$$

$a = (R_l + r_{\pi K})/2$—the first type of the scheme, $a = (R_l + r_{\alpha K})/2$ the second type of the scheme, where K_i, R_i, r_i, $(i = 1, \ldots, l)$ are the gravitational parameter, the radius of the orbit, and the radius of the target satellite.

Condition (5.8) is written for the case of capturing into a highly-extended elliptic orbit of an artificial satellite of the natural planetary satellite ($r_{\pi f} \geqslant r_l$, $e_f \approx 1$).

If any other orbit is to be entered (e.g. a circular orbit $r_{\pi f} \geqslant r_l$) the necessary deceleration impulse ΔV_Σ is greater.

In this formulation ΔV_Σ will depend both on the parameters of the initial elliptic orbit $r_{\pi 0}$, T_0, and, accordingly, ΔV_1, and on the number N of flybys being performed (this will determine ΔV_2). In a more general case, the criteria function can have a more complex form incorporating a temporal parameter—the total time of the mission accomplishment, bringing about restrictions on the number N of the flybys of natural planetary satellites included in the trajectory.

In the computation algorithm for the optimization of the trajectory to reach the given natural planetary satellite in terms of energy criterion (5.6) the inverse solution approach was used, that is, the final orbit of apsidal tangency was taken as the initial one (with an arbitrarily chosen semi-major axis). After that, the energy of the orbit under consideration is increased. In the moment at which the spacecraft passes the pericenters of the orbits of the chain (5.5), impulse ΔV_1 is evaluated such that is necessary for the transfer from the planetocentric hyperbolic orbit with the specified energy. Thus, the minimum of the total characteristic velocity can be found:

$$\Delta V_\Sigma\{r_{\pi k}, (N, m_j)_l\} \Rightarrow \min \quad j=1, \ldots, N \tag{5.9}$$

at $T_\Sigma \leqslant T_{\text{lim}}$, for the first type of the flight scheme;

$$\Delta V_\Sigma\{r_{\alpha k}, (N, m_j)_l\} \Rightarrow \min \quad j=1, \ldots, N \tag{5.10}$$

l is the number of the selected flyby satellite at $T_\Sigma \leqslant T_{\text{lim}}$, for the second type of the flight scheme.

Here, $r_{\pi k}$ (or $r_{\alpha k}$) are the parameters of the final orbit of the spacecraft apsidal approach to the target satellite orbit, $(N, m_j)_l$ is the optimal number of the spacecraft flybys of the lth natural satellite of the planet with arbitrarily selected relative flyby altitude m_j over the satellite surface at each of the jth flybys, T_{lim} is the specified limiting time of maneuvering in the orbits with multiple flybys of the lth natural satellite of the planet.

When assessing the total time T_Σ for the entire operation (from the moment the spacecraft leaves the hyperbolic incoming orbit to the moment it encounters the target satellite), it was assumed that at the first and last revolutions of the chain of orbits being formed, gravity assist flybys are performed, whereas the revolutions between them may contain no flybys (in accordance with the condition (5.3)). Then

$$T_\Sigma = t_{01} + \sum_j (x_j+1) T_j + t_l, \quad j=1, \ldots, N, \tag{5.11}$$

where t_{01} is the time required for the spacecraft to pass the trajectory leg from leaving the hyperbolic orbit to the moment of the first gravity assist maneuver; T_j is the period of the corresponding resonant orbit after jth flyby was performed, $j=1, \ldots, N$; x_i is the number of revolutions ("idle") made by the spacecraft until the next gravity assist maneuver is performed; t_l is the time for passing the trajectory leg from the moment of the last gravity assist maneuver to the moment of encounter with the target satellite.

Thus, the algorithm for optimizing the energy expenditure for the flight to the target satellite with multiple gravity assist flybys at natural planetary satellites can be represented as follows:

$$\Delta V_\Sigma^{\text{opt}} = \min_{r_{\pi k}(r_{\alpha k})} \min_{(N, m_j)_l} \Delta V_\Sigma, \quad j=1, \ldots, N \tag{5.12}$$

at $T_\Sigma \leqslant T_{\text{lim}}$.

If the planet has a system of natural satellites (as is the case with Jupiter), the best trajectory for reaching the target natural satellite of the planet can be selected:

$$\Delta V_{\Sigma}^{opt} = \min_{r_{\alpha K}(r_{\alpha K})} \min_{l} \min_{N, m_j} \quad \begin{matrix} j=1,\ldots,N \\ l=1,\ldots,M \end{matrix} \tag{5.13}$$

at $T_{\Sigma} \leqslant T_{\lim}$.

Here, M is the number of attracting natural satellites of the planet under consideration. The suggested algorithm can be used for numerical examination of the best (in terms of energy expenditure) trajectory options for reaching a natural satellite of a planet in any planetary satellite system.

Table 5.8 presents a numerical solution for one such option—the results of calculating the trajectory for the spacecraft flight to the orbit of a natural satellite of Europa with multiple flybys of Ganymede (the first type of flight scheme).

The trajectory was assumed to begin at the moment of the pericentral single-impulse departure of the spacecraft from a hyperbolic incoming orbit ($V_{\infty} = 6$ km/s) and transfer to an elliptic orbit for the first stage of maneuvering.

The first spacecraft encounter with Ganymede was assumed to occur at the first revolution of the spacecraft orbit. The period of each of the subsequent revolutions was selected so as to provide an encounter with the flyby natural satellite of the planet. The spacecraft encounter with Europa was assumed to take place at the last revolution of the trajectory with subsequent transfer to a highly-extended orbit of an artificial satellite of Europa ($r_{\pi} \approx r_{e}$, $e \approx 1$).

Each line of Table 5.8 represents an operation being performed and gives the parameters of the newly formed spacecraft orbit (r_{π}, T). In the obtained optimal

Table 5.8

Current time [days]	Operation performed	Number of the orbital revolution	Relative altitude over the natural satellite of the planet m	Expenditure [km/s] ΔV	Orbital period [day]	Pericentral distance of the spacecraft orbit $r_{\pi} \cdot 10^6$ [km]
0	Spacecraft entering the orbit of an artificial satellite of Jupiter	0	—	2.198	78.738	1.0705
1.12	Flyby of Ganymede	1	1.00	0	50.152	1.0513
51.272	Flyby of Ganymede	2	1.00	0	21.471	0.992
72.743	Flyby of Ganymede	3	1.0312	0	10.727	0.8696
94.197	Flyby of Ganymede	5	1.1562	0	6.119	0.6714
98.237	Entering the orbit of an artificial satellite of Europa	6		0.971	6.119	0.6714

solution (Figure 5.12) a flyby of Ganymede is made at each revolution of the trajectory except for the last but one, when an additional "idle" revolution is necessary (to meet the condition of divisibility of the periods of the natural planetary satellite and the spacecraft).

The results obtained show that the successive gravity assist maneuvers yield notable energy savings when entering the orbit of an artificial satellite of Europa $\Delta V_{\Sigma}=3.169$ km/s, which is well below the expenditure in the schemes with powered deceleration.

The suggested algorithm was also applied to the numerical examination of all possible trajectories for flights to the Galilean satellites of Jupiter with multiple gravity assists. The optimal options of the flights in terms of total energy requirements are presented in Tables 5.9–5.12.

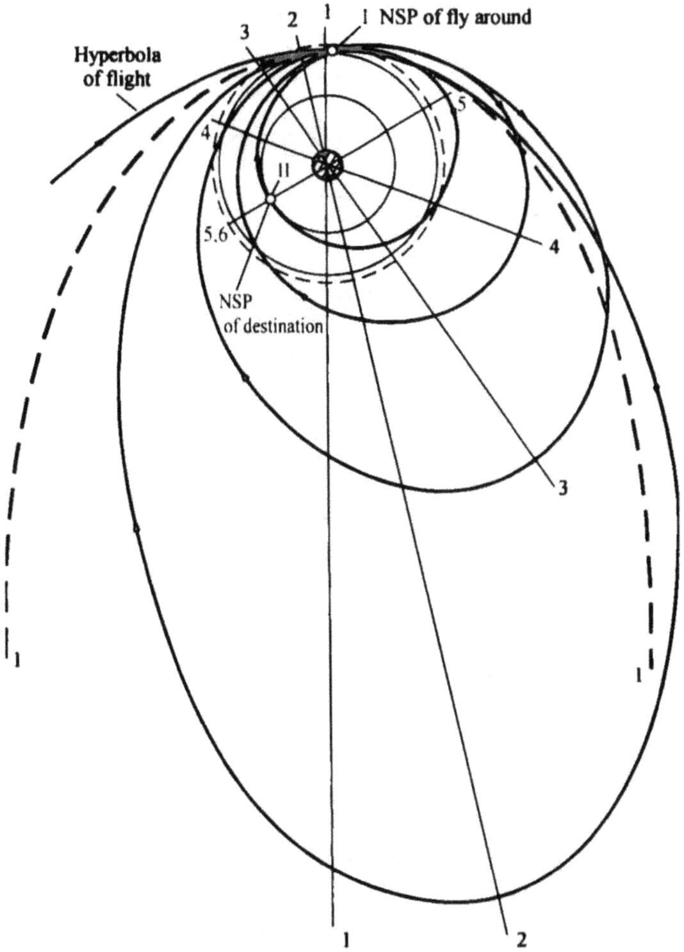

Figure 5.12 Diagram of spacecraft flight to the orbit of an artificial satellite of Europa via Ganymede (4 flybys).

Table 5.9 Paths via Io.

Parameters	Target natural satellite of the planet														
	Europa					Ganymede					Callisto				
T_{lim} [days]	50	100	150	200	250	50	100	150	200	250	50	100	150	200	250
ΔV_1 [km/s]	3.271	2.633	2.633	2.633	2.633	2.869	1.959	1.703	1.440	1.440	2.041	1.503	1.346	1.185	1.123
ΔV_2 [km/s]	1.057	1.269	1.269	1.269	1.269	1.285	1.425	1.352	1.425	1.425	1.788	1.935	1.879	1.935	1.851
ΔV_Σ [km/s]	4.329	3.902	3.902	3.902	3.902	4.154	3.384	3.055	2.865	2.865	3.829	3.438	3.225	3.121	2.975
N_{opt}	6	9	9	9	9	4	7	9	9	9	3	5	5	6	6
T_Σ [day]	46.0	74.3	74.3	74.4	74.3	47.8	36.7	141.5	159.2	159.2	49.6	90.2	129.1	166.3	222.8

Table 5.10 Paths via Europa.

Parameters of the scheme	Target natural satellite of the planet					
	Io		Ganymede		Callisto	
T_{lim} [days]	50	100	50	100	50	100
ΔV_1 [km/s]	3.968	3.198	3.732	2.912	2.606	2.147
ΔV_2 [km/s]	1.771	1.478	0.425	0.564	1.040	1.056
ΔV_Σ [km/s]	5.471	4.676	4.157	3.476	3.646	3.203
N_{opt}	5	8	4	7	4	6
T_Σ [day]	49.695	99.419	42.567	92.366	42.676	92.293

Table 5.11 Paths via Ganymede.

Parameters of the scheme	Target natural satellite of the planet					
	Io		Europa		Callisto	
T_{lim} [days]	50	100	50	100	50	100
ΔV_1 [km/s]	3.227	2.519	3.045	2.198	2.635	2.498
ΔV_2 [km/s]	2.291	2.289	0.811	0.972	0.616	0.473
ΔV_Σ [km/s]	5.519	4.809	3.856	3.169	3.251	2.971
N_{opt}	4	5	3	4	3	3
T_Σ [day]	42.938	71.498	42.953	93.078	43.056	57.321

Table 5.12 Paths via Callisto.

Parameters of the scheme	Target natural satellite of the planet					
	Io		Europa		Ganymede	
T_{lim} [days]	50	100	50	100	50	100
ΔV_1 [km/s]	2.779	2.779	3.325	3.289	3.093	3.696
ΔV_2 [km/s]	3.276	3.276	1.832	1.589	0.484	0.482
ΔV_Σ [km/s]	6.056	6.056	5.158	4.878	4.377	4.178
N_{opt}	3	3	1	4	1	2
T_Σ [day]	33.48	33.48	13.356	83.669	22.210	66.922

As can be seen from the results presented, an increase in the time of spacecraft maneuvering in the resonant orbits (i.e. the number of flybys) favorably influences the energy characteristic of the flights to various satellites of Jupiter. However, an indefinite increase in T_{lim} cannot be justified. The most efficient is the rise in T_{lim} until the first 100 days, and a further increase in the number of flybys (and, accordingly, T_Σ) is not very efficient. This can be explained primarily by the fact that by selecting more extended initial spacecraft orbits (noting that the transfer to these orbits from the incoming hyperbola is less energy-consuming), we will obtain greater values of $v_{\infty NSP}$, at the point of arrival at the target satellite. Therefore, in such schemes it appears expedient to restrict oneself to consideration of orbits whose

T_Σ is not greater than three months; this appears to be justified by the considerations of the total duration of the mission to a natural satellite of Jupiter.

It should be noted that in most solutions, a rather significant portion of ΔV_Σ is due to the impulse ΔV_1. Therefore, to reduce energy requirements it appears advantageous to examine variants incorporating the deceleration leg of the spacecraft in Jupiter's atmosphere to place it in an orbit of an artificial satellite of Jupiter.

Analysis of the results presented allows a series of interesting observations. Thus, when forming paths in the planetary satellite system of Jupiter, it turns out that the most efficient in terms of energy requirements are the paths via the satellite nearest to the target satellite (in most cases, this is an "outer" natural satellite of the planet). In the flight to Io, it is most expedient to perform a flyby of Europa; in the flight to Europa, a flyby of Ganymede is best; in the flight to Ganymede, a flyby of Io; and in the flight to Callisto, a flyby of Ganymede. In terms of minimum time for maneuvers, the most expedient are the schemes of flight via Callisto ($13.56 \leqslant T_\Sigma \leqslant 83.66$ days).

The greatest number of flybys ($N=9$) is required in the paths including maneuvers at Io—the closest to Jupiter, and the least ($N=1$ to 2), in the flyby of Callisto (in this case the required number of spacecraft revolutions is even more, because the spacecraft must make "idle" revolutions to meet the resonance condition).

The least energy consuming among all those considered ($\Delta V_\Sigma < 3$ km/s) are flights to Ganymede via Io, to Callisto via Io, and to Callisto via Ganymede.

Table 5.13 gives the characteristics of the best flight options to the natural satellites of Jupiter. The table also shows single- and double-impulse options of the schemes of powered entry to the orbits of artificial satellites of Galilean satellites ($r_{\pi j} = r_{NSPj}, e_j \approx 1$).

The results shown indicate that all schemes with gravity assist maneuvers yield considerable energy savings as compared with flights with powered deceleration (thus, the energy requirements for these paths are more than 2–2.5 times less than those for the common double-impulse scheme).

Similar studies were also made for flight paths to satellites of Saturn (see Table 5.14).

The use of multiple-flyby trajectories in some cases allows a substantial reduction in the energy requirements for the flight to a natural satellite of a planet. For comparison, the energy required by the schemes of single- and double-impulse flights to natural satellites of Saturn are also given in Table 5.14 (for the incoming hyperbola to Saturn it equals 6 km/s).

As one would expect, the most expedient is the utilization of the gravity potential of Titan. Flyby of these satellite allows a significant (by a factor of 1.5–2) reduction in the energy expenditure for the flight (as compared with powered deceleration schemes). Therefore, schemes for flights to Tethys, Diona, Rhea, and Japetus can be formed based on flybys of Titan (no more than 3 of them are necessary).

Flybys of the satellites Japetus and Rhea can be used for the flight to Titan, though their effect is not too large. In this case Japetus should be preferred, because this scheme requires fewer flybys of the satellite and yields an acceptable reduction in the energy required for the flight.

Table 5.13 Flights to satellites of Jupiter.

Target satellite	ΔV [km/s] 1 impulse	ΔV [km/s] 2 impulses	Flyby of Io	Flyby of Europa	Flyby of Ganymede	Flyby of Callisto
Io	5.70	10.5	—	$\Delta V_\Sigma = 4.676$ km/s $T_\Sigma = 99.419$ day $N = 8$	$\Delta V_\Sigma = 4.809$ km/s $T_\Sigma = 71.498$ day $N = 5$	$\Delta V_\Sigma = 6.056$ km/s $T_\Sigma = 33.48$ day $N = 3$
Europa	4.85	9.0	$\Delta V_\Sigma = 3.902$ km/s $T_\Sigma = 74.3$ day $N = 9$	—	$\Delta V_\Sigma = 3.169$ km/s $T_\Sigma = 93.078$ day $N = 4$	$\Delta V_\Sigma = 4.878$ km/s $T_\Sigma = 83.569$ day $N = 4$
Ganymede	3.60	7.0	$\Delta V_\Sigma = 2.865$ km/s $T_\Sigma = 159.2$ day $N = 9$	$\Delta V_\Sigma = 3.476$ km/s $T_\Sigma = 92.366$ day $N = 7$	—	$\Delta V_\Sigma = 4.178$ km/s $T_\Sigma = 66.922$ day $N = 2$
Callisto	3.10	6.0	$\Delta V_\Sigma = 2.975$ km/s $T_\Sigma = 222.8$ day $N = 6$	$\Delta V_\Sigma = 3.202$ km/s $T_\Sigma = 92.299$ day $N = 6$	$\Delta V_\Sigma = 2.971$ km/s $T_\Sigma = 57.321$ day $N = 3$	—

Table 5.14 Flights to satellites of Saturn.

Target satellite	ΔV_Σ [km/s] 1 impulse	ΔV_Σ [km/s] 2 impulses	Flyby of Titan	Flyby of Japetus	Flyby of Rhea
Tethys	5.4	8.1	$\Delta V_\Sigma = 4.06$ km/s $N = 3$ $T_\Sigma = 50$ days	$\Delta V_\Sigma = 5.85$ km/s $N = 5$ $T_\Sigma = 284$ days	$\Delta V_\Sigma = 4.8$ km/s $N = 12$
Diona	4.8	7.4	$\Delta V_\Sigma = 3.25$ km/s $N = 3$ $T_\Sigma = 60$–70 days	$\Delta V_\Sigma = 4.75$ km/s $N = 5$ $T_\Sigma = 370$ days	$\Delta V_\Sigma = 4.05$ km/s $N = 12$
Rhea	4.4	6.5	$\Delta V_\Sigma = 2.6$ km/s $N = 2$ $T_\Sigma = 40$ days	$\Delta V_\Sigma = 3.72$–4.1 km/s $N = 6-5$ $T_\Sigma = 320$–435 days	—
Titan	2.4	3.4	—	$\Delta V_\Sigma = 1.5$–2.11 km/s $N = 4-5$ $T_\Sigma = 254$–440 days	$\Delta V_\Sigma = 1.6$ km/s $N = 20$
Japetus	3.8	4.4	$\Delta V_\Sigma = 1.7$ km/s $N = 1$ $T_\Sigma = 57$ days	—	$\Delta V_\Sigma = 2.6$ km/s $N = 5$ $T_\Sigma = 242$ days

It is interesting that in this case the above mentioned regularity also holds: in multi-flyby schemes it is most expedient to perform a flyby of the nearest outer satellite (provided that the planet has a system of satellites).

By and large, the studies conducted on flights to the Galilean satellites of Jupiter show that it is beneficial to accomplish schemes with multiple flybys of one of the natural satellites of the planet, as these paths are more efficient in terms of energy and provide additional possibilities for studying the near-planetary regions of Jupiter, Saturn, and their satellites.

6. FLIGHTS TO SMALL BODIES IN THE SOLAR SYSTEM

Exploration of small bodies in the solar system (comets and asteroids) is among the most promising lines of space studies. The flight schemes to these bodies have received much consideration in recent publications [185–231]. In this chapter we present the results of numerical analysis of different flight scheme options to comets and asteroids based on the approaches developed, and assess the feasibility of the accomplishment of different types of missions.

The selection of the reference orbits for the international project Earth–Venus–Halley's comet, accomplished in the mid-1980s, is used here as an example to illustrate some features of mission design in the case of multi-purpose flights to small bodies.

6.1 Examination of flight schemes to comets

Nowadays, B. Marsden's catalogue [24] contains the orbital elements of 710 comets determined from the results of land-based astrometric measurements in different transits. Among these, the number of comets that were observed twice or more and whose orbital elements were reliably estimated, is 80. Therefore, analysis of the optimal trajectory characteristics for the selected year of the spacecraft launch as well as the selection of the comets to become potential mission targets is a computationally difficult problem. To reduce the time required for studies at the early stages of design works it is reasonable to compile specialized catalogues of comets, containing the dates at which the comet passes through ascending (descending) nodes or through perihelions (aphelions) of their orbits, because for most comets, these lines are close to the line of apsides. Tentative calculations show that optimal trajectories provide the spacecraft encounter with the comet near the node of its orbit. Therefore, given possible spacecraft performance and the planned interval of the spacecraft launch date (starting years), it is possible to determine the preliminary intervals of the spacecraft's arrival at a comet and then to use the catalogue to select the comets meeting this condition.

6.1.1 Direct flights to comets

Possible schemes of a direct spacecraft flight to the comet can be divided into two types: flights leading to an encounter with the comet and its exploration from a flyby trajectory, and flights where the spacecraft passes to an orbit to follow the comet and to perform long-term exploration.

The search for optimal trajectories for the Earth–comet flight is based on the criterion of minimal energy requirements:

- for single-impulse trajectories opt $V = \min_{T_1, T_2} V_p$
- for double-impulse trajectories opt $V = \min_{T_1, T_2}(V_p + V_{\infty 2})$.

To assess the energy required and to study the features of optimal trajectories, the following comets were selected: Grigg–Skjellerupp, Tempel-2, Tempel-1, Clark, Virtanen, Kopf, Jakobini–Zinner, Wolf–Harington, Curyumov–Gerasimenko, Borelly, Hunn, Arend–Rigo, Finley, Ashbroock–Jackson, Halley. These comets have been repeatedly observed in recent decades, they are very bright at perihelion, and their movement can be forecast with acceptable reliability. The selection of comets for the given starting year is made using the above-mentioned catalogue.

The orbits of comets have some peculiarities, making them different from the planetary orbits. Thus, most of the comets have their aphelions near the orbit of Jupiter. The inclination of their orbits to the plane of the ecliptic varies from 4° to 32° and is far greater than the orbital inclinations of most planets. One more feature of the motion of comets is that their lines of apsides are close to the lines of nodes. Considering the large inclinations of the orbits of comets, it can be suggested that it is optimal in terms of energy to encounter the comet near its orbital node. This allows a division of optimal flight trajectories into two types: trajectories encountering the comet near the perihelion of its orbit, and trajectories flying to the zone of its aphelion.

Consider first the flights to perihelion. It should be mentioned that an encounter with the comet near its perihelion provides most favorable conditions for its exploration, because in this portion of the trajectory, large amounts of gases are released under the effect of solar radiation to form the coma and tail of the comet.

Table 6.1 gives the characteristics of single-impulse trajectories, which are optimal in terms of energy, for the flight from Earth to the comet to encounter it near its perihelion.

The presented data suggests that within the start date intervals for most comets (Grigg–Skjellerupe, Finley, Tempel-1, Tempel-2, Clark, Kopf, Arend–Rigo, Wolf–Harrington, Jacobiny–Zinner, Churyumov–Gerasimenko), the values of boost velocities range from 3.41 to 4.23 km/s and never exceed the boost values required for the flights to Venus and Mars. For other comets (Borelly, Hunn, Virtanen, Ashbroock–Jackson), the boost velocities vary from 4.39 to 5.56 km/s.

Note that the energy required to boost the spacecraft for the flight from the orbit of an artificial satellite of Earth to the comet is determined by two main factors: perihelion altitude and the date of the comet passing through the orbital perihelion. The calculations performed showed the second factor to exert a greater effect on the energy requirements.

The velocities at infinity of the encounter with comets vary from 8.5 to 31.8 km/s and are determined by the aphelion radius, orbital inclination, and the date of passing through perihelion.

Table 6.2 gives the results of calculating optimal double-impulse trajectories for the flight to the comet in the zone of its perihelion. It can be seen that optimal double-impulse trajectories typically have higher boost velocities at Earth (V_p= 3.75 to 10.1 km/s) and lower arrival velocity at the comet ($V_{\infty 2}$=5.95 to 12.97 km/s) as compared to single-impulse trajectories. However, even optimal double-impulse trajectories fail to guarantee the spacecraft passing to the trajectory accompanying the comet using a high-thrust engine.

Table 6.1 Optimal trajectories of Earth–comet flight in terms of V_p.

	The comet	T_f, days	$V_{\infty 1}$, km/s	V_p, km/s	$V_{\infty 2}$, km/s
1	Ashbroock–Jackson	319	5.61	4.58	8.89
2	Grigg–Skjellerup	428	2.04	3.41	14.50
3	Enke	337	5.17	4.39	30.92
		242	6.29	4.90	31.82
4	Finley	541	3.93	3.91	15.00
5	Borelly	213	7.02	5.27	17.50
6	Tempel-1	465	4.04	3.94	11.00
7	Tempel-2	206	2.31	3.47	10.70
8	Clark	368	3.63	3.81	8.50
9	Kopf	434	3.95	3.91	11.20
10	Hunn	245	7.55	5.56	11.70
11	Arend–Rigo	552	4.81	4.23	17.60
12	Wolf–Harrington	301	3.47	3.76	11.50
13	Jakobini–Zinner	517	3.34	3.72	21.30
14	Virtanen	233	6.77	5.14	13.30
15	Curyumov–Gerasimenko	469	3.68	3.84	9.46

Table 6.2 Optimal trajectories of Earth–comet flight in terms of $V_\Sigma = V_p + V_{\infty 2}$ (encounter near perihelion).

	The comet	T_f, days	$V_{\infty 1}$, km/s	V_p, km/s	$V_{\infty 2}$, km/s	V_Σ, km/s
1	Ashbroock-Jackson	324	5.67	4.61	8.86	13.47
2	Enke	475	6.81	5.17	19.16	24.33
		429	25.50	19.97	5.78	25.75
3	Grigg–Skjellerup	434	7.61	5.60	8.33	13.93
4	Finley	542	4.44	4.09	5.95	10.04
5	Borelly	200	11.74	8.31	11.02	19.33
6	Tempel-1	454	5.79	4.66	8.42	13.08
7	Tempel-2	224	3.43	3.75	9.72	13.43
8	Curyumov–Gerasimenko	460	4.30	4.05	7.87	11.92
9	Kopf	412	4.71	4.19	8.23	12.42
10	Clark	370	3.82	3.87	8.31	12.18
11	Hunn	297	9.11	6.50	8.70	15.20
12	Arend–Rigo	533	11.66	8.25	7.45	15.70
13	Wolf–Harrington	299	5.59	4.56	9.55	14.11
14	Jakobini–Zinner	527	14.05	10.10	7.64	17.74
15	Virtanen	241	7.09	5.31	12.97	18.28

Whereas the boost velocities at Earth required to fly to comets in the zone of their perihelion are the same as those to fly to Mars and Venus, the boost velocities required to reach comets near their aphelion are comparable with the velocities needed to fly to Jupiter. This is due to the fact that the orbital aphelions of most short-period comets are located near Jupiter's orbit. Table 6.3 gives the characteristics of orbits which are optimal in terms of energy (criterion V_{Σ}), reaching the comet near its aphelion, and accompanying it in its subsequent motion. The data show that the boost velocities required to fly into the zone of aphelion vary from 6.22 to 8.81 km/s. The relative velocities of approach to the comet lie within 0.61 to 3.94 km/s.

6.1.2 Earth–Venus–comet flights

Consider the potential of planetary gravity assists in flights to comets. Let us take Venus as the flyby planet, because the orbital perihelions of many comets lie between the orbits of Earth and Venus. Considering the potential of the planet's gravitational field, the spacecraft post-flyby orbit for typical Earth–Venus trajectories ($V_p = 4$ km/s) can lie within the range 0.5–2.5 AU. Therefore, considering the above recommendations, it is beneficial to locate the encounter with the orbit near the perihelion. As is shown in Section 4.1, the use of single flyby of Venus to reach Jupiter is not good in terms of energy. Therefore, there is no use in considering a flyby of Venus for the flight to the zone of the comet's aphelion, which for most comets lies near the orbit of Jupiter. The comets that can be studied using spacecraft were selected based on a catalogue containing dates at which the comets pass through their perihelions. Once the window for launch to Venus is selected (it occurs every 1.6 years), the list of comets that will pass through their orbital perihelions at proper times can be determined, and optimal trajectories can be calculated for them. The calculation results presented in Table 6.4 show that for each starting window, as a rule, several comets can be regarded as possible targets of the Earth–Venus–comet flight.

Table 6.3 Optimal trajectories of Earth–comet flight in terms of $V_{\Sigma} = V_p + V_{\infty 2}$ (encounter near the aphelion).

	The comet	T_f, days	$V_{\infty 1}$, km/s	V_p, km/s	$V_{\infty 2}$, km/s	V_{Σ}, km/s
1	Jakobini–Zinner	1417	9.57	6.81	3.21	10.02
2	Virtanen	1309	9.15	6.53	1.57	8.10
3	Arend–Rigo	1432	10.44	7.38	2.55	9.93
4	Grigg–Skjellerup	1048	9.20	6.56	2.39	8.95
5	Borelly	1431	9.43	6.71	3.44	10.45
6	Tempel-2	864	8.74	6.27	2.76	9.03
7	Finley	1940	9.24	6.59	0.61	7.20
8	Tempel-1	905	8.98	6.42	1.73	8.15
9	Clark	1045	12.42	8.81	1.76	10.57
10	Hunn	761	8.66	6.22	3.58	9.80
11	Kopf	1319	9.00	6.43	1.68	8.11
12	Wolf–Harrington	1235	9.10	6.50	3.94	10.44

6.1.3 Ballistic design of "Vega" mission

The international project "Venus–Halley's comet" [210] can be considered an example of an accomplished multi-purpose mission. The project included exploration of Venus and Halleys comet using a single spacecraft. The general view of the spacecraft "Vega" is shown in Figure 6.1. Consider the design and ballistic characteristics of this mission.

Figure 6.2 shows the trajectory of Vega in the projection to the ecliptic plane.

Table 6.4 Trajectories of Earth–Venus–comet.

The comet	T_{E-V}, days	T_{V-C} days	V_P, km/s	$V_{\infty 2}$, km/s	$V_{\infty 1}$, km/s	r_π, thous. km	V_m, km/s
Tzuchinchana-1	143	615	4.12	7.2	11.5	7.4	1.15
Arend–Rigo	95	415	3.86	5.9	13.9	6.5	1.32
Deviko–Swift	149	290	4.01	6.76	12.4	6.6	0.65
Tailor	170	415	3.82	6.18	11.9	6.1	0.77
Barnard-3	155	365	3.98	6.4	18.3	7.8	0.14
Virtanen	107	415	3.95	3.3	14.4	4.1	1.29
Forbes	82	418	4.17	9.7	8.8	4.2	1.7

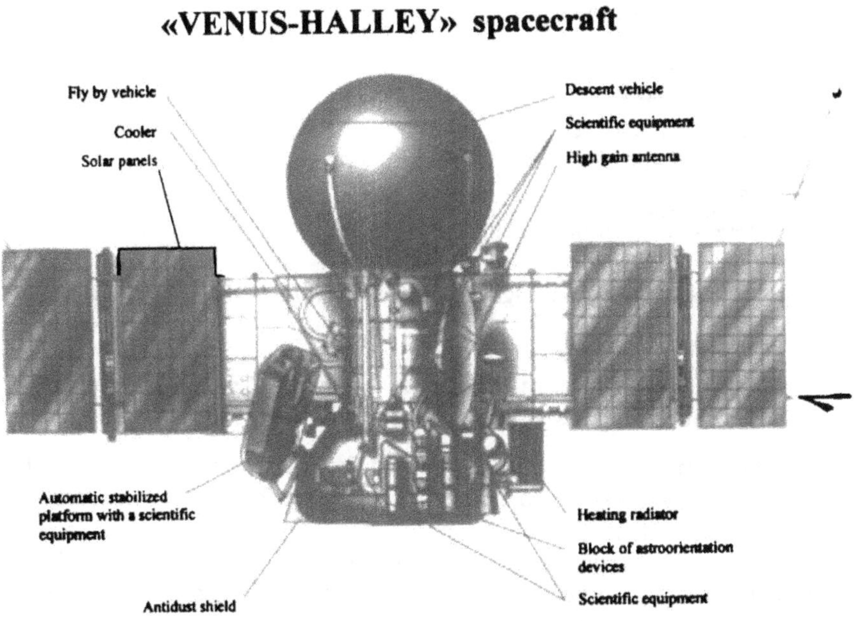

Figure 6.1 "Venera–Halley" space vehicle.

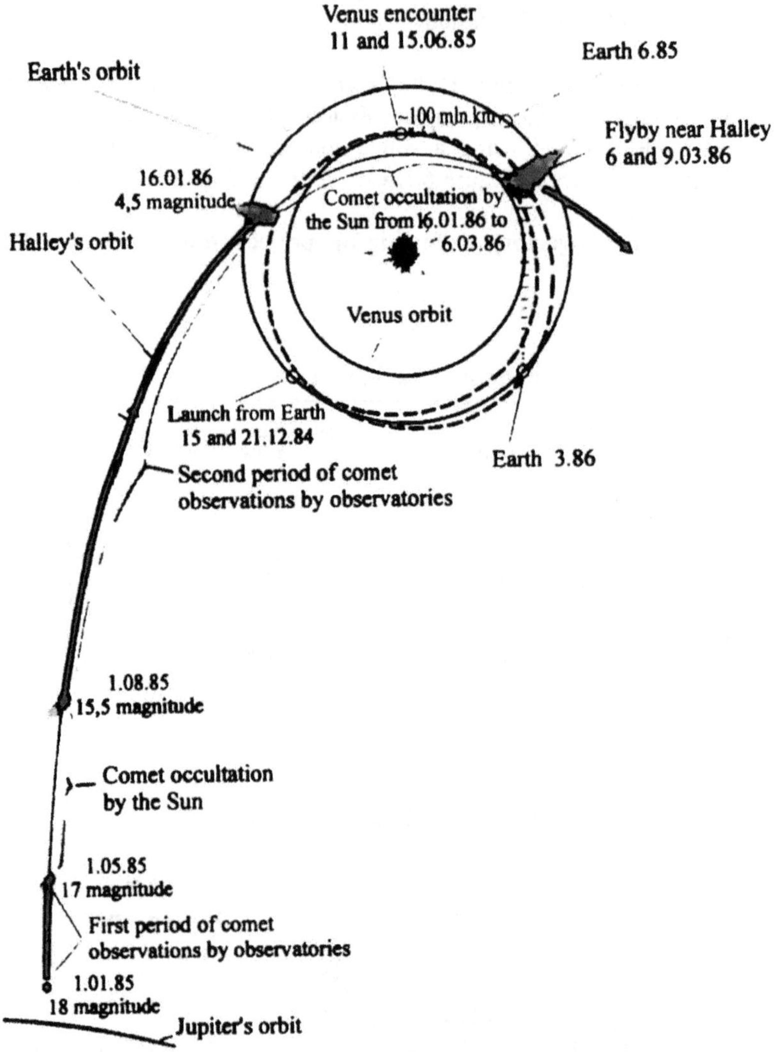

Figure 6.2 Schematic diagram of flight of "Venera–Halley" space vehicle.

When selecting reference trajectories for the Earth–Venus–Halley's Comet flight (E–V–H), the following design restrictions should be taken into account:

- an atmospheric probe is to be introduced on the side of the planet visible from Earth, and conditions are to be provided for interferometric measurements from the planetary surface;
- the velocity of the lander entering Venus' atmosphere must not exceed 11.0 km/s;
- the spacecraft boost velocity from the orbit of Earth's artificial satellite must not exceed 3.89 km/s in order to provide the required initial mass of the spacecraft;

- intervals between the launch dates two spacecrafts are not to be less than 4 days;
- intervals between the dates of arrival of two spacecrafts at Venus are not to be less than 3 days;
- radio communication between the lander and the module flying by ("bus") is to be maintained throughout the descent into the atmosphere, and the duration of radio communication between the lander and the orbiter after landing is to be not less than 14 minutes with a probability of 0.8.

Analysis of E–V–H trajectories was performed in the isoline fields of ballistic parameters presented in Figures 6.3–6.5. For start dates in 1984–1985, three domains of optimal trajectories in terms of the outgoing velocity at Earth are seen to exist. The first domain corresponds to Earth–Venus flights along trajectories of the first half-revolution (Earth–Venus flight time equals 120 to 130 days); the second and third domains correspond to the encounter with Venus at the second half-revolution of heliocentric orbit (the flight time is 170 days). In all these cases, the Venus–Halley's Comet flight is made along the second-revolution trajectories. Table 6.5

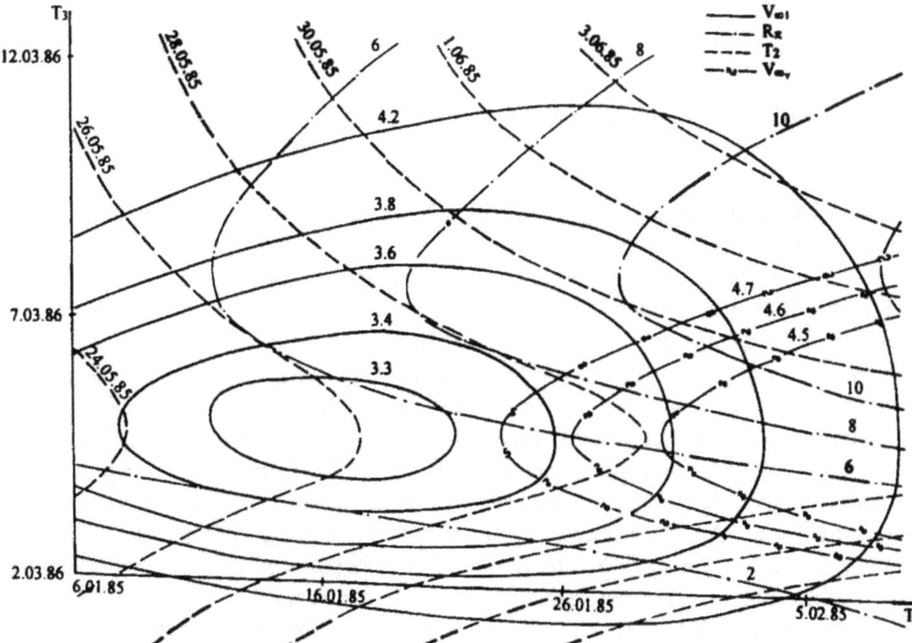

Figure 6.3 Isolines of parameters of Earth–Venus–Halley's comet trajectories (1st domain), constructed on a plane: the date of launching from Earth (T_1)—the date of arrival to the Halley's comet (T_3); the velocity at infinity of departure from Earth ($V_{\infty 1}$); the velocity at infinity of approaching Venus (V_{∞_v}); the pericentral radius of planetocentric trajectory of flyby of Venus ($R\pi$); the date of arrival to Venus (T_2).

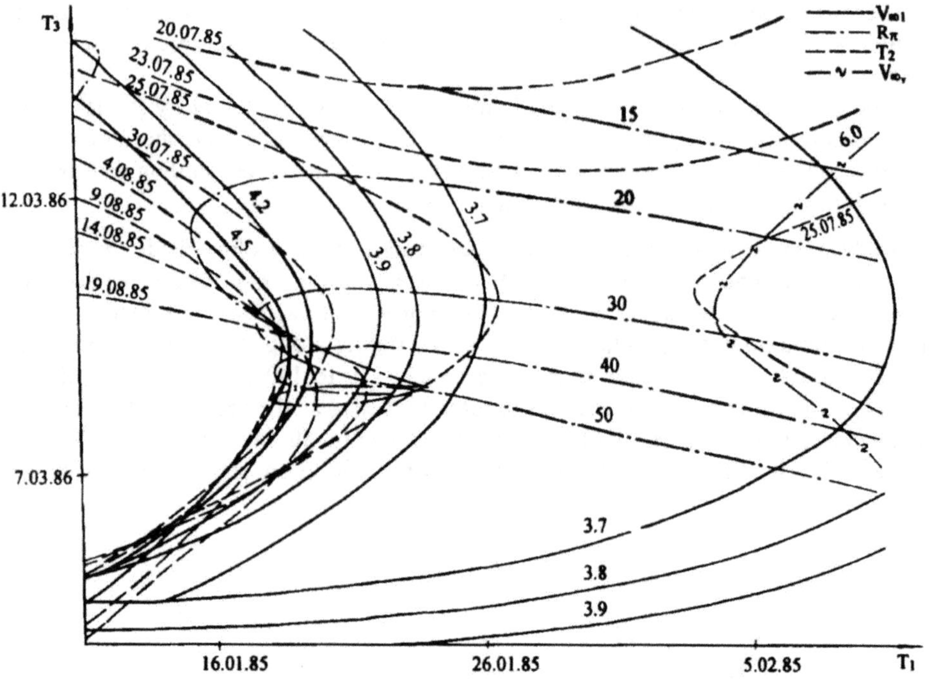

Figure 6.4 Isolines of parameters of Earth–Venus–Halley's comet trajectories (2nd domain).

gives the characteristics of E–V–H trajectories which are optimal in terms of V_p for the three domains.

The E–V–H trajectories belonging to the first and second domains are seen to have lower departure velocities at Earth, as compared with the trajectories of the third domain; however they fail to meet the above conditions concerning the entry velocity, the duration of radio communication, and the conditions of the probe entering the planetary atmosphere. Therefore, they were excluded from our consideration.

The trajectories belonging to the third domain meet all the necessary requirements, so they were used to select the reference orbits for the E–V–H flight. The procedure of selecting reference trajectories was iterative and included the following principal operations. First, isolines of ballistic parameters of Venus gravity assist flyby were constructed for the selected domain of E–V–H orbits (the condition of gravity assist flyby implies that the modules of ingoing $V_{\infty i}$ and outgoing $V_{\infty o}$ velocities at Venus at infinity are equal). Next, the characteristics of E–V–H reference trajectories satisfying all the design requirements were determined. Conditions of radio communication between the lander and the orbiter were determined.

Calculations show that the condition on the necessary duration of lander–orbiter radio communication during the lowering of the spacecraft and after its landing leads to a 50–100 m/s increase in the departure velocity at Venus relative to the calculated value obtained from a purely ballistic flyby of Venus. Therefore, a powered

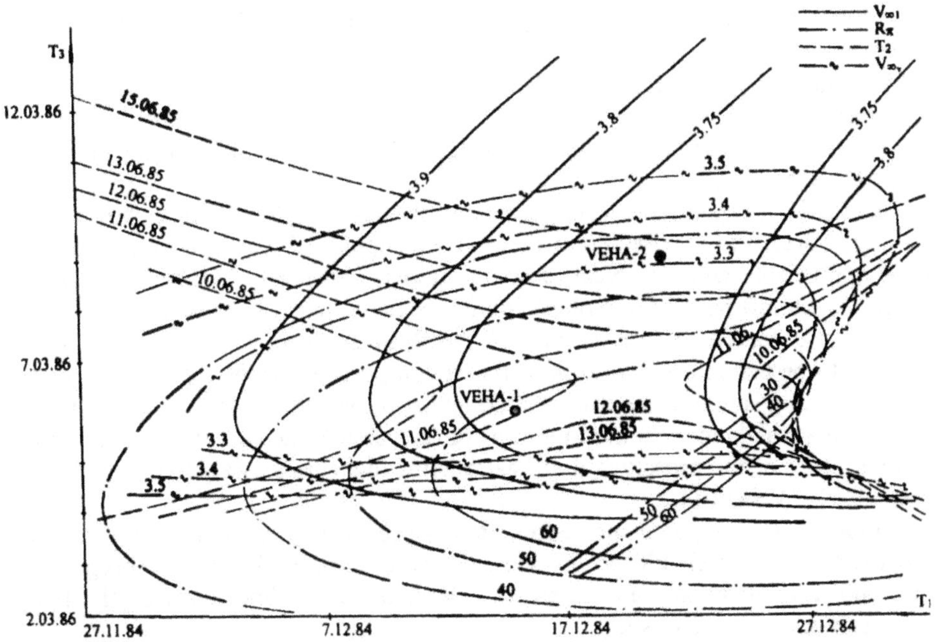

Figure 6.5 Isolines of parameters of Earth–Venus–Halley's Comet trajectories (3rd domain). Black dots mark the trajectories selected for the "Venus–Halley's Comet" project.

Table 6.5

Domain of trajectories E–V–H	Date of start from Earth	T_{E-V}, days	T_{V-H}, days	$V_{\infty 1}$, km/s	$V_{\infty 2}$, km/s	V_H, km/s	r_π, thous. km
1st domain 1st half-revol. +2nd half-revol.	17.01.85	129	283	3.30	5.1	90.4	4
2nd domain 2nd half-revol. + 2nd half-revol.	2.02.85	173	223	3.61	6.0	81.0	28
3rd domain 2nd half-revol. + 2nd half-revol.	21.12.84	174	267	3.75	3.3	78.9	40

The following designations are used in Table 6.5:

- T_{E-V}, T_{V-H} denote the flight time from Earth to Venus and from Venus to Halley's Comet, respectively;
- $V_{\infty 1}$, $V_{\infty 2}$, V_H are the velocities at infinity of departure from Earth, arrival at Venus and Halley's Comet, respectively;
- r_π is the pericentral radius of the planetocentric flyby trajectory.

maneuver is required to correct the flyby trajectory within 10–20 days after the flyby of Venus.

At the next stage, the reference orbits are selected so as to minimize the value of departure maneuver velocity. To do this, isolines of parameters were constructed for E–V–H trajectories satisfying the conditions of powered–gravitation flyby of Venus ($V_{\infty i} - V_{\infty o} = 50$ m/s).

Examination of trajectories using powered–gravitation flyby of Venus showed that the use of these trajectories allows the velocity impulse of the departure maneuver to be appreciably reduced. Next, after the design and mass characteristics of the "Vega" spacecraft have been refined, isolines of the mass of the spacecraft flying by ("bus") were constructed. The mass of the flyby spacecraft was selected as the criterion to be maximized, because of the design peculiarities of the "Vega" spacecraft. The thing is that in the course of designing the spacecraft the requirement was imposed that the lander should inherit as far as possible the construction and mass characteristics of the prototype that has satisfied flight tests as a part of spacecraft "Venera-9"–"Venera-14".

In practice, this requirement implied that the spacecraft mass and entry velocity into the atmosphere should be the same as those accepted for the prototype. The final choice of nominal trajectories was based on the isoline fields of the flyby spacecraft mass characteristics.

The performed analysis allowed the selection of reference trajectories for flights of two spacecrafts to reach Venus with a 4-day interval and Halley's Comet with a 3-day interval; this made it possible to successively interact with the two spacecrafts in the course of flying by Venus and encountering the comet.

The selected E–V–H trajectories provided reception of data throughout the descent into the atmosphere and within 14 minutes after the spacecraft landing.

The E–V–H reference trajectories have some interesting features to distinguish them from the trajectories commonly used for Earth-to-comet flights. In particular, the velocity vector at infinity of departure from Venus ($\vec{V}_{\infty o}$) demonstrates high sensitivity to changes in the dates of planetary flybys (T_2) and arrival at the comet (T_3). Thus, a 1-day change in the date T_2 (T_3) can cause a 160 m/s (210 m/s) change in the departure velocity at infinity $V_{\infty o}$, and the direction of vector $\vec{V}_{\infty o}$ can move by 7°–10°. The corresponding change in the flyby-spacecraft encounter with the comet (V_e) does not exceed 800 m/s, and its direction changes by not more than 0.5°. The small variations in the direction of $\vec{V}_e$ due to variations in T_2 (T_3) can be accounted for by the fact that its module reaches 80 km/s because of the spacecraft and comet moving towards one another.

The above analysis shows that the parameters of planetocentric trajectories of Venus flybys are strongly dependent on the moment T_3 the comet is encountered, and therefore, on the actual parameters the orbit of Halley's Comet, whose refined estimates were obtained based on the astrometric observations from Earth after the spacecraft had already started from it. On the other hand, virtually no dependence was found between the ballistic conditions of the spacecraft encounter with the comet and the date of this event. The principal ballistic characteristics of the selected trajectories are given in Table 6.6.

Table 6.6 Principal design and ballistic characteristics of the trajectories selected for the "Vega" project.

Start from Earth	15.12.84	21.12.84
Flight time from Earth to Venus, days	178	176
Date of arrival at Venus	11.06.85	15.06.85
Flight time from Venus to Halley's Comet, day	268	267
Date of arrival at Halley's Comet	06.03.86	09.03.86
Total mission duration, days	446	443
Characteristic boost velocity from the orbit of artificial satellite of Earth, km/s	3.89	3.89
Velocity at infinity of departure from Earth, km/s	3.73	3.72
Velocity of entry into Venus' atmosphere, km/s	10.83	10.83
Angle of the spacecraft entry into Venus' atmosphere, degrees	19 ± 5	19 ± 5
Relative velocity of spacecraft arrival at Halley's Comet, km/s	79.7	77.2
Probability of providing conditions for radio communication with the lander: lander–"bus" communication throughout the stage of descent in the atmosphere	0.9	0.9
Guaranteed time of orbiter–lander communication after landing ($P = 0.8$); minutes	14	14
Velocity at infinity of arrival at Venus, km/s	3.18	3.36
Velocity at infinity of departure from Venus, km/s	3.14	3.31
Angle of rotation of velocity vector at infinity due to flyby of Venus, degree	43.6	54.4
Pericentral radius of the flyby trajectory at Venus, thousands km	47.4	31.0
Velocity impulse of maneuver when approaching Venus, m/s	284	182
Velocity impulse of maneuver after flyby of Venus (within 10 days after the flyby), m/s	31	133

6.2 Examination of flight schemes to asteroids

Up-to-date catalogues [6] contain the orbital parameters of about 3500 asteroids, whose diameters are estimated to vary from 1 to 1000 km. All asteroids can be conventionally subdivided into three large groups according to the ballistic peculiarities of their orbits.

The first group is the largest, incorporating about 3000 small bodies, and including the asteroids of the Main Belt, whose semi-major orbital axes vary from 2.2 to 3.2 AU. The eccentricities of these orbits are not large, ranging from 0 to 0.2 and rarely exceeding 0.4. Inclinations of the orbits to the ecliptic plane commonly amount to 5°–10°. Some asteroids move along orbits with inclinations up to 35–40°. The orbits of the asteroids of the Main Belt lie within a ring located between the orbits of Mars and Jupiter.

The second group consists of about 80 small bodies and includes the asteroids of Earth's group. Among these, the orbital elements of 36 asteroids were reliably determined and included in catalogues. The diameters of the asteroids are estimated to vary from 200 m to 40 km. The common feature allowing them to be included in a single group is the fact that their orbit lies between the orbit of Venus and the Main Asteroid Belt.

The third group consists of about 20 small bodies and incorporates asteroids moving along the orbit of Jupiter in the triangular points of libration. These asteroids are rather large, their diameters are about 150 km. In their motion they form two groups: one group moves ahead of Jupiter (asteroids-Greeks) and the other group moves within 60° behind Jupiter (asteroids-Trojans) in an angular motion around the sun. Estimates obtained recently by researchers show that about 700 weak asteroids with a stellar magnitude of up to +20.6 move near each Lagrangian point.

Consider the options for a mission to explore asteroids. At the first (reconnaissance) stage of the studies, it is expedient to use the flight schemes allowing the spacecraft to quickly (with relative velocity of 5 to 12 km/s) pass by one or several asteroids. Considering the long duration of the flight to asteroids and relatively short time the spacecraft moves in the nearest vicinity of each asteroid, it is advantageous to perform flybys of planets or several asteroids to increase the scientific efficiency of the mission.

At the second stage, when more detailed exploration of the asteroids is performed, it is advantageous to use the mission schemes allowing special-purpose landers to be delivered to the asteroid surface. Of most interest in this case are the trajectories providing minimum possible velocities of encounter with the asteroid. Tentative design studies showed that to provide a landing on an asteroid, the approach velocity must not exceed 4 km/s.

6.2.1 Approximate method for selecting asteroids to be explored by different types of missions

Ballistic and design studies of missions to asteroids differ in some ways from those concerning the flights to planets and, therefore, they require special methods and algorithms. The primary distinction is that at the stage of preliminary flight scheme analysis, the designer deals with a large set (more than 3000) of asteroids in order to find the target asteroids that satisfy selected design criteria and restrictions. Therefore, the conventional methods of design and ballistic analysis incorporating the *a priori* construction of calendars of optimal flights and fields of isolines of principal ballistic characteristics for a wide range of launch and arrival dates are virtually inapplicable at the early stages of studying the trajectories to asteroids, because of the great number of these and due to the necessary calculations being too involved. Therefore, a procedure is required that allows a preliminary selection of asteroids meeting design criteria and to be considered at the later stages.

To select trajectories to fly by several asteroids, the following calculation scheme, named "principal target method", can be used. First, for the selected year of mission implementation, trajectories for flights to the largest asteroids are selected so as to optimize a certain criterion, e.g. the boost velocity at Earth. Thus, the number of asteroids whose diameters are estimated to exceed 200 km is as large as 31, and all of them are located in the Main Belt. Asteroids meeting the restriction on the boost velocity are selected, and isolines of the principal ballistic parameters are constructed for these. Once the reference trajectories have been selected, elements of their orbits are calculated, and minimum distances from each of the asteroids in the catalogue are determined for the spacecraft moving along the reference trajectories. The asteroids that are closest to the spacecraft trajectory are considered to be the most probable targets for a multi-asteroid flyby mission. As the number of asteroids in the catalogue is large, it is usually possible to find a number of asteroids the spacecraft will fly by near the aphelion of its orbit with maneuvers which are acceptable in terms of velocity impulses. Such schemes will be referred to as multiple-flyby asteroid missions.

The following method was developed for selecting asteroids that can be approached at the minimum possible velocity. Assuming that the Earth's orbit is circular and coplanar with the orbits of the asteroids, let us determine the characteristics of the orbit (velocities at infinity of departure from Earth ($V_{\infty 1}$) and arrival at the asteroid ($V_{\infty 2}$)) to reach the aphelion and perihelion of the asteroids orbit.

The parameters $V_{\infty 1}$ and $V_{\infty 2}$ can be calculated from the following formulae:

$$\left.\begin{aligned} V_{\infty 1} &= \sqrt{\frac{2\mu_c r_{\alpha A}}{r_{\oplus}(r_{\oplus}+r_{\alpha A})}} - \sqrt{\frac{\mu_c}{r_{\oplus}}} \\ V_{\infty 2} &= \sqrt{\frac{2\mu_c r_{\oplus}}{r_{\alpha A}(r_{\oplus}+r_{\alpha A})}} - \sqrt{\frac{2\mu_c r_{\pi A}}{r_{\alpha A}(r_{\pi A}+r_{\alpha A})}} \end{aligned}\right\} \text{for the flight to the aphelion of the asteroid's orbit}$$

$$\left.\begin{aligned} V_{\infty 1} &= \sqrt{\frac{2\mu_c r_{\pi A}}{r_{\oplus}(r_{\oplus}+r_{\pi A})}} - \sqrt{\frac{\mu_c}{r_{\oplus}}} \\ V_{\infty 2} &= \sqrt{\frac{2\mu_c r_{\oplus}}{r_{\pi A}(r_{\oplus}+r_{\pi A})}} - \sqrt{\frac{2\mu_c r_{\alpha A}}{r_{\pi A}(r_{\pi A}+r_{\alpha A})}} \end{aligned}\right\} \text{for the flight to the perihelion of the asteroid's orbit}$$

$$r_{\pi A} = a(1-e) \quad r_{\alpha A} = a(1+e)$$

where $r_{\oplus}$ is the radius of Earth's orbit; $r_{\alpha A}(r_{\pi A})$ is the altitude of aphelion (perihelion) of the asteroid's orbit; μ_c is the gravitational parameter of the sun; e is the eccentricity of the asteroid's orbit; a is the semi-major axis of the asteroid's orbit.

It can easily be seen that $V_{\infty 1}$, $V_{\infty 2}$ are functions of two variables, e.g. the semi-major axis (a) and the eccentricity of the asteroid's orbit (e). Therefore, the method presented in Section 3.1 was used to construct the isolines of parameters $V_{\infty 1}$ and $V_{\infty 2}$ in coordinates a and e; the values of eccentricity varied from 0 to 0.85, and the range of variations in the semi-major axis covered all possible values for the orbits of the asteroids (from 0.7 to 4 AU). Thus, we obtain the isolines of departure velocities from Earth and arrival at the asteroid in the a–e plane. Real phasing of the Earth and the asteroids in their orbits taken into account in the calculations will generally result, in an increase in the arrival and departure velocities, therefore the isolines should be considered the lower bounds of the energy requirements.

The use of the isolines at the preliminary mission analysis stage allows one to:

- determine minimum values of the departure and arrival velocities for each of the above groups of asteroids;
- determine which type of flight is the most advantageous: to aphelion or to perihelion;
- select the interval of values of a and e, to provide acceptable departure and arrival velocities;
- determine the list of asteroids for which further mission analysis is expedient, with real orbital characteristics taken into account.

Among the asteroids selected using the above procedure, a further selection can be made based on some additional restrictions, e.g. the orbital inclination and size of the asteroid.

One more advantage of the suggested procedure is the possibility of assessing different flight schemes to asteroids with the use of gravity assist flybys of planets. To do this, isolines of parameters $V_{\infty 1}$ and $V_{\infty 2}$ were constructed for Venus–asteroid, Earth–asteroid, Mars–asteroid, and Jupiter–asteroid flights. They allow *a priori* estimates of the potential of different flight schemes to asteroids, for example, Earth–Venus–asteroid, Earth–Venus–Earth–asteroid, Earth–Mars–asteroid, Earth–Venus–Mars–asteroid in terms of providing minimum encounter velocities in the flights to different groups of asteroids.

Figures 6.6 and 6.7 show isolines of velocities $V_{\infty 2}$ for Mars–asteroid flights. An analysis of these isolines with the aim of minimizing the velocities of the spacecraft encounter with asteroids allows the following recommendations:

(a) for asteroids of the Main Belt:

- minimum approach velocities can be attained when the asteroids are encountered near the aphelion of their orbits; in this case, the orbits of the asteroids, considered as possible targets, should have minimum semi-major axes and maximum values of eccentricity;
- flights to the pericenter of the orbit of an asteroid result in an increase in the encounter velocity by ~2 km/s as compared to flights to aphelions;
- the use of trajectories with a gravity assist maneuver at Mars makes it possible to obtain minimum approach velocities to asteroids (less than 4 km/s);
- the use of Earth–asteroid trajectories results in an increase in the encounter velocity by ~2 km/s as compared with flights via Mars;
- the use of Venus–asteroid trajectories leads to an increase in the approach velocities by ~2 km/s relative to Earth–asteroid flights;

(b) for asteroids of the near-Earth group:

- minimum approach velocities can be provided when the asteroids are encountered near the aphelions of their orbits; in most of such cases it is advantageous to use Earth–asteroid and Venus–asteroid trajectories;
- minimum boost velocities from Earth can be provided when the encounters of asteroids take place near the perihelions of their orbits.

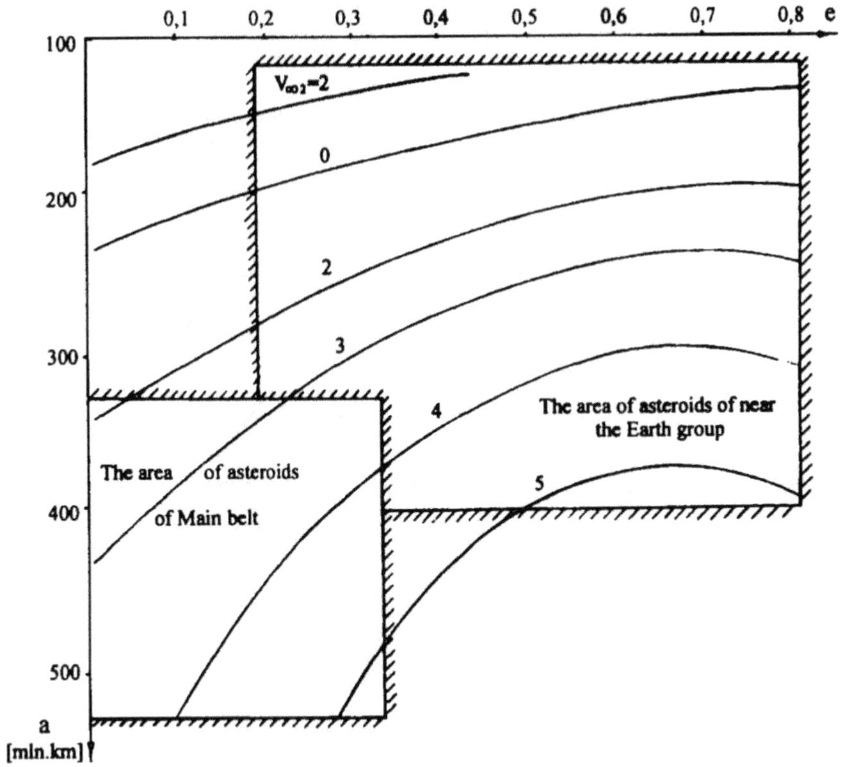

Figure 6.6 Isolines of velocities of approaching the asteroid in Mars–asteroid flight (encounter in perihelion).

6.2.2 Direct flights to asteroids

To assess the energy requirements for direct flights, the characteristics of flight trajectories to asteroids of different groups were examined.

Table 6.7 gives the characteristics of optimal double-impulse trajectories for flights to asteroids of the near-Earth group for a launch window in 2000–2001.

The flight time to the near-Earth group asteroids amounts to 150–900 days.

Energy requirements for flights to the asteroids of the Jupiter group are the same as those for the flight to Jupiter. Parameters of Earth–Jupiter trajectories have been closely studied. The boost velocity required to reach asteroids of the Jupiter group is 6.2–6.8 km/s, and the encounter velocity in this case will be 6–8 km/s.

6.2.3 The use of inner planetary flybys to reach asteroids

Results of the examination of Earth–Venus–asteroid trajectories for flights to asteroids of the Earth group and Main Belt are given in Table 6.8. The boost

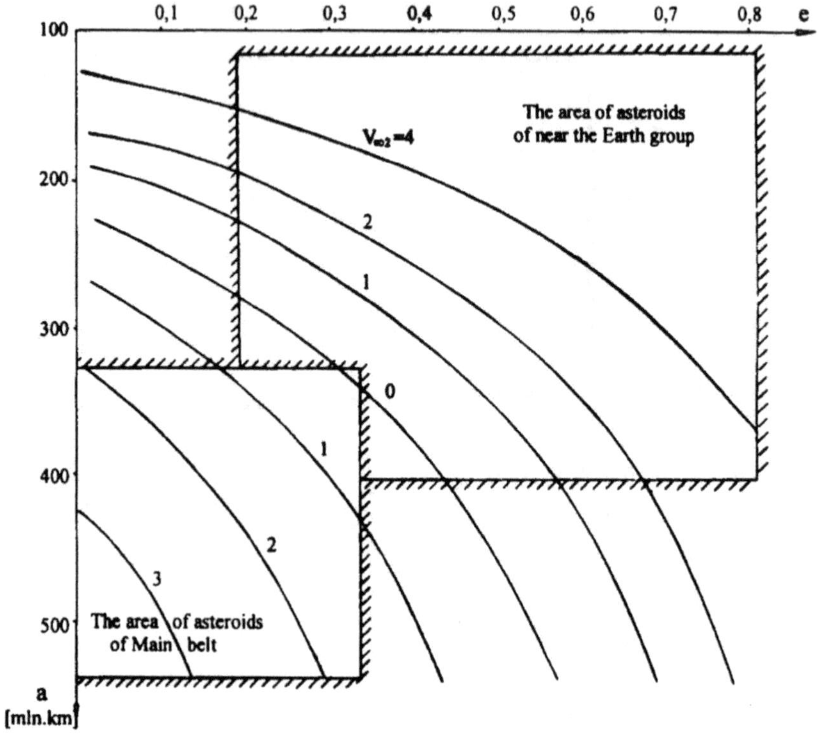

Figure 6.7 Isolines of velocities of approaching the asteroid in Mars–asteroid flight (encounter in aphelion).

Table 6.7 Earth–asteroid trajectories optimal in terms of criterion: min ($dV_p + V_{as2}$) (options for missions with soft landing). Trajectories of types 1 and 2 (angular distance of the flight is less than 360°).

Name	Family	Launch date	Time of flight from Earth to asteroid	dV_p, km/s	V_{as1}, km/s	V_{as2} km/s
433 Eros	Am	31/01/2000	380.0	5.750	7.873	2.188
1620 Geographos	Ap	17/09/2000	280.0	5.794	7.947	3.317
1685 Toro	Ap	01/01/2000	150.0	6.860	9.656	1.790
1943 Anteros	Am	07/06/2001	340.0	4.271	4.910	3.877
3102 1981 QA	Am	08/08/2000	680.0	5.807	7.969	2.515
3288 Seleucus	Am	05/05/2000	680.0	5.183	6.851	1.757
4486 Mithra	Ap	09/06/2000	720.0	5.624	7.654	1.720
5189 1990 UQ	Ap	27/06/2001	790.0	4.995	6.487	2.431
1989 QF	Ap	26/03/2000	380.0	4.652	5.782	2.525
1991 OA	Am	04/07/2000	890.0	5.335	7.134	2.684

Note: dV_p is the characteristic boost velocity from the parking circular orbit of an artificial satellite of Earth with an altitude of 200 km (gravitational losses are neglected); V_{as1} is the asymptotic departure velocity from Earth; V_{as2} is the asymptotic velocity of arrival to asteroid.

Table 6.8 Characteristics of Earth–Venus–asteroid trajectories.

Flight scheme	T_{st}	T_{E-V}, days	T_{E-V-A}, days	$V_{\infty 1}$, km/s	V_{p}, km/s	$V_{\bullet}$ km/s	V_{m}, km/s	$V_{\infty A}$, km/s	r_{J1}
E–V–Sizif	06.10.89	112	495	4.76	4.21	13.0	0	30.5	6.4
E–V–Cerber	26.10.89	110	420	3.75	3.85	11.8	0.27	8.0	11.6
E–V–Bakhus	05.11.89	220	220	3.90	3.89	12.4	0	11.7	8.4
E–V–Ivar	11.10.89	110	395	4.52	4.12	13.1	0	15.1	6.1
E–V–Eros	10.11.89	170	370	3.13	3.66	11.5	0	10.5	12.8
E–V–Bakhus	12.07.91	159	390	3.99	3.93	12.3	0	4.5	26.4
E–V–Anteros	07.06.91	105	420	3.48	3.76	11.4	0	14.6	6.1
E–V–Midas	12.07.91	143	330	5.12	4.36	12.0	0	33.7	16.6
E–V–Hungary	09.05.91	125	390	4.16	3.98	12.3	0.69	9.2	6.2
E–V–Clio	08.06.91	184	435	3.62	3.81	11.9	0.73	9.1	8.9
E–V–Victory	16.05.91	156	455	5.01	4.31	11.9	0	7.9	6.8
E–V–Germia	12.05.91	158	500	4.83	4.23	12.1	0	6.7	6.3

velocities at Earth for these trajectories amount to 3.7–4.4 km/s, and the velocity of encounter reaches 6.7–33 km/s. The flight time varies from 220 to 500 days. Note that the trajectory for flight to the Earth group asteroids can be accomplished with a purely ballistic flyby of Venus, and in the flight to the asteroids of the Main Belt, the maneuver required at the planet does not exceed 1 km/s.

Thus, the flight schemes involving a flyby of Venus provide acceptable departure velocities at Earth, but fail to provide low velocities of approaching to the asteroids. The recurrence of the launch windows for this flight scheme coincides with the periodicity of launches to Venus and equals ~1.6 years.

It should be noted that the Earth–Venus–asteroid flight schemes are most advantageous for exploring the near-Earth group asteroids.

As can be seen from the data presented above, flights involving a flyby of Mars provide minimum possible velocities of approach the Main Belt asteroids, and the minimum approach velocities are attained when the encounter takes place near the aphelion of their orbits. Table 6.9 presents the characteristics of Earth–Mars–asteroid trajectories for the launch window in 2001.

The chief disadvantage of the E–M–asteroid flight scheme is that, as a rule, it requires a powered maneuver in the flyby of Mars.

Among the merits of these schemes is the possibility of increasing the scientific efficiency of missions to asteroids by delivering a lander to the Martian surface or orbiting a probe as an artificial satellite of Mars with the use of powered or powered-atmospheric deceleration.

In the case where trajectories with two gravity assist maneuvers at Mars are used (E–M–M–asteroid), the first gravity assist maneuver is purely ballistic, and the maneuver velocity during the second flyby is notably less than that for trajectories with a single flyby of Mars (see Table 6.9).

A promising scheme for flights to asteroids is the Earth–Venus–Earth–asteroid (E–V–E–A) flight. It allows the exploration of Venus and asteroids to be conducted

Table 6.9 Trajectories of flights to asteroids of the Main Belt with a single gravity assist maneuver at Mars. Optimization in terms of criterion $V_{sum} = dV_1 + dV_2 + V_3$.

Asteroid (number and name in catalogue	Radius, km	Start from Earth	Flight time from Earth to Mars	Date of flyby of Mars	Flight time Mars–asteroid, days	Date of arrivals asteroid	Total flight time, years	Boost velocity from the orbit of artificial Earth satellite, dV_1, km/s	Velocity of powered maneuver at Mars, dV_2, km/s	Velocity of arrival at asteroid, V_3, km/s	Entry velocity to the Martian atmosphere
12 Victoria	68	27.03.2001	270	22.12.2001	580	25.07.2003	2.3	3.70	0.88	4.27	6.87
27 Euterpe	58	10.06.2001	375	20.06.2002	910	16.12.2004	3.5	3.97	1.44	3.85	7.34
40 Harmonia	58	27.03.2001	285	06.01.2002	580	09.08.2003	2.4	3.60	1.91	2.93	6.80
47 Aglaja	79	26.05.2001	360	21.05.2002	610	21.01.2004	2.7	3.91	1.61	4.04	7.35
84 Kljo	44	11.05.2001	330	06.04.2002	520	08.09.2003	2.3	3.76	0.33	4.93	7.25
111 Ate	78	11.05.2001	345	21.04.2002	580	22.11.2003	2.5	3.85	2.50	2.75	7.34
203 Pompeja	53	12.03.2001	165	24.08.2001	400	28.09.2002	1.6	3.65	2.04	3.86	7.61

Trajectories of flights to asteroids of the Main Belt with two gravity assist maneuvers at Mars. Optimization in terms of criterion. $V_{sum} = dV_1 + dV_2 + V_3$.

Asteroid (number and name in catalogue	Radius, km	Start from Earth	Flight time from Earth to Mars	Date of the first flyby of Mars	Flight time Mars–asteroid, days	Date of arrival at asteroid	Total flight time, years	Boost velocity from the orbit of artificial Earth satellite, dV_1, km/s	velocity of powered maneuver in the second fly-by of Mars dV_2, km/s	Velocity of arrival at asteroid, V_3, km/s	Entry velocity to the Martian atmosphere
7 Iris	104	10.06.2001	285	06.02.2004	780	27.03.2006	4.8	4.26	0.38	1.75	7.69
46 Hestia	82	26.04.2001	300	07.01.2004	690	27.11.2005	4.6	3.61	0.09	2.83	7.04
89 Vibilia	66	26.05.2001	315	21.02.2004	630	12.11.2005	4.5	3.89	0.00	3.59	7.28
146 Lucina	70	26.05.2001	315	21.02.2004	540	14.08.2005	4.2	3.89	0.22	3.86	7.28
554 Peraga	50	11.05.2001	330	21.02.2004	690	11.01.2006	4.7	3.76	0.36	1.68	7.25

within a single mission. The ballistic potential of the gravity assist maneuver at Earth make it possible to reach asteroids of the Main Belt, near-Earth and Jupiter groups.

At the first stage, the following procedure was used to search for optimal E–V–E–asteroid trajectories. First, isolines for different types of E–V–E trajectories (1 half-revolution + 1 half-revolution), (1 half-revolution + 2 half-revolution), (2 half-revolution + 2 half-revolution) were constructed. Next, domains of E–V–E trajectories, meeting the principal design and ballistic requirements, were found in the fields of isolines. E–V–E trajectories for flights to asteroids selected as possible targets which are optimal in terms of the criterion min $V_{\Sigma} = \min(V_p + V_{m1} + V_{m2})$ where $V_{m1}(V_{m2})$ is the velocity of flyby of Earth (Venus), were sought for within each domain of admissible trajectories.

Table 6.10 gives the characteristics of trajectories for flights to the asteroids Iris, Gigeya, Tisbe, Daphna, Euterpia, Lutecia, Metida, etc., which have been selected by the above procedure and which are optimal in terms of V_{Σ}. It can be seen that the boost velocities at Earth for flights to different asteroids are approximately equal and vary within 3.79 to 3.86 km/s. Almost all the trajectories allow purely ballistic maneuvers at Venus and Earth. The entry velocity to the atmosphere of Venus ranges from 11.4 to 11.6 km/s and provide acceptable conditions for entering a lander into the planetary atmosphere. The velocities at which the asteroids are encountered are rather high (6.3–9.7 km/s), except for the asteroid Iris, where this velocity amounts to 4 km/s. The results of calculations confirm the recommendations concerning selection of different flight schemes based on the analysis of isoline fields constructed with the coordinates: semi-major axis—eccentricity of the asteroid's orbit.

At the second stage of the studies, isolines of the of E–V–E–A trajectory parameters were constructed for the selected asteroids using the methods and algorithms discussed in Section 3.1. The use of isoline fields makes it possible to perform mission analysis of multi-purpose trajectories for Earth–Venus–Earth–asteroid flight. It should be noted that E–V–E–asteroid flights lead, as a rule, to an increase in the velocity of encounter with asteroids relative to Earth–Mars–asteroid trajectories. Therefore, it is interesting to consider the possibility of combining the flights E–V–A and E–V–E–A in a single mission. As an example of such a mission, Table 6.11 gives the characteristics of multipurpose E–V–E–asteroid trajectories for a 1991 launch window. These trajectories are remarkable in that their Earth–Venus leg allows the aiming parameters for the flyby of Venus to be selected so as to provide either a flight to Earth or an encounter with the asteroid Anteros.

6.2.4 Flights to asteroids with orbit of an artificial satellite of an intermediate planet

In [196], a promising scheme was suggested for flights to asteroids with a delay on the orbit of artificial satellite of an intermediate planet. The use of delay time as an additional parameter for minimizing V_{Σ} in some cases allows a reduction in the total energy expenditure. The periodicity of such trajectories is related to the optimal dates for launch to Mars (the recurrence is $\sim$2.135 years).

Table 6.10 Characteristics of Earth–Venus–Earth–asteroid trajectories.

Name of the asteroid	T_{st}	T_{E-V}, days	T_{V-E}, days	T_{E-A}, days	T_{Σ}, days	V_{p}, km/s	V_{e}, km/s	V_{M1}, km/s	$r_{\pi V}$, thous. km/s	V_{s}, km/s	V_{M2}, km/s	$r_{\pi E}$, thous. km	$V_{\infty A}$, km/s
Iris	16.05.91	194	331	642	1167	3.8	11.6	0	7.19	9.3	0	8.1	3.9
Metida	20.05.91	175	626	548	1309	3.9	11.6	0	11.04	8.3	0.1	6.9	8.2
Gigea	14.05.91	195	320	820	1335	3.8	11.6	0	12.38	9.1	0	6.9	9.6
Fortuna	14.04.91	180	631	724	1535	3.9	11.5	0	16.0	8.2	0.5	6.9	9.4
Lutecia	15.05.91	193	679	897	1679	3.8	11.6	0	14.6	8.9	0.2	6.9	8.9
Evternia	16.04.91	180	642	549	1371	3.8	11.4	0	26.9	8.1	0	9.2	9.2
Gisbo	12.05.91	196	321	921	1438	3.8	11.6	0	11.8	9.1	0.1	6.9	9.7
Daphna	16.04.91	180	273	667	1120	3.8	11.4	0	24.6	7.9	0	7.7	8.7
Urania	07.05.91	198	676	735	1609	3.9	11.6	0	15.7	8.5	0	6.9	6.3
Eva	09.05.91	196	680	870	1746	3.8	11.5	0	14.1	8.5	0	6.9	8.0

Table 6.11 Trajectories of Earth–Venus–Earth–asteroid flight.

Name of the asteroid, number in the catalogue	Iris 7	Gigea 10	Tisbe 88	Daphna 41	Evterpia 27	Lutecia 21	Metida 9	Urania 30	Eva 164
Diameter, km	230	440	215	205	118	114	168	95	111
Date of start from Earth	16.05.91	14.05.91	12.05.91	16.04.91	15.05.91	15.05.91	20.05.91	9.05.91	14.04.91
Flight time Earth–Venus, days	194	195	196	180	180	193	175	198	196
Flight time Venus–Earth, days	331	320	321	273	642	679	626	676	680
Flight time Earth–asteroid, days	642	820	921	667	549	897	548	735	870
Total flight time, days	1167	1335	1438	1120	1371	1679	1309	1609	1746
Boost velocity from the orbit of Earth's artificial satellite, km/s	3.82	3.83	3.84	3.80	3.80	3.79	3.86	3.85	3.82
Velocity of maneuver at Venus, km/s	0	0	0	0	0	0	0	0	0
Pericentral radius of Venus flyby, thous. km	7.19	12.38	11.83	24.61	26.9	14.55	11.04	15.70	14.1
Velocity of maneuver at Earth, km/s	9	0.04	0.127	0	0	0.177	0.131	0	0.02
Pericentral radius of Earth flyby, thous. km	8.12	6.87	0.87	7.69	9.21	6.87	6.87	6.92	6.87
Velocity of arrival to the asteroid, km/s	3.94	9.61	9.68	8.74	9.23	8.98	8.17	6.31	7.96

Table 6.11 (*continued*) Parameters of Earth–Venus–Anteros trajectories combined with E–V–E–A trajectories along the E–V leg.

Name of the asteroid, number in the catalogue	Iris 7	Gigea 10	Tisbe 88	Daphna 41	Evterpia 27	Lutecia 21	Metida 9	Urania 30	Eva 164
Flight time Venus–Anteros, days	122	127	129	165	165	127	171	112/133	132/114
Velocity of maneuver at Venus, km/s	0.02	0	0.014	0.09	0.09	0.007	0.015	0	0.02
Pericentral radius of Venus flyby, thous. km	42	25	21	21	21	28	19	174/15	17/182
Velocity of flyby to Anteros, km/s	9.3	9.7	9.8	8.2	8.2	9.6	8.1	8.4/10.1	10/8.4

In all the cases considered, the energy required to reach an asteroid along direct trajectories $V_\Sigma = V_p$ is less than that for trajectories with a delay on the orbit of an artificial satellite of Mars, where $V_\Sigma = V_p + V_d + V_m$.

However, in the case of trajectories with a delay on the orbit of an artificial satellite of Mars, the velocities of encounter with asteroids can be considerably reduced relative to direct flights by optimizing the parameters of the Mars–asteroid leg.

This means that in missions where it is intended that spacecraft should accompany the asteroid for a long time, trajectories with a delay can yield an advantage in the payload as compared with direct flights.

6.2.5 The use of flyby of large planets for flights to asteroids

Jupiter is the heaviest planet in the solar system, therefore gravity assist during a flyby is very efficient and allows the spacecraft to be aimed at virtually any asteroid without additional fuel expenditures. However, the energy requirements for the accomplishment of trajectories to Jupiter are greater than those for flights to asteroids (because the orbits of most asteroids of the Main Belt lie within Jupiter's orbit). Therefore, a flight to an asteroid after a flyby of Jupiter can be considered a means of enhancing the scientific efficiency of the mission to Jupiter.

The characteristics of optimal trajectories for flights to the largest asteroids (Ceres, Pallada, Junona, and Vesta) were examined. The boost velocities at Earth for all the asteroids were virtually equal to those for the optimal flight from Earth to Jupiter ($V_p = 6.37$ km/s). The maneuver during flyby of Jupiter is purely ballistic (the engine is not turned on). The only exception is the asteroid Junona, for which the required velocity impulse is less than 0.5 km/s.

The total time for flights to asteroids is rather high: it amounts to 4.7 years for the flight to Vesta and varies from 5.7 to 6.6 years for other asteroids. An increase in the boost impulse at the parking orbit of an artificial Earth satellite up to 6.5 km/s allows a significant reduction in the flight time to Junona (4.9 years instead of 6.44 years). The arrival velocities at asteroids range from 7.2 to 20 km/s. Thus, the scheme makes it possible, for example, after a probe was delivered into Jupiter's atmosphere, to send the spacecraft to a selected asteroid with virtually no energy expenditure.

Note that the use of a Jupiter flyby for a mission involving a landing on an asteroid is not expedient, because of the large arrival velocities at the asteroid.

7. SOLUTION OF NAVIGATIONAL PROBLEMS FOR MULTI-PURPOSE MISSIONS AT NEAR-PLANET SEGMENTS OF THE FLIGHT TRAJECTORY

The examination of possible trajectories and the choice of target point in the B-plane of the target planet or flyby planet are the second stages of the problem of designing spacecraft trajectories.

As was mentioned in Chapters 1 and 3, once the interplanetary leg of the trajectory has been determined, i.e. the reference dates of the start and approach to the planet have been selected, we obtain the velocity vector at "infinity" $\bar{V}_{\infty n}$ of approach to the planet. By specifying the vector of aiming distance $\bar{d}(\xi, \eta)$ in the B-plane we obtain unique elements of the planetocentric orbit. Modulus of the aiming distance vector is the semi-minor axis of the hyperbolic orbit in the sphere of activity of the planet.

By mapping isolines of the design parameters into the B-plane

$$q_i(\xi, \eta) = q_i^*$$

we obtain the domain of admissible trajectories at the near-planet flight segment.

Since the aiming distance vector $\bar{d}$ uniquely determines the near-planet trajectory, the velocity vector at infinity is also determined for the departure from the planet $\bar{V}_{\infty 0}$. Then, within the framework of the method of spheres of activity, the heliocentric orbit that results from the gravitational flyby is also determined. Note that $|\bar{V}_{\infty n}| = |\bar{V}_{\infty 0}|$.

If such a trajectory never reaches any specific celestial body, we call it an *interplanetary probe trajectory*. In this case, any parameter of the interplanetary probe Q also can be expressed as a function of aiming parameters ξ, η and can be represented as a level line in B-plane

$$Q_i(\xi, \eta) = Q_i^*.$$

To construct the isolines one can use the general method of isoline construction described in Section 3.1. However, as will be shown below, most problems allow analytical solutions, which makes the computation much simpler and makes it possible to reveal a series of interesting regularities and to obtain qualitative results.

This method is efficient for analysis of both simple trajectories (flyby, descent) and complex (combined) flight schemes (flyby-descent, orbital-descent, descent with formation of an interplanetary probe, etc.).

It should be noted that the B-plane method is also widely used for correcting interplanetary trajectories. It is from here that the notion of aiming parameters has originated. Elements of scatter of trajectories and the ellipse of the resulting accuracy of aim are also projected onto this plane. The admissible plane of trajectories, together with the domain of resulting accuracy, yields an estimate of the flight feasibility.

On the other hand, the dimensions and shape of the admissible domain determine the requirements on the radius of the resulting aiming accuracy. The admissible domain can have a rather complicated form, consisting of a number of isoline segments of different parameters. Let D be the closed boundary of the domain, ξ_a, η_a are the aiming parameters that are to be selected within the admissible domain, then, the maximum possible radius of the resulting aiming neighborhood ρ^* can be found from the solution of the following problem:

$$\rho^*(\xi_a, \eta_a) = \max_{\xi, \eta} \min_{D} \rho$$

where ρ is the distance from the aiming points to the domain boundary.

Thus, the method of parameter mapping into the B-plane can be used not only in design studies, but also to accomplish space flights. In this case, the flight time to the planet and, accordingly, $V_{\infty n}$ are determined based on actual interplanetary orbit.

By this time, a rather wide variety of flight schemes in the near-planetary part of the trajectory have been implemented in the flights of Russian and American interplanetary spacecraft. Some cases of selection of the aiming parameters in the B-plane for some flight schemes have been published. In the present chapter we systematically discuss the entire range of problems associated with the parameter mapping into the B-plane for different types of trajectories.

7.1 Formation of descent and flyby trajectories

Consider the basic criteria and restrictions essential for the analysis and selection of trajectories at the near-planet flight segment.

1. Parameters of hitting trajectories:
   - the angles at which the planetary atmosphere is entered;
   - the maximum g-loading during motion in the atmosphere as a function of the entry angle and velocity;
   - the angles between the local vertical direction at the entry point and the direction to the Earth for the analysis of characteristics of direct radio communication with the Earth;
   - the angles between the local vertical direction and the direction to the sun for determining the conditions of functioning of scientific equipment, in particular, photographic and television equipment;
   - the planetographic coordinates of the zone of entry into the atmosphere;
   - the planetographic longitudes of the landing in order to coordinate landing time with visibility from observation points.

2. Parameters of flyby trajectories:
   - the range of pericentral altitudes (bounded below by the height of atmosphere, its radiation belts, and the accuracy of guiding, and bounded above by the requirements of scientific experiments for studying the planet, restrictions imposed by the system of radio communications between the "bus" and lander, and the admissible energy regime of entering the orbit of an artificial satellite of the planet);
   - the angles between the directions to the sun and Earth and the edge of the planet, enabling the celestial navigation system to function, because the apparatus entering the zone of the planet's shadow with respect to the sun or Earth can be either inadmissible or planned maneuver;
   - the angular elements i, Ω of flyby trajectories, which in the case of coplanar (i, Ω) and apsidal (i, Ω, ω) transfers become the elements of the orbits of natural planetary satellites;
   - the conditions of planet visibility for the devices mounted on board;
   - the parameters of the heliocentric trajectories after the planetary flyby.

Usually, the coordinate system in the B-plane is constructed in the following way:

$$\bar{\xi}^0 = \bar{\eta}^0 \times \bar{\zeta}^0, \qquad \bar{\eta}^0 = \frac{\bar{V}_\infty \times \bar{r}}{|\bar{V}_\infty \times \bar{r}|}, \quad \bar{\zeta}^0 = \frac{-\bar{V}_\infty}{|\bar{V}_\infty|}.$$

Here, $\bar{V}_\infty$ is the vector of approaching velocity at "infinity" with respect to the planet, $\bar{\xi}^0, \bar{\eta}^0$ are unit vectors in the B-plane, $\bar{r}$ is the direction vector.

For the selected interplanetary trajectory, the planetocentric orbit is completely determined by the aiming distance vector $\bar{d}$, lying in the plane $\bar{\xi}^0, \bar{\eta}^0$.

It may be convenient to select the individual direction of the axes $\bar{\xi}_i^0, \bar{\eta}_i^0$, i.e. vector $\bar{r}_i$ for each mapping, and only after that to convert all isolines to a certain base system determined by vector r. The transition between the two systems ($\bar{r}$ and $\bar{r}_i$) is determined by the formulas:

$$\begin{Vmatrix} \xi \\ \eta \end{Vmatrix} = \begin{Vmatrix} \bar{\xi}^0 \cdot \bar{\xi}_i^0, & \bar{\xi}^0 \cdot \bar{\eta}_i^0 \\ \eta^0 \cdot \bar{\xi}_i^0, & \bar{\eta}^0 \cdot \eta_i^0 \end{Vmatrix} \begin{Vmatrix} \xi_i \\ \eta_i \end{Vmatrix}$$

and the angle of rotation between the systems in the B-plane will be

$$\bar{r} \to \bar{r}_i \qquad \mathscr{H} = -\operatorname{sign}(\bar{\xi}^0 \cdot \bar{\eta}_i^0) \arccos(\bar{\eta}^0 \cdot \bar{\eta}_i^0).$$

It is convenient to take as the base vector r, the sun–planet direction at the moment of approach to the planet, because a series of restrictions is associated with this direction.

7.1.1 Mapping of the parameters of incoming orbits into the B-plane

Isolines of the trajectory parameters, which depend only on the semi-major axis $a = \mu / V_\infty^2$ and the aiming distance d, are represented in the B-plane by concentric circles. The pericentral altitude r_π of the planetocentric orbit and the angle at which the atmosphere is entered θ_{ent} are the most important among these values:

$$d=\sqrt{r_\pi^2+2ar_\pi},$$

$$d=\sqrt{R_h^2+2aR_h}\cos\theta_{ent};$$

$R_h=R_{pl}+h$ is the distance from the center of the planet; where R_{pl} is the radius of the planet, h is the design altitude.

In some cases, the restrictions are imposed not on the entry angle, but on the value of the maximum admissible load factor n_{max} in the course of motion in the atmosphere, which is a function of the entry angle and the velocity at which the atmosphere is entered. This means that each circle θ_{ent} has its own value of $n_{max}=const$.

Consider now the temporal relationships that are also represented by circles.

The time of motion along a hyperbolic orbit from a sphere of radius r to the pericenter r_π is [9]:

$$t=n(e\,\mathrm{sh}\,H-H)$$

where

$$n=a^{3/2}\mu^{-1/2},\qquad \mathrm{ch}\,H=\frac{1}{e}\left(\frac{r}{a}+1\right).$$

Applying the formulas of hyperbolic functions we have

$$t=n\left[\sqrt{\mathscr{H}^2-e^2}-\ln\frac{\mathscr{H}+\sqrt{\mathscr{H}^2-e^2}}{e}\right], \tag{7.1}$$

$$\mathscr{H}=\frac{r+a}{a}.$$

Suppose that we have two orbits with equal a, but with different eccentricities e_1, e_2. We will find the difference between the motion times

$$\Delta t=t_2-t_1, \tag{7.2}$$

where $t_1=t_1(r_0)-t_1(r_1)$, $t_2=t_2(r_0)-t_2(r_2)$ are the times it will take the spacecraft to move from the sphere of radius r_0 to the spheres r_1, r_2 along the first and second orbits, respectively.

Substituting (7.1) and (7.2) we obtain

$$\Delta t=n[N_1+N_2+N_3+N_4], \tag{7.3}$$

$$N_1=\sqrt{\mathscr{H}_0^2-e_2^2}-\sqrt{\mathscr{H}_0^2-e_1^2};$$

$$N_2=\sqrt{\mathscr{H}_1^2-e_1^2}-\sqrt{\mathscr{H}_2^2-e_2^2};$$

$$N_3=\ln\frac{\mathscr{H}_0+\sqrt{\mathscr{H}_0^2-e_1^2}}{\mathscr{H}_0+\sqrt{\mathscr{H}_0^2-e_2^2}}\cdot\frac{e_2}{e_1};$$

$$N_4=\ln\frac{\mathscr{H}_2+\sqrt{\mathscr{H}_2^2-e_2^2}}{\mathscr{H}_1+\sqrt{\mathscr{H}_1^2-e_1^2}}\cdot\frac{e_1}{e_2};$$

$$\mathscr{H}_0=\frac{r_0+a}{a};\quad \mathscr{H}_1=\frac{r_1+a}{a};\quad \mathscr{H}_2=\frac{r_2+a}{a}.$$

Let us consider particular cases of Formula (7.3).

1. The difference between the times of arrival at the planet. Let us assume that $r_0\to\infty$, then $N_3\to\ln(e_2/e_1)$; $N_1\to\infty\cdot 0$.

Transforming the uncertainty we obtain that $N_1\to 0$. Next $r_1=r_2=R_h$, $\mathscr{H}_h=(R_h+a)/a$, then

$$\Delta t=n\left[\sqrt{\mathscr{H}_h^2-e_1^2}-\sqrt{\mathscr{H}_h^2-e_2^2}+\ln\frac{\mathscr{H}_h+\sqrt{\mathscr{H}_h^2-e_2^2}}{\mathscr{H}_h+\sqrt{\mathscr{H}_h^2-e_1^2}}\right],\tag{7.4}$$

$$e_{1,2}=\sqrt{1+\frac{d_{1,2}^2}{a^2}}.$$

Formula (7.4) gives the difference between the times of entering the planetary atmosphere in the orbits with d_1, d_2. If we set $d_1=0$, which corresponds to $\theta_{ent}=\pi/2$, then, each circle of radius d in the B-plane will be associated with a value Δt, which is equal to the difference between the entry time in the given orbit and that in the vertical orbit.

2. Difference between the times of reaching the pericenters. Setting $r_0\to\infty$, $r_1=r_{\pi 1}$, $r_2=r_{\pi 2}$, we get $N_1=N_2=N_4=0$, then

$$\Delta t=n\ln\frac{e_2}{e_1}=\frac{n}{2}\ln\frac{a^2+d_2^2}{a^2+d_1^2}.\tag{7.5}$$

If we set $d_1=0$, then

$$\Delta t=\frac{n}{2}\ln\left(1+\frac{d_2^2}{a^2}\right)=n\ln e_2.\tag{7.6}$$

Formula (7.5) gives the difference between the times of passing the pericenters in two arbitrary orbits, while (7.6) gives the same difference for an arbitrary orbit and the vertical one. In the latter case it is convenient to directly express d in terms of Δt,

$$d = a\sqrt{e^{2\Delta t/h} - 1}$$

where e is the base of natural logarithms.

The above formulae can be also used to assess the errors in the moment at which the engine turned on, the possible time of entry into the atmosphere, as well as errors in the time of turning on scientific equipment, etc. In this case, $d_2 = d_1 + \Delta d$, where Δd is the expected error in the aiming distance.

7.1.2 Method of aiming

Let us pose the following problem: to find the parameters of a hyperbolic orbit given the vector of velocity at "infinity" $\bar{V}_\infty$ and radius vector $\bar{r}$, pertaining to a certain point M of the orbit to be found (Figure 7.1).

Based on the well-known relationships of conic sections [9] we can express all the variables in terms of aiming distance d and semi-major axis a, thus obtaining the equation

$$d^2 - r\sin\upsilon \cdot d - ar(1 - \cos\upsilon) = 0, \tag{7.7}$$

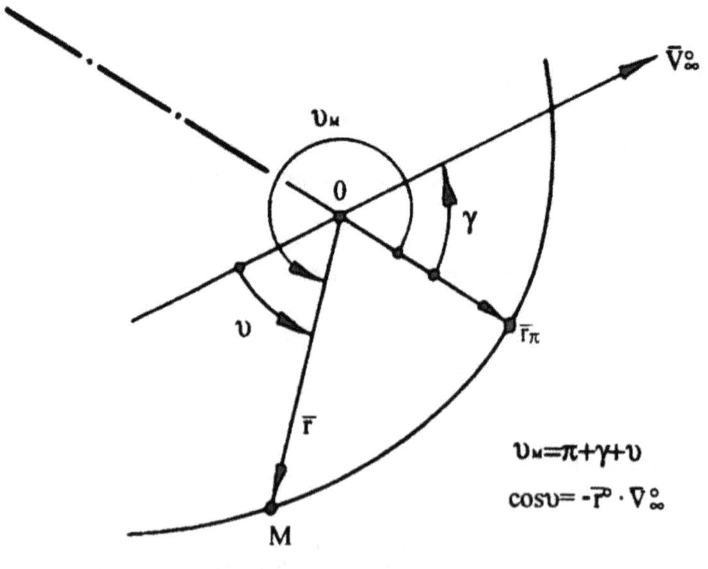

Figure 7.1 Trajectory with a specified vector and passing through given point M.

the solution of which has the form

$$d_{1,2}=\frac{r}{2}\sin\upsilon\pm\sqrt{\frac{r^2}{4}\sin^2\upsilon+ar(1-\cos\upsilon)}. \tag{7.8}$$

From Solution (7.8) it follows that there are two types of trajectories meeting the point M, and $d_1\geqslant 0$, $d_2\leqslant 0$. Different signs of d_1 and d_2 correspond to different directions of flying by the planet. Let us analyze possible orbits depending on the value of υ. A typical dependence $d=d(\upsilon)$ is shown in Figure 7.2. Let us consider the boundary points $\upsilon=0$, $\upsilon=\upsilon^*$, $\upsilon=\pi$ separately as they correspond to limiting types of orbits:

1. $\upsilon=0$. Then $d_{1,2}=0$; the orbit is rectilinear; the encounter with the point M takes place on the incoming leg of the orbit (Figure 7.3a).
2. $\upsilon=\upsilon^*$. From the expression $\partial d/d\upsilon=0$ we find

$$\cos\upsilon^*=\frac{-\mu}{rV_\infty^2+\mu};\qquad d_{1\max}=\sqrt{r^2+2ar}.$$

   That is, when $\upsilon=\upsilon^*$, the point M became the pericenter of the first orbit. The encounter with the point M in the second trajectory (d_2) takes place on the outgoing leg (Figure 7.3b).
3. $\upsilon=\pi$. We get two orbits with equal aiming distances $d_{1,2}=\sqrt{2ar}$ that are symmetrical with respect to the vector $\bar{V}_\infty$. The encounter takes place on the outgoing legs. Formally, the vertical orbit also belongs to this class (Figure 7.3c).
4. $0<\upsilon<\upsilon^*$. Figure 7.3d presents two trajectories passing through the point M on the incoming leg (solution d_1) and on the outgoing one (d_2).
5. $\upsilon^*<\upsilon<\pi$. On two types of orbits (d_1 and d_2), the encounter with the point M takes place on the outgoing leg (Figure 7.3e).

Thus, the encounter with the point M on the incoming leg of the orbit takes place when $0<\upsilon<\upsilon^*$ for the first solutions d of Equation (7.7). For all other options of the orbit, the encounter with the given point takes place after passing the pericenter.

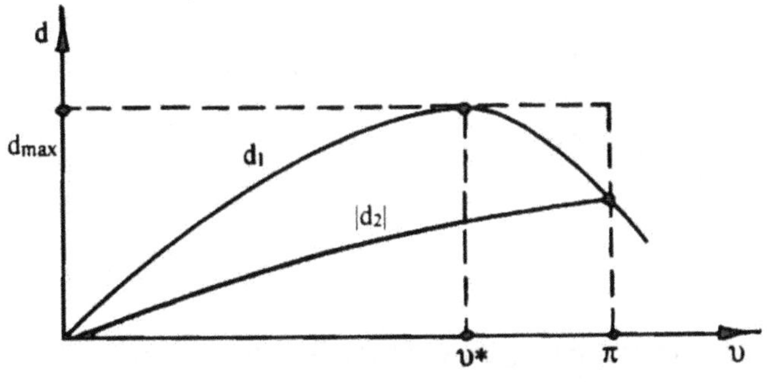

Figure 7.2 Dependence of aiming distance d on angle.

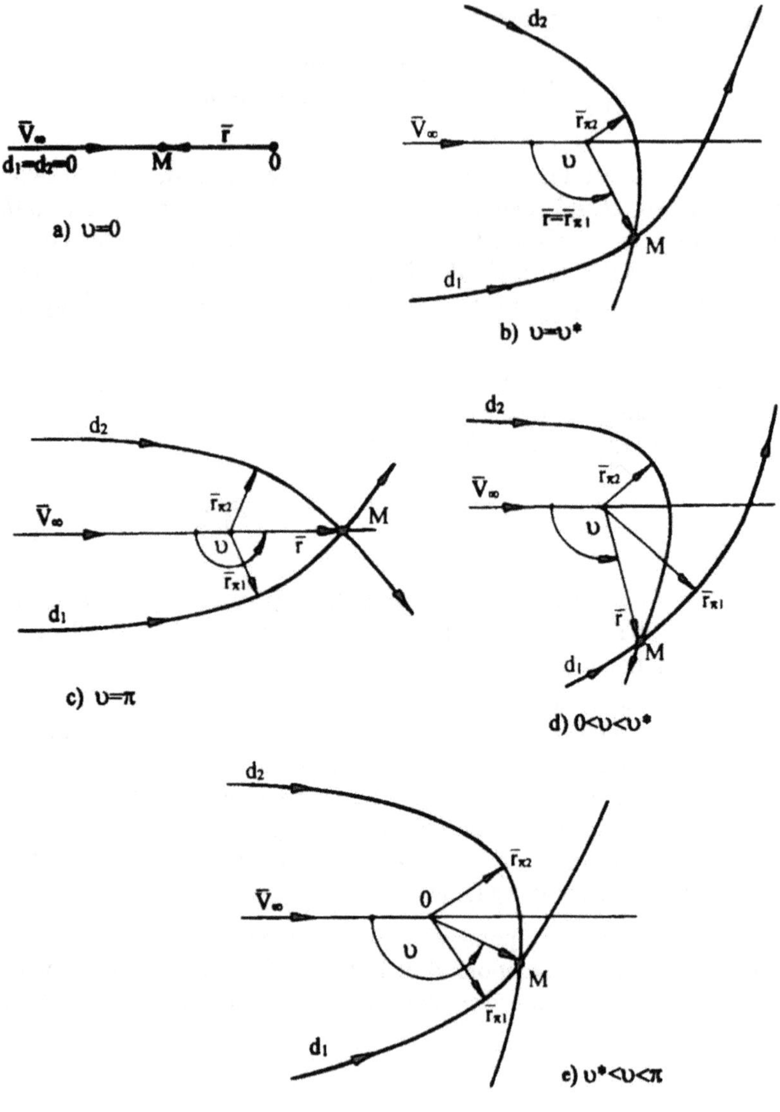

Figure 7.3 Possible near-planet trajectories.

Now the space elements of the orbit can easily be found from the vectors of kinetic momentum and aiming distance

$$\bar{C}^0_{1,2} = \pm \frac{\bar{r} \times \bar{V}_\infty}{|\bar{r} \times \bar{V}_\infty|}; \quad \bar{d}_{1,2} = |d_{1,2}| \bar{d}^0_{1,2}; \quad \bar{d}^0_{1,2} = \bar{V}^0_\infty \times \bar{C}^0_{1,2} \tag{7.9}$$

and the coordinates of the aiming point in the B-plane will be

$$\xi_{1,2} = \bar{\xi}^0 \cdot \bar{d}_{1,2}, \qquad \eta_{1,2} = \bar{\eta}^0 \cdot \bar{d}_{1,2}. \tag{7.10}$$

It is common to use the first solution of the problem (d_1), since it provides the encounter in the incoming leg of the hyperbola. As will be shown below, all types of trajectories should be considered for some problems.

In a general case, if the end of a vector describes some spatial curve $\bar{r}=\bar{r}(t)$, then formulae (7.8–7.10) allow us to determine its mapping into the B-plane, which in this case consists of two planar curves $\bar{d}_1=\bar{d}_1(t), \bar{d}_2=\bar{d}_2(t)$.

In solving some problems it is convenient to use the following parametrization of the vector $\bar{r}$. Let ρ^0 be a certain central direction, with respect to which the vector $\bar{r}$ describes a cone with an angle of λ, and the position of the vector $\bar{r}$ on the cone is specified by the angle γ (Figure 7.4).

Introduce the B-plane associated with a unit vector $\bar{\rho}^0$

$$\bar{\xi}_\rho^0=\bar{\eta}^0\times(-\bar{V}_\infty^0); \qquad \bar{\eta}_\rho^0=\frac{\bar{\rho}^0\times\bar{V}_\infty}{|\bar{\rho}_0\times\bar{V}_\infty|}. \tag{7.11}$$

In this system, the image of the circle described by the vector $\bar{r}=\bar{r}(\gamma)$ can be found by simple formulae

$$\xi_{1,2}=d_{1,2}\cos\mathscr{H}_{1,2}; \qquad \eta_{1,2}=d_{1,2}\sin\mathscr{H}_{1,2};$$

$$\operatorname{tg}\mathscr{H}=\frac{\sin\gamma}{\sin\upsilon_0\operatorname{ctg}\lambda-\cos\upsilon_0\cos\gamma}; \tag{7.12}$$

$$\cos\upsilon=\cos\lambda\cos\upsilon_0+\sin\lambda\sin\upsilon_0\cos\gamma$$

$$\cos\upsilon_0=-\bar{V}_\infty^0\,\bar{\rho}^0.$$

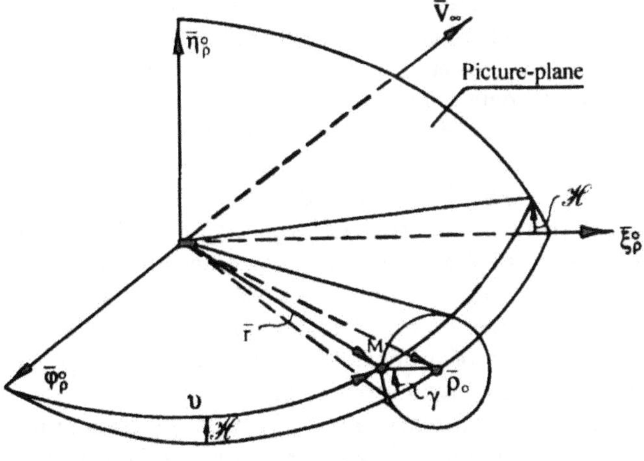

Figure 7.4 Spatial parameterization of vector.

Here $d_{1,2}$ are calculated from (7.8), and in view of the symmetry of the curves (7.12) with respect to the plane determined by the vectors $\bar{V}_\infty$ and $\bar{\rho}^0$ as well as with respect to the axis ξ in the B-plane, it is sufficient to vary the parameter γ within the range $0 \leqslant \gamma \leqslant \pi$.

The above method allows the solution of the following basic problems of mapping into the B-plane.

1. Construction of the latitude and longitude network on the planetary surface. For construction of parallels, it is assumed that $\bar{\rho}^0 = \bar{P}_N^0$, $\lambda_i = \varphi_i$; where $\bar{P}_N^0$ is the direction to the planetary north pole, φ_i are latitudes, and the values of φ equal to $\pi/2, 0, \pi$ yield the equator, and north and south poles, respectively. To construct meridians, it is assumed that

$$\lambda = \frac{\pi}{2} \quad \text{and} \quad \bar{\rho}_i^0 = \left\{\cos\left(\psi_i + \frac{\pi}{2}\right), \sin\left(\psi_i + \frac{\pi}{2}\right), 0\right\},$$

i.e. the vectors $\bar{\rho}_i^0$ lie within the planetary equatorial plane and are perpendicular to the meridian of longitude ψ_i.

2. Construction of the angles of elevation for directions to celestial bodies. The direction to the celestial body (e.g. the sun or Earth) is accepted as $\bar{\rho}^0$, and the parameter λ_i takes the values $\pi/2 - \delta$, where δ is the angle of elevation. If $\lambda_i = 0$, we obtain the point, where either the sun or Earth is at the zenith, if $\lambda = \pi/2$, we have a terminator, or the boundary of the Earth visibility.

In the first two problems, trajectories of the first solutions in the range $0 \leqslant \upsilon \leqslant \upsilon^*$ are used, which corresponds to physical feasibility.

3. Mapping of the planetary satellite orbits into the B-plane. In this case, the vector $\bar{\rho}^0 = \bar{C}_{\text{sat}}^0$, $\bar{C}_{\text{sat}}^0$ is the unit vector of the kinetic momentum of the satellite orbital motion, $\lambda = \pi/2$ and γ correspond to the true anomaly of the satellite position in its orbit. Each value of γ corresponds to a quite definite time when satellite passes through this point. These mappings allow one to analyze the encounter with a planetary satellite, or passage near it in terms of the requirements regarding the moment of arrival at the planet. This has importance for the organization of a rendezvous in a specified point. One can also analyze the encounter in terms of the requirements for the guiding accuracy, with allowance made for the errors in determining the satellite location in its orbit.

Note that the satellite orbit is represented by two closed curves in the B-plane, which correspond to two types of orbits (d_1, d_2). Two possible ways of forming the scheme of action in the near-planet segment follow from here: an encounter with the satellite after passing the pericenter (all the second solutions and part of first ones for $\upsilon^* < \upsilon \leqslant \pi$), or an encounter on the incoming leg (first solutions for $0 \leqslant \upsilon \leqslant \upsilon^*$).

If the planet has a system of artificial satellites, their mappings can be used for planning the orbits that pass near a series of the satellites.

4. Mapping of the shadowed zones into the B-plane. In solving this problem, the vector $\bar{\rho}^0$ is oriented along the direction to the celestial body, with respect to which the entrance to the planet's shadow is considered, and the parameter λ_i is determined

by the distance at which the spacecraft crosses the boundary of the shadowed zone. All the considered types of orbits are used in the analysis. The method for determining the boundaries of the shadowed zones is discussed in the following section.

7.1.3 Determination of the shadowed zones

Let $\bar{\rho}^0$ be a certain fixed direction in space, and $\bar{r}$ be a radius vector of the points for which the conditions

$$\alpha-\alpha_{sh}=\alpha^*,$$

$$\cos\alpha=-\bar{\rho}^0\cdot\bar{r}^0, \tag{7.13}$$

$$\sin\alpha_{sh}=\frac{R_{sh}}{r}$$

are satisfied. Here $R_{sh}=R_{pl}+h_{sh}$ is the effective radius of the planet, determining the size of shadow; h_{sh} is the height, in particular, the height of the atmosphere or ionosphere of the planet; α^* is a specified value.

After transformations, Equation (7.13) in the coordinate system where $\bar{X}^0=\bar{\rho}^0$, can be written as

$$-(r_x-r_{x0})^2+\mathrm{ctg}^2\,\alpha^*\cdot r_y^2+\mathrm{ctg}^2\,\alpha\cdot r_z^2=0, \tag{7.14}$$

which determines a circular cone with the vertex $r_x=r_{xo}=R_{sh}/\sin\alpha^*$, $r_y=0$, $r_z=0$. At $r_x=const$ we have the circle

$$r_y^2+r_z^2=\mathrm{tg}^2\alpha(r_x-r_{x0})^2$$

in the plane y, z, and in sections $r_y=0$, $r_z=0$, we obtain a pair of straight lines $r_z=\pm r_x\,\mathrm{tg}\,\alpha^*-(R/\cos\alpha^*)$. Note that at $\alpha^*=0$ we have a shadow cylinder $r_y^2-r_z^2=R_{sh}^2$ whose axis is parallel to the direction ρ^0 (Figure 7.5).

The geometric locus of the points where the spacecraft crosses the shadow cone at the distance of r from the center of the planet is a circle, whose mapping on the B-plane can be found from (7.8) and (7.12). When r varies from R_{sh} to infinity, we get a set of mappings, whose envelope will be the boundary of the shadowed zone we are searching for. Examples of determining the shadowed zones are published elsewhere. In these cases, the problem was solved either by way of graphical construction of the envelope, or through directed individual examination of points in the B-plane, for each of which the intersection of the orbit with the shadow cone was found by analyzing the roots of a fourth-degree algebraic equation.

Based on Expressions (7.8), (7.12), let us directly find the envelope equations. The set of curves that cross the shadow cone at a given distance r can be specified in the

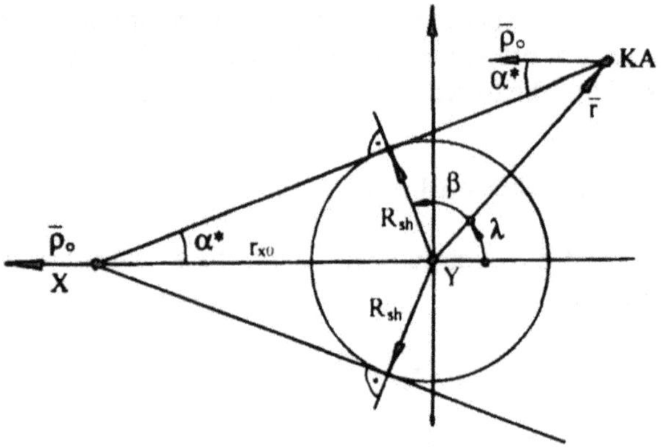

Figure 7.5 The cone of shadow.

parametric form:

$$\xi = d\cos\mathscr{H};\quad \eta = d\sin\mathscr{H}; \tag{7.15}$$

$$d = d(\lambda, \upsilon);\quad \mathscr{H} = \mathscr{H}(\lambda, \upsilon)$$

Here υ is the parameter determining the point on the curve Figure 7.4, and λ is the parameter specifying the curve from the set and equivalent to specifying r. From Figure 7.5 it follows

$$\sin(\lambda - \alpha^*) = \frac{R_{sh}}{r}. \tag{7.16}$$

When r varies from R_{sh} to ∞, the value of λ varies from $(\alpha^* + \pi/2)$ to α^*.
Let us write the equation

$$\frac{\partial\xi}{\partial\lambda}\cdot\frac{\partial h}{\partial\upsilon} - \frac{\partial\xi}{\partial\upsilon}\cdot\frac{\partial\eta}{\partial\lambda} = 0, \tag{7.17}$$

which together with (7.15) determines the envelope of the family of curves (7.15) with arbitrary λ. After transformation we shall obtain:

$$d = \frac{1-\cos\upsilon}{\sin\upsilon}\cdot\frac{\sin^2(\lambda-\alpha^*)+\cos\upsilon-\cos\upsilon^0\cos(\lambda-\alpha^*)}{\cos\upsilon_0\cos(\lambda-\alpha^*)-\cos\upsilon}\cdot a. \tag{7.18}$$

Thus, each specified value of r is associated with a certain value of λ, and the point at the curve (7.15), where condition (7.18) is satisfied, belongs to the envelope.

Since the angle γ from the Expressions (7.12) is convenient to use as a parameter, Equation (7.18) after transformations can be presented in the form

$$\cos\gamma = \frac{P}{Q}, \tag{7.19}$$

where

$$P = 2\cos\upsilon_0 \sin\upsilon \sin\left(\lambda - \frac{\alpha^*}{2}\right)\sin\frac{\alpha^*}{2} d + (1-\cos\upsilon) \times\left[2\cos\upsilon_0 \sin\left(\lambda - \frac{\alpha^*}{2}\right)\sin\frac{\alpha^*}{2} - \sin^2(\lambda-\alpha^*)\right]a, \tag{7.20}$$

$$Q = \sin\upsilon_0 \sin\lambda\,[d\sin\upsilon + a(1-\cos\upsilon)].$$

For the shadow cylinder we have $\alpha^* = 0$, and the relationships 7.18–7.20 take the simple form

$$d = \frac{1-\cos\upsilon}{\sin\upsilon}\cdot\frac{\sin\lambda + \sin\upsilon_0\cos\gamma}{\sin\upsilon_0\cos\gamma} \tag{7.21}$$

$$\cos\gamma = -\frac{a\sin\lambda(1-\cos\upsilon)}{[a(1-\cos\upsilon)+d\sin\upsilon]\sin\upsilon_0}. \tag{7.22}$$

Thus, the scheme for determining the envelope reduces to the standard scheme for calculating the aiming parameters 7.8 and 7.12 with a simultaneous search for the value of γ satisfying either of Equations 7.18 or 7.22. In view of the symmetry of the envelopes with respect to the plane $\bar{V}_\infty$, $\bar{\rho}^0$, the range $0 \leqslant \gamma \leqslant \pi$ is considered.

We search for the boundary between the shadowed zones for the first and second solutions of d_1, d_2. Note that the shadowed zone for the first solutions is always unlimited, whereas for the second solutions, it can be closed. Finally, note that the search can be based on a one-dimensional iteration process using Equations (7.19), (7.22) in the form $\gamma = f(\gamma)$.

7.1.4 Angular elements of planetocentric orbits

Consider the problem of mapping angular elements i, Ω, ω of planetocentric orbits onto the B-plane. As mentioned above, all these elements also persist for the orbit of an artificial satellite of the planet in the case of coplanar apsidal transfer.

Let $\bar{x}^0, \bar{y}^0, \bar{z}^0$ be the unit vectors of the planetocentric coordinate system, with respect to which the variables i, Ω, ω are determined, and angles λ_∞ and δ_∞

coordinate vector $\bar{V}_\infty$ in this system. We introduce the coordinate system in the B-plane in the following way:

$$\bar{\xi}^0 = \frac{\bar{V}_\infty \times \bar{Z}^0}{|\bar{V}_\infty \times \bar{Z}^0|}; \quad \bar{\eta}^0 = \zeta^0 \times \bar{\xi}^0; \quad \bar{\zeta}_0 = -\frac{\bar{V}_\infty}{|\bar{V}_\infty|}.$$

Figure 7.6 presents possible orbits in unit sphere, circumscribed about the planetary center. Angles λ_∞, Ω are measured from the $\bar{X}^0$ axis, the positive direction being counterclockwise; $\mathscr{H}$ is the angle in the B-plane between the ξ axis and the trajectory plane with the positive direction from the axis ξ to η. Subscripts 1, 2 denote direct orbits $0 \leqslant i \leqslant \pi/2$, and subscripts 3, 4 denote reverse orbits $\pi/2 < i \leqslant \pi$.

Introduce variables

$$\mathscr{H}' = \arccos\frac{\cos i}{\cos\delta_\infty},$$

$$\psi = \arcsin\frac{\operatorname{tg}\delta_\infty}{\operatorname{tg} i}, \quad -\frac{\pi}{2} \leqslant \psi \leqslant \frac{\pi}{2},$$

$$|\delta_\infty| \leqslant i \leqslant \pi - |\delta_\infty|.$$

Then, the formulae for determining Ω and $\mathscr{H}$ as functions of i in accordance with numeration of orbits in Figure 7.6 take the form

$$\left.\begin{array}{ll} \mathscr{H}_1 = -\mathscr{H}' & \Omega = \lambda_\infty - \psi \\ \mathscr{H}_2 = \mathscr{H}' & \Omega = \lambda_\infty + \psi + \pi \end{array}\right\} 0 \leqslant i \leqslant \frac{\pi}{2}$$

$$\left.\begin{array}{ll} \mathscr{H}_3 = \mathscr{H}' & \Omega = \lambda_\infty + \psi + \pi \\ \mathscr{H}_4 = -\mathscr{H}' & \Omega = \lambda_\infty - \psi \end{array}\right\} \frac{\pi}{2} < i \leqslant \pi. \tag{7.23}$$

These expressions hold also for $\delta_\infty < 0$ in accordance with numeration of orbits in Figure 7.7.

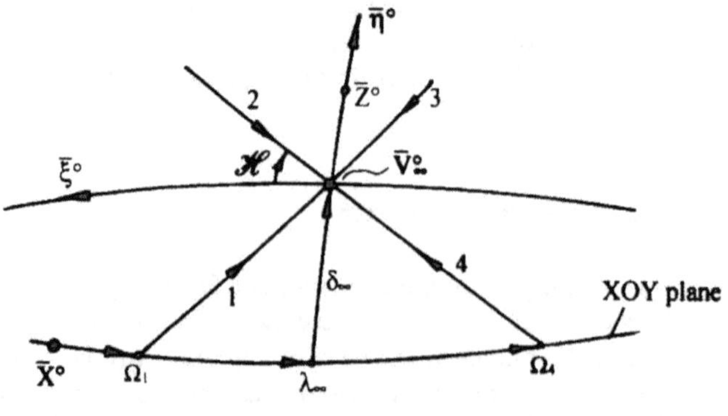

Figure 7.6 The angular elements of planetocentric orbits, the case of $\delta_\infty > 0$.

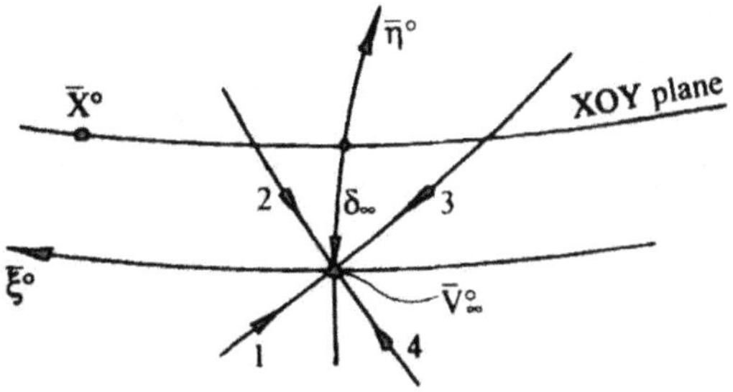

Figure 7.7 The angular elements of planetocentric orbits, the case of $\delta_\infty < 0$.

Thus, the values of Ω and $\mathscr{H}$ are two-valued functions of inclination i, and there is a unique correspondence between each ray in the B-plane, which forms in the center an angle $\mathscr{H}$ with the ξ axis, and the elements i, Ω for which this ray is an isoline.

Let us pass now to determining the isolines of the argument of the pericenter ω. γ_∞ denotes the angle between $\bar{V}_\infty$ and the orbital pericenter, u_∞ denotes the argument of the latitude of the vector $\bar{V}_\infty$, then

$$\gamma_\infty = u_\infty - \omega; \quad \operatorname{tg}\gamma_\infty = \frac{\operatorname{tg} u_\infty - \operatorname{tg}\omega}{1 + \operatorname{tg}\omega \operatorname{tg} u_\infty}.$$

The value of u_∞ can be expressed in terms of δ_∞ and $\mathscr{H}$ as follows: $\operatorname{tg} u_\infty = -\operatorname{tg}\delta_\infty / \sin\mathscr{H}$.

Substituting this into the previous expression and considering that $\operatorname{tg}\gamma_\infty = d/a$, we obtain two forms for representing the isoline $\omega = const$ in polar coordinates in the B-plane

$$d = \frac{\operatorname{tg}\delta_\infty + \operatorname{tg}\omega \sin\mathscr{H}}{\operatorname{tg}\delta_\infty \operatorname{tg}\omega - \sin\mathscr{H}} \cdot a \tag{7.24a}$$

$$\sin\mathscr{H} = \frac{d \operatorname{tg}\omega - a}{d + a \operatorname{tg}\omega} \operatorname{tg}\delta_\infty. \tag{7.24b}$$

From (7.24b) it follows that the isolines $\omega = const$ are symmetrical with respect to the η axis.

Possible positions of the argument of pericenter ω lie within the circle of radius $\pi/2$ with the center in $\bar{V}^0_\infty$ (Figure 7.8). Therefore, the range of possible values of ω is $\delta_\infty - \pi/2 \leqslant \omega \leqslant \pi - \delta_\infty$. Hence, the value of ω can lie in the I, II, and IV quadrants for $\delta_\infty > 0$, and in the II, III, and IV quadrants for $\delta_\infty < 0$.

Let us examine the behavior of curves $\omega = const$ in the B-plane using expressions (7.24). To do this, let us consider some limiting cases.

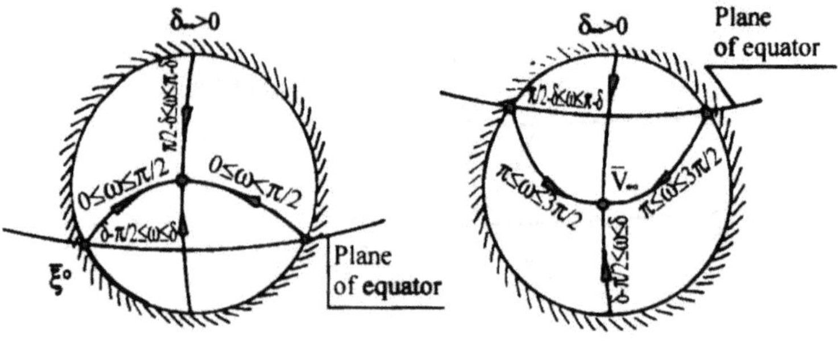

Figure 7.8 Possible position of the argument of pericenter.

1. $\omega=0$ for $\delta_\infty>0$, $\omega=\pi$ for $\delta_\infty<0$ then $d=-a\,\mathrm{tg}\,\delta_\infty/\sin\mathscr{H}$ or $d_\xi=a\,\mathrm{tg}\,\delta\,\mathrm{ctg}\,\mathscr{H}$, $d_\eta=-a\,\mathrm{tg}\,\delta_\infty$ we obtain a straight line parallel to ξ axis and located within $-a\,\mathrm{tg}\,\delta_\infty$ from it.

2. $\omega=\pi/2$ for $\delta_\infty>0$, $\omega=-\pi/2$ for $\delta_\infty<0$ then

$$d=a\frac{\sin\mathscr{H}}{\mathrm{tg}\,\delta_\infty}.$$

Let us represent this equation in coordinate form

$$d_\xi^2+\left(d_\eta-\frac{a}{2\,\mathrm{tg}\,\delta_\infty}\right)^2=\left(\frac{a}{2\,\mathrm{tg}\,\delta_\infty}\right)^2$$

i.e. we obtain the circle centered at $d_\xi=0$, $d_\eta=a/(2\,\mathrm{tg}\,\delta_\infty)$ of radius $a/(2\,\mathrm{tg}\,\delta_\infty)$.

From the cases considered it follows that the isolines can be either closed or nonclosed.

3. $d\to\infty$. From Equation (7.24b) it follows that $\sin\mathscr{H}=\mathrm{tg}\,\omega\,\mathrm{tg}\,\delta_\infty$, i.e. for all possible values of ω that satisfy the inequality $|\mathrm{tg}\,\omega\,\mathrm{tg}\,\delta_\infty|\leqslant 1$ the isolines are not closed. In particular, at $\omega=\pi/2-\delta_\infty$ the value $d\to\infty$ along the η axis.

4. $d=0$. We get from this the condition of the line $\omega=const$ passing through the origin $\sin\mathscr{H}=\mathrm{tg}\,\delta_\infty/\mathrm{tg}\,\omega$; $|\mathrm{tg}\,\omega|\leqslant|\mathrm{tg}\,\delta_\infty|$.

To illustrate the form of isolines of argument of pericenter, we show curves $\omega=const$ in Figures 7.9–7.11 for $\delta_\infty=30$, 60, and $-45°$. The hatching shows the isoline that tends to close up on the η axis at infinity. This line separates the families of closed and open curves, and the line with double hatching separates the families of curves passing through the origin (these curves run together at $\delta_\infty=\pm45°$).

The table below shows the quadrant where ω is located depending on where the point is located in the B-plane.

	Domain within the circle	Domain between the circle and straight line	Domain below the straight line
	$\omega=\frac{\pi}{2}$ $(\delta_\infty>0)$	$\omega=\frac{\pi}{2}$ and $\omega=0$ $(\delta_\infty>0)$	$\omega=0$ $(\delta_\infty>0)$ above line
	$\omega=\frac{3\pi}{2}$ $(\delta_\infty<0)$	$\omega=\frac{3\pi}{2}$ and $\omega=\pi$ $(\delta_\infty<0)$	$\omega=\pi$ $(\delta_\infty<0)$
$\delta_\infty>0$	$\omega\in$ II quadr.	$\omega\in$ I quadr.	$\omega\in$ IV quadr.
$\delta_\infty<0$	$\omega\in$ IV quadr.	$\omega\in$ III quadr.	$\omega\in$ II quadr.

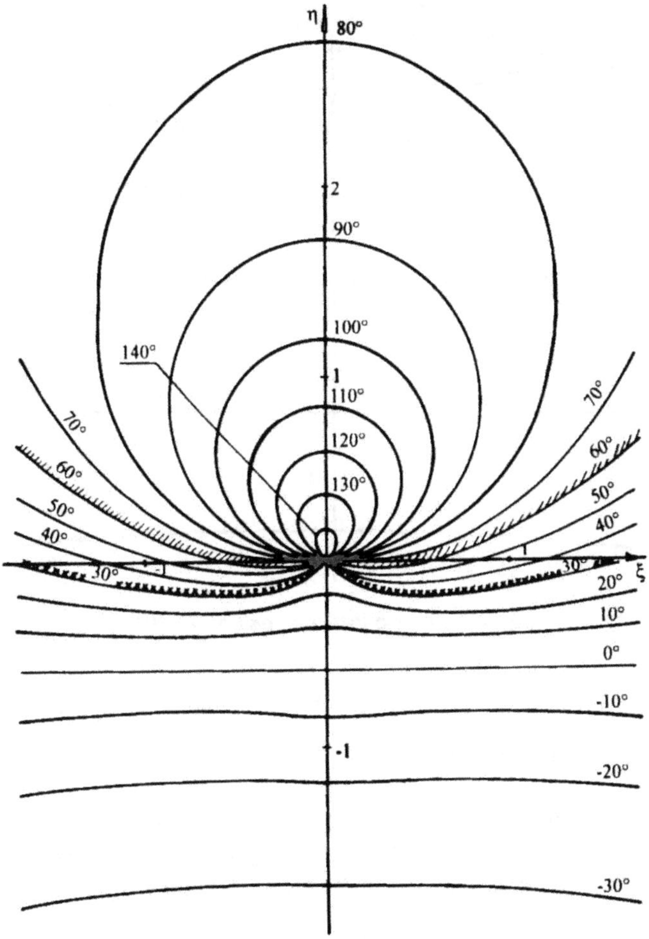

Figure 7.9 Isolines $\omega=const$ in B-plane, $\delta_\infty=30°$.

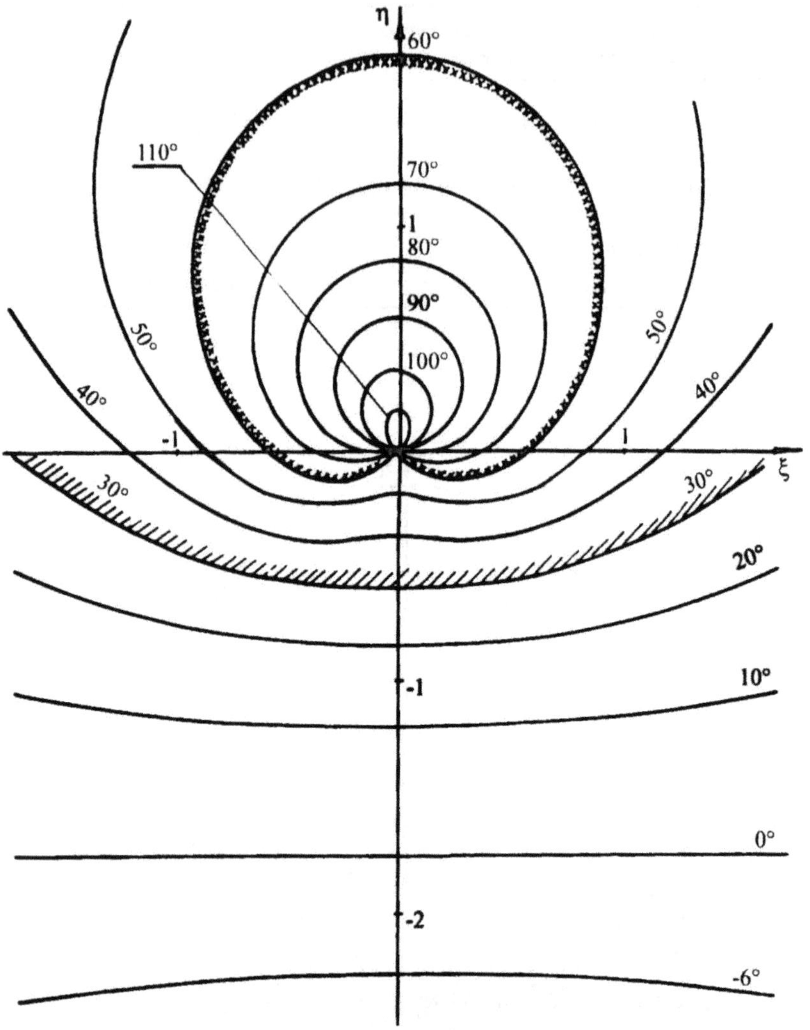

Figure 7.10 Isolines $\omega = const$ at $\delta = 60°$.

7.1.5 Landing on a specified point

Once the planetographic coordinates of the landing point and the angle of entry into the atmosphere have been specified, the aiming point (ξ, η) and the moment of landing within the day of arrival at the planet are completely determined. Consider this problem.

Let $\varphi^*, \theta^*_{\text{ent}}$ be the specified values of the landing latitude and the angle of entry into the planetary atmosphere. Then, the radius vector in the stationary coordinate

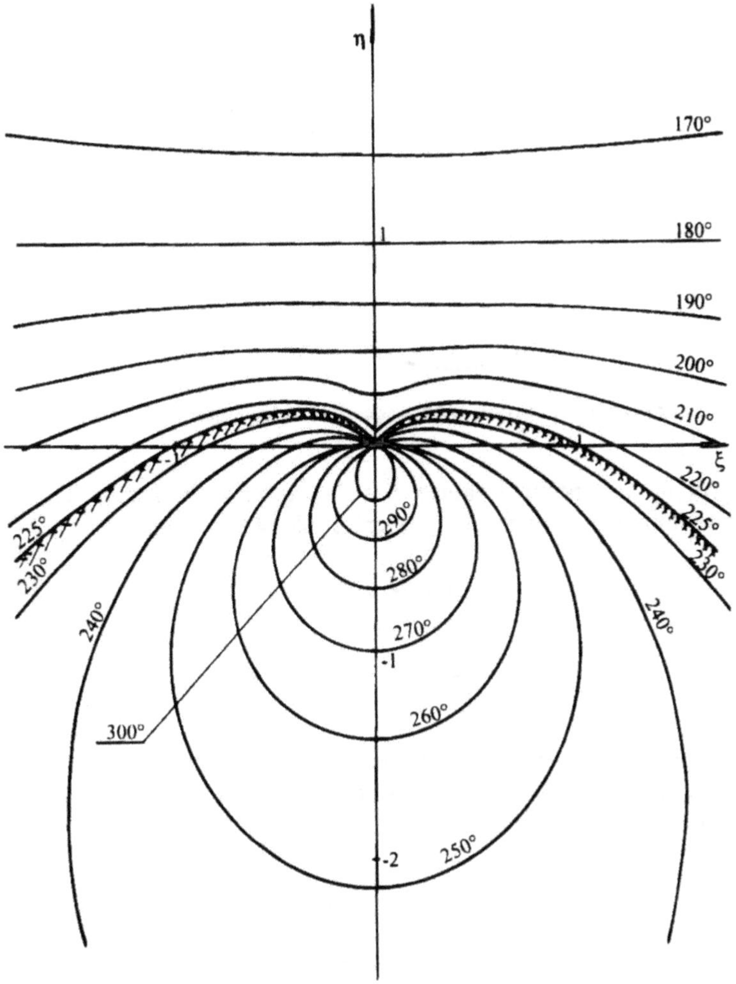

Figure 7.11 Isolines $\omega = const$ at $\delta_\infty = -45°$.

system can be found as the intersection of two circles on the sphere (Figure 7.12). Since the entry angle uniquely determines the aiming distance d, the problem reduces to determining the unit vector of landing ρ_n^0, which together with the vector $\bar{V}_\infty$ determines the angular elements of the planetocentric trajectory and aiming parameters ξ, η.

Let us solve an auxiliary system of vector equations

$$\left.\begin{aligned} \bar{b}_1^0 \cdot \bar{r} &= b_1, \\ \bar{b}_2^0 \cdot \bar{r} &= b_2, \\ \bar{r} \cdot \bar{r} &= 1. \end{aligned}\right\} \tag{7.25}$$

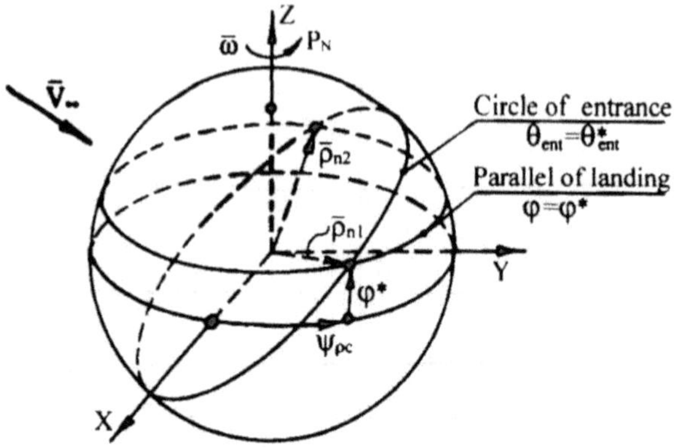

Figure 7.12 Landing on the surface with a specified entry angle to the specified latitude.

Here, $\bar{b}_1^0, \bar{b}_2^0, b_1, b_2$ are specified values, $\bar{r}$ is the vector to be found. The first two equations determine two planes, which intersect along the straight line

$$\bar{r}=\bar{\rho}+q\bar{\sigma}^0,$$

$$\bar{\sigma}^0=\frac{\bar{b}_1\times\bar{b}_2}{\sin\alpha};\quad \cos\alpha=\bar{b}_1^0\cdot\bar{b}_2^0 \tag{7.26}$$

where $\bar{\rho}$ is a vector satisfying the first two equations (it is convenient to assume that $\bar{\rho}\perp\bar{\sigma}$); and q is the parameter of the point on the straight line; this parameter must satisfy the third Equation. Then

$$q_{1,2}=\pm\sqrt{1-\rho^2}. \tag{7.27}$$

After transformations we obtain an expression for $\bar{\rho}$

$$\bar{\rho}=\frac{1}{\sin^2\alpha}[(b_1-b_2\cos\alpha)\bar{b}_1^0+(b_2-b_1\cos\alpha)\bar{b}_2^0]. \tag{7.28}$$

Thus, taking into account (7.26)–(7.28), the solution of the system (7.25) in vector form is as follows:

$$\bar{r}_{1,2}=\bar{\rho}\pm\sqrt{1-\rho^2}\,\frac{\bar{b}_1^0\times\bar{b}_2^0}{\sqrt{1-(\bar{b}_1^0\times\bar{b}_2^0)}}. \tag{7.29}$$

Geometric interpretation of the solution is presented in Figure 7.13. Solutions $\bar{r}_{1,2}$ are symmetrical with respect to the plane determined by vectors $\bar{b}_1^0, \bar{b}_2^0$ and are the points of intersection of the circles of radii $z_1=\arccos b_1$ and $z_2=\arccos b_2$ centered at b_1^0 and b_2^0, respectively.

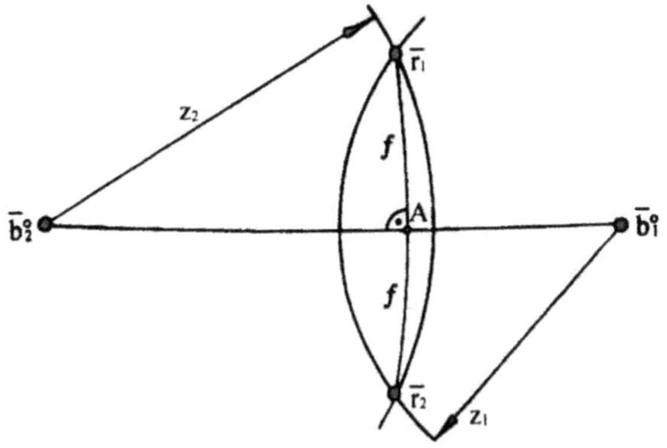

Figure 7.13 Geometric interpretation of the solution of vector equation on sphere.

Solutions $\bar{r}_{1,2}$ can also be determined in terms of angles $\mathscr{K}=\bar{b}_1^0 A$ and f, where $\cos f=\rho$; $\operatorname{tg}\mathscr{K}=(b_2-b_1\cos\alpha)/(b_1\sin\alpha)$.

The value $2f$ is the angular distance between the solutions.

Modulus of the vector ρ can be found from the relationship (7.28)

$$\rho=\frac{1}{\sin\alpha}\sqrt{b_1^2+b_2^2-2b_1b_2\cos\alpha}. \tag{7.30}$$

Note that vector $\bar{\sigma}^0$ is always directed toward the first solution of system (7.25).

The problem of determining the landing vector $\bar{\rho}_{\text{л}}^0$ is reduced to solving system (7.25), which will be written in the form

$$\bar{V}_\infty^0\cdot\bar{\rho}_{\text{л}}^0=\cos\beta_{\text{ent}},$$
$$\bar{\rho}_{\text{N}}^0\cdot\bar{\rho}_{\text{л}}^0=\sin\varphi^*,$$
$$\bar{\rho}_{\text{л}}^0\cdot\bar{\rho}_{\text{л}}^0=1.$$

The angle β_{ent} is formed as follows:

$$\beta=\upsilon_\infty+\upsilon_{\text{ent}}-\Delta(\theta_{\text{ent}})$$

where

$$\cos\upsilon_\infty=\frac{1}{e};\quad \cos\upsilon_{\text{ent}}=\frac{1}{e}\left(\frac{P}{r_{\text{ent}}}-1\right);\quad e=\sqrt{1+\frac{d^2}{a^2}};$$

$$P=\frac{d^2}{a};\quad a=\frac{\mu}{V_\infty^2};\quad d=\sqrt{r_{\text{ent}}^2+2ar_{\text{ent}}\cos\theta_{\text{ent}}}.$$

The variable $\Delta(\theta_{\text{ent}})$ is the angular distance of the flight segment within the atmosphere; this distance can be specified in the form of a table $\Delta=\Delta(\theta_{\text{ent}})$.

Once the vector $\bar{\rho}_\pi^0$ has been determined, the coordinates of the aiming point ξ_π, η_π can be easily found (7.9), (7.10). The time of arrival at the planet, i.e. the landing time, is determined by the planetocentric longitude ψ_{pc} of the vector ρ_π^0 (Figure 7.12) and the specified planetographic longitude of the landing point ψ_{pg}^*.

$$t_{lan} = [\psi_{pc} - (\psi_{p_0} + \psi_{pg}^*)]\omega_{pl}$$

where ψ_{p_0} is the planetocentric longitude of the zero meridian at the beginning of a day. ω_{pl} is the angular velocity of the planet's own rotation.

If the latitude of the landing point is specified in the form of the belt $\varphi_{min} \div \varphi_{max}$, then, varying φ^* within this range, we will obtain the curve of landing points in the B-plane $\xi = \xi(\varphi^*)$, $\eta = \eta(\varphi^*)$, each point of this curve being associated with $t_{lan} = t_{lan}(\varphi^*)$.

In conclusion we note that a wide range of orientation and navigational problems can be reduced to the solution of the system (7.25), e.g. [131].

7.2 Determining the trajectories of interplanetary probes

Interplanetary probes can be used for the following purposes:

- studying near-solar space (solar probes);
- studying extraecliptic space (extraecliptic probes);
- studying distant regions of the solar system and galactic space (galactic probes).

These problems impose requirements on the parameters Q_i of heliocentric orbits after the planetary flyby. The following parameters can be used as Q_i: minimum distance from the sun, inclination, period of rotation, maximum deviation from the ecliptic plane, the quickest route to the sphere of activity of the sun, etc.

As mentioned earlier, once the interplanetary trajectory of approach to the planet has been specified, the parameters Q_i are the function of only the aiming parameters ξ, η.

Two problems are possible:

1. The objective of the flyby trajectory is exclusively to form an interplanetary probe trajectory. In this case, the aiming parameters in the B-plane are determined by the following variants:

(i) $Q_{opt} = \text{extr}\, Q$ for $Q_i\{\geqslant; =; \leqslant\}Q_i^*$

(ii) $Q_1(\xi, \eta) = Q_1^*$; $Q_2(\xi, \eta) = Q_2^*$.

From here it follows that the flyby trajectories are determined uniquely.

Parameters Q_i become functions of the positions of the planets $Q_i = Q_i(T_1, T_2)$ and their isolines are constructed on the plane of dates, using the method described in Section 3.1.

However, in this formulation of the problem, it is also necessary to map isolines Q_i in the B-plane, both in order to solve the problem of guiding accuracy, and to study the domain of trajectories lying within the vicinity of the solutions obtained.

2. The main objective of the planetocentric trajectory is to examine the planet (e.g. landing of a lander, photographic and television surveying of specified regions, etc.).

If these conditions are not rigid, i.e. leave room for choice, the problem arises of how to create a space probe using flyby apparatus. To do this, the parameters of both planetocentric trajectories q_j and heliocentric orbits Q_i are mapped into the B-plane. Complex analysis of isolines $q_j = q_j^*, Q_i = Q_i^*$ is used to find the best compromise trajectories between all the criteria and restrictions.

7.2.1 The main features of mapping for flyby trajectories

Let us derive the relationships between the incoming $\bar{V}_{\infty\text{н}}$ and departure $\bar{V}_{\infty 0}$ velocity vectors at "infinity", aiming distance $\bar{d}$, and pericenter $\bar{\rho}_\pi$. The relationship between the unit vectors of these vectors is shown in Figure 7.14.

The following equalities hold for the unit vectors

$$\bar{V}^0_{\infty 0} = -(\cos 2\gamma \bar{V}^0_{\infty\text{н}} + \sin 2\gamma \bar{d}^0), \tag{7.31}$$

$$\bar{\rho}^0_\pi = \cos\gamma \bar{V}^0_{\infty\text{н}} + \sin\gamma \bar{d}^0. \tag{7.32}$$

The value of angle γ is determined in terms of $a = \mu / V^2_\infty ud$

$$\left.\begin{aligned} \cos\gamma &= \frac{a}{\sqrt{a^2+d^2}} \\ \sin\gamma &= \frac{d}{\sqrt{a^2+d^2}} \end{aligned}\right\} \quad \begin{aligned} \cos 2\gamma &= \frac{a^2-d^2}{a^2+d^2}, \\ \sin 2\gamma &= \frac{2ad}{a^2+d^2}. \end{aligned} \tag{7.33}$$

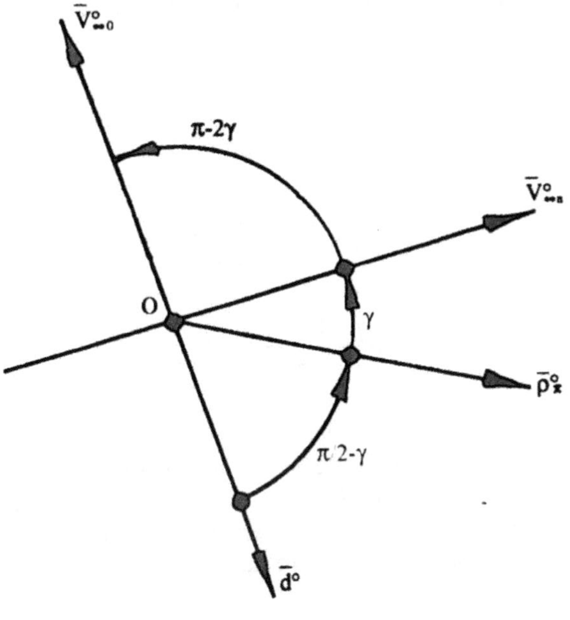

Figure 7.14 Relationship between vectors for solving the flyby problem.

Coordinates $\bar{V}^0_{\infty н}$ and $\bar{d}^0$ in the coordinate system ξ, η, ζ associated with the B-plane will be

$$\bar{V}^0_{\infty н} = \begin{Vmatrix} 0 \\ 0 \\ -1 \end{Vmatrix}, \quad \bar{d}^0 = \begin{Vmatrix} d_\xi/d \\ d_\eta/d \\ 0 \end{Vmatrix}.$$

Using the above relationships, we have

$$\bar{V}^0_{\infty 0} = \begin{Vmatrix} -\dfrac{2ad_\xi}{a^2+d^2} \\ -\dfrac{2ad_\eta}{a^2+d^2} \\ \dfrac{a^2-d^2}{a^2+d^2} \end{Vmatrix}; \quad \bar{\rho}^0_\pi = \begin{Vmatrix} \dfrac{d_\xi}{\sqrt{a^2+d^2}} \\ \dfrac{d_\eta}{\sqrt{a^2+d^2}} \\ -\dfrac{a}{\sqrt{a^2+d^2}} \end{Vmatrix}; \tag{7.34}$$

$$\bar{V}_{\infty 0} = V_\infty \bar{V}^0_{\infty 0}; \quad V_\infty = |\bar{V}_{\infty н}|; \quad \bar{\rho}_\pi = \rho_\pi \rho^0_\pi;$$

$$\rho_\pi = \sqrt{a^2+d^2} - a; \quad d = \sqrt{d_\xi^2 + d_\eta^2}. \tag{7.35}$$

In the case where the vector $\bar{V}^0_{\infty 0}$ is specified, whose coordinates will be denoted by $\dot{X}^0_{\infty 0}, \dot{Y}^0_{\infty 0}, \dot{Z}^0_{\infty 0}$, the aiming point in the B-plane can be found by the inverse mapping

$$d_\xi = \frac{\dot{X}^0_{\infty 0}}{1+\dot{Z}^0_{\infty 0}} \cdot a,$$

$$d_\eta = -\frac{\dot{Y}^0_{\infty 0}}{1+\dot{Z}^0_{\infty 0}} \cdot a, \tag{7.36}$$

$$d^2 = \frac{1-\dot{Z}^0_{\infty 0}}{1+\dot{Z}^0_{\infty 0}} \cdot a^2.$$

Formulae (7.34)–(7.36) determine a one-to-one correspondence between the points of the B-plane and points on the unit sphere of departure velocities.

Consider the following problem. Suppose that a circle is specified on the unit sphere $\bar{V}^0_{\infty 0}$. The center of the circle is determined by the unit vector $\bar{r}^0\{X^0, Y^0, Z^0\}$, and the radius, by the angle β, which corresponds to the condition

$$\bar{r}^0 \cdot \bar{V}^0_{\infty 0} = \cos\beta = K. \tag{7.37}$$

Using (7.34) we get

$$(K+Z^0)d^2+2aX^0d_\xi+2aY^0d_\eta+a^2(K-Z^0)=0 \tag{7.38}$$

and, if $K+Z^0\neq 0$, we arrive at the expression

$$\left(d_\xi+\frac{aX^0}{K+Z^0}\right)^2+\left(d_\eta+\frac{aY^0}{K+Z^0}\right)^2=a^2\frac{1-K^2}{(K+Z^0)^2} \tag{7.39}$$

which determines a circle with a shifted center.

Equations (7.38) and (7.39) suggest the basic properties of the mapping of the sphere of departure velocities into the B-plane:

1. $K=-Z^0$, i.e. circles on the sphere pass through the pole with the coordinates $\{0,0,-1\}$. This family of circles is associated with a family of straight lines in the B-plane:

$$X^0d_\xi+Y^0d_\eta+aK=0. \tag{7.40}$$

   For $X^0=0$ or $Y^0=0$ we obtain straight lines parallel to the coordinate axes $\bar{\xi}^0, \bar{\eta}^0$: $d_\eta=-aK/Y^0$ or $d_\xi=-aK/X^0$, i.e. the correspondence between the coordinate grids on the plane and sphere.
2. $K=Z^0=0$—circles on the sphere are meridians with respect to the poles $\{0,0,+1\}$, i.e., the vector $\bar{r}^0$ lies in the B-plane, and $\beta=\pi/2$. Images of these circles will be straight lines passing through the center of the B-plane:

$$X^0d_\xi+Y^0d_\eta=0.$$

3. $K+Z^0\neq 0$—an arbitrary circle on the sphere that is mapped on the plane (7.39) into the circle of radius ρ and centered at ξ, η_c

$$\xi=-\frac{X^0}{K+Z^0}\cdot a;\quad \eta_c=-\frac{Y^0}{K+Z^0}\cdot a;\quad \rho=\left|\frac{a}{K+Z^0}\sqrt{1-K^2}\right|. \tag{7.41}$$

   When $X^0=Y^0=0$, we obtain a family of concentric circles with respect to the origin.
4. $K=\pm 1$ is a point on the sphere, corresponding to the vector $\bar{r}_0$; this point is mapped into the point

$$d_\xi=-\frac{X^0}{\pm 1+Z^0}\cdot a,\quad d_\eta=-\frac{Y^0}{\pm 1+Z^0}\cdot a \tag{7.42}$$

   in the B-plane. These relationships are equivalent to (7.36). The image of the pole $\{0,0,-1\}$ is a circle of infinite radius.

Thus, all the circles on the sphere of departure velocities are mapped into circles or straight lines in the B-plane. The inverse statement is also valid.

The stereographic projection of points on a sphere to a plane [27] is known to possess similar properties. Consider the relation between the stereographic projection and the mappings (7.34) and (7.36). Suppose that we have a sphere with the center located in the origin and of radius $R=a/2$; a point on the sphere is specified by the vector $\bar{r}_1=R\bar{r}_1^0$, where r_1 is the unit vector $\{X_1^0, Y_1^0, Z_1^0\}$, and d_{x_1}, d_{y_1} is the stereographic projection of the point $\bar{r}_1$ into the plane $Z=-a/2$. Then

$$X_1^0=\frac{2ad_{x_1}}{a^2+d^2}, \qquad Y_1^0=\frac{2ad_{y_1}}{a^2+d^2}, \qquad Z_1^0=\frac{d^2-a^2}{a^2+d^2}, \tag{7.43}$$

$$d_{x_1}=\frac{X_1^0}{1-Z_1^0}\cdot a, \qquad d_{y_1}=\frac{Y_1^0}{1-Z_1^0}\cdot a, \qquad d_1^2=\frac{1+Z_1^0}{1-Z_1^0}\cdot a^2. \tag{7.44}$$

We will refer to the relations (7.36) and (7.44) as forward projection and to (7.34) and (7.43) as backward projection. Figure 7.15 shows stereographic projections of points M and M_1, that are specified by vectors $\bar{r}^0$ and $\bar{r}_1^0$ in the plane of the meridian containing the vectors $\bar{r}_1^0=-\bar{r}^0$.

The relationships (7.43), (7.44) and (7.34), (7.36) specify the following correspondence between the points of plane and sphere:

$$M_1 \underset{(7.43)}{\overset{(7.44)}{\rightleftharpoons}} N_1 \qquad M_1 \underset{(7.34)}{\overset{(7.36)}{\rightleftharpoons}} N$$

$$M \underset{(7.43)}{\overset{(7.44)}{\rightleftharpoons}} N \qquad M \underset{(7.34)}{\overset{(7.36)}{\rightleftharpoons}} N_1$$

If we denote $SN=d$ and $SN_1=d_1$, then $d\cdot d_1=4R^2=a^2$, i.e. we found that the point N passes into N_1 (and vice versa) by inversion with respect to the circle and reflection with respect to the point S.

From this analogy it follows that:

- mapping the sphere of departure velocities onto the B-plane is a forward stereographic projection with a subsequent inversion with respect to a circle of radius $a=\mu/V_\infty^2$ and a reflection with respect to the origin;
- mapping the B-plane on to the unit sphere of departure velocities is a backward stereographic projection with a subsequent centrally-symmetric reflection.

In the general case, the problem of constructing isolines of the parameters of heliocentric trajectories in the B-plane is reduced to the mapping of the line of intersection of the surfaces

$$\begin{gathered} F(\bar{V},\bar{r},Q)=0, \\ \bar{V}_{\infty 0}\cdot\bar{V}_{\infty 0}=V_\infty^2, \end{gathered} \tag{7.45}$$

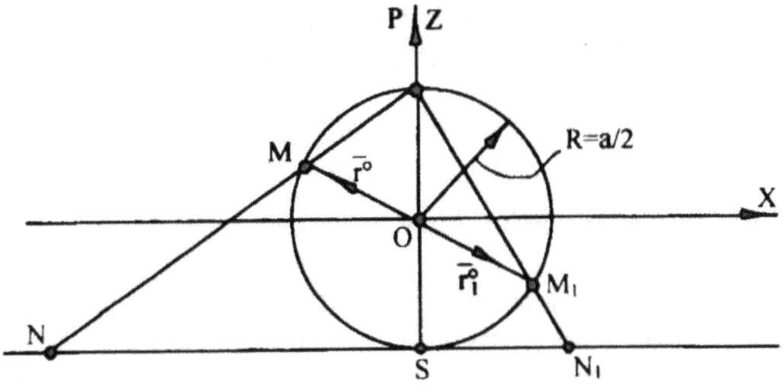

Figure 7.15 Stereographic projection of points of the departure velocity sphere onto B-plane.

where $\bar{V}$ is the heliocentric velocity of the spacecraft after the flyby of the planet, and

$$\bar{V}=\bar{V}_n+\bar{V}_{\infty 0} \tag{7.46}$$

$\bar{r}$, $\bar{V}_n$ are the heliocentric radius vector and velocity of the planet in the flyby point, $Q=const$ is a parameter of the heliocentric trajectory.

It is convenient to interpret (7.45) as an equation of a curve on the sphere of departure velocities. If this curve is a circle with a specified center $\bar{r}^0$, then its image in the B-plane can easily be found from expressions (7.40) and (7.41). In the cases where the curves are more complicated, it is convenient to represent Equations (7.45) in a parametric form, e.g. $\bar{V}_{\infty 0X}=f_1(V_{\infty 0Z}), V_{\infty 0Y}=f_2(V_{\infty 0Z})$, and to perform the mapping using relationships (7.36). Finally, in the most general case, where the function $F=0$ fails to be expressed analytically, the line of intersection of the planes is constructed by the method presented in Section 3.1 with subsequent mapping of each point by the formulae (7.36). However, as will be shown below, most problems allow an analytical form of solution or parametric representation of the line.

It is convenient to express the function F in terms of the components of the velocity vector V in the planetocentric coordinate system r, m, b for the flyby moment; the unit vectors of this system are:

$$\bar{r}^0=\frac{\bar{r}}{|\bar{r}|};\quad \bar{m}^0=\bar{b}^0\times\bar{r}^0;\quad \bar{b}^0=\frac{\bar{r}\times\bar{V}_n}{|\bar{r}\times\bar{V}_n|}.$$

To simplify notation we will denote the departure velocity at infinity by $\bar{V}_\infty$.

After the flyby of the planet, the spacecraft trajectory is determined by $\bar{r}_a$ and $\bar{V}_a=\bar{V}_n+\bar{V}_\infty$, which in the planetocentric coordinate system will have the form

$$\bar{r}_a=\begin{Vmatrix} r \\ 0 \\ 0 \end{Vmatrix},\qquad \bar{V}_a=\begin{Vmatrix} V_{nr}+V_{\infty r} \\ V_{nm}+V_{\infty m} \\ V_{\infty b} \end{Vmatrix}. \tag{7.47}$$

We denote the complete transversal component of the velocity V_{am1} as $\bar{V}_a$; this component determines the planar parameters of the post-flyby orbit, and we have

$$V_{am1}^2 = V_{am}^2 + V_{ab}^2 = (V_{nm} + V_{\infty m})^2 + V_{\infty b}^2.$$

7.2.2 Constant-energy characteristics of the interplanetary probe trajectories

Let us determine the geometric locus of aiming points in the B-plane that correspond to specified values of the components of departure heliocentric velocity. There is a series of problems that require such a formulation. For example, the construction of orbits with a period divisible by the period of the orbit of the flyby planet or by any other value, isolation of the domain of trajectories leaving the solar system, determination of the orbits that after the flyby will be directed towards the sun or from it, etc.

(i) The specified value of the total heliocentric velocity of the spacecraft after the planetary flyby $V_a = V_a^*$. Using the relationships (7.47), we write

$$\bar{V}_a^2 = (\bar{V}_n + \bar{V}_\infty)^2 = V_a^{*2},$$

$$\bar{V}_n^0 \cdot \bar{V}_\infty^0 = \frac{V_a^{*2} - V_n^2 - V_\infty^2}{2V_n V_\infty} = K_v, \tag{7.48}$$

$$V_n - V_\infty \leqslant V_a^* \leqslant V_n + V_\infty.$$

Note that specifying V_a^* is equivalent to specifying the semi-major axis or the period of revolution in the heliocentric orbit (for ellipses).

(ii) The specified value of the radial component of heliocentric velocity of the spacecraft $V_{ar} = V_{ar}^*$. We have $\bar{r}^0 \cdot \bar{V}_a = V_{ar}^*$ or

$$\bar{r}^0 \cdot \bar{V}_\infty^0 = \frac{V_{ar}^* - V_{nr}}{V_\infty} = K_r,$$

$$V_{nr} - V_\infty \leqslant V_{ar}^* \leqslant V_\infty + V_{nr}. \tag{7.49}$$

(iii) The specified value of the transversal component $V_{am} = V_{am}^*$

$$\bar{m}^0 \cdot \bar{V}_\infty^0 = \frac{V_{am}^* - V_{am}}{V_\infty} = K_m,$$

$$V_{nm} - V_\infty \leqslant V_{am}^* \leqslant V_{nm} + V_\infty. \tag{7.50}$$

(iv) The specified value of the binormal component $V_{ab} = V_{ab}^*$

$$\bar{b}^0 \cdot \bar{V}_\infty^0 = \frac{V_{ab}^*}{V_\infty} = K_b; \quad -V_\infty \leqslant V_{ab}^* \leqslant V_\infty. \tag{7.51}$$

Equations (7.48)–(7.51) determine the planes perpendicular to the vectors $\bar{V}_n^0, \bar{r}^0, \bar{m}^0, \bar{b}^0$, and spaced from the center by K_v, K_r, K_m, K_b, respectively. Adding the condition $\bar{V}^0_\infty \cdot \bar{V}^0_\infty = 1$ to each equation, we will get circles on the sphere of departure velocities of radius β, where $\cos\beta = K$. By varying the specified value of the corresponding velocity we will get concentric circles on the sphere.

Conditions (7.48)–(7.51) are equivalent to the problem solved in the previous Tsection (7.21). Hence, the isolines of velocity $V_a, V_{ar}, V_{am}, V_{ab}$ in the B-plane are either circles or straight lines that can be found from simple relationships (7.39)–(7.42) under the condition that the unit vectors $\bar{V}^0_n, \bar{r}^0, \bar{m}^0, \bar{b}^0$ are transformed into the coordinate system of the B-plane ξ, η, φ.

(v) The specified value of the total transversal component of the spacecraft post-flyby heliocentric velocity $V_{am1} = V^*_{am1}$. Using (7.47) we obtain

$$(V_{nm} + V_{\infty m})^2 + V^2_{\infty b} = V^{*2}_{am1},$$
$$V^2_{\infty r} + V^2_{\infty m} + V^2_{\infty b} = V^2_\infty.$$

These equations determine the curve of intersection of the cylinder, whose axis is parallel to the axis $\bar{r}^0$, with the sphere of departure velocities. It is convenient to present this curve in the parametric form.

$$V^2_{\infty r} = V^2_{nm} + 2V_{nm}V_{\infty m} + V^2_\infty - V^{*2}_{am1},$$
$$V^2_{\infty b} = V^{*2}_{am1} - (V_{nm} + V_{\infty m})^2. \tag{7.52}$$

Projection of the curve (7.52) onto the plane r, m is a parabola which is symmetrical with respect to the axis $\bar{m}^0$, and its projection onto the plane m, b is a circle shifted along the axis $\bar{m}^0$ to value V_{nm}, i.e. the curve is closed and symmetric with respect to these planes.

The conditions of existence of the solution are: $V_{nm} - V_\infty \leqslant V^*_{am1} \leqslant V_{nm} + V_\infty$, and the domain of variations of $V_{\infty m}$ for construction of the curve (7.52) are

$$\frac{V^{*2}_{am1} - V^2_{nm} - V_\infty}{2V_{nm}} \leqslant V_{\infty m} \leqslant V^*_{am1} - V_{nm}. \tag{7.53}$$

When $V_{\infty m}$ is equal to the left limit, we have $V_{\infty r} = 0$; when it is equal to the right limit, $V_{\infty b} = 0$.

Mapping the line $V_{am1} = const$ onto the B-plane reduces to application of transformation (7.36) to each point of the curve (7.52) within the limits (7.53).

(vi) Suppose that any two conditions from (i)–(iv) are to be satisfied. This is equivalent to solving a system of vector equations of the type

$$\bar{\rho}_1 \cdot \bar{V}^0_\infty = \cos\beta_1,$$
$$\bar{\rho}_2 \cdot \bar{V}^0_\infty = \cos\beta_2,$$
$$\bar{V}^0_\infty \cdot \bar{V}^0_\infty = 1$$

from where two solutions $V^0_{\infty 1,2}$ can be found (Section 7.1). These points are then transformed into points on the B-plane using transformation (7.36).

The second method consists in directly determining the points in the B-plane where the circles or straight lines obtained by the above-considered transformations intersect.

Consider the classes of trajectories that can be implemented using constant-energy mapping:

- $V_a^* = V_{apar} = \sqrt{2\mu_s/r}$: post-flyby heliocentric parabolic orbits. Boundary of the domain within which the spacecraft leaves the solar system and becomes a galactic probe;
- $V_a^* = V_c = \sqrt{\mu_s/r}$: post-flyby circular heliocentric velocities;
- $V_a^* = V_{amax} = V_n + V_\infty$: trajectory with maximum heliocentric energy.

If $V_{amax} \geqslant V_{par}$, then $V_{\infty a} = \sqrt{V^2_{amax} - (2\mu_s/r)}$ is the maximum velocity at infinity at the moment of leaving the sun's sphere of activity.

$V_a^* = V_{amin} = V_n - V_\infty$: if $V_\infty < V_n$ we obtain the minimum-energy trajectory with a minimum semi-major axis, and, therefore, with a minimum period of revolution around the sun. If $V_{nr} = 0$, this condition is equivalent to that of the minimum pericentral altitude of heliocentric orbit. Here we have

$$r_{\pi\,\min} \approx \frac{V_\alpha^2}{(2\mu/r_\alpha) - V_\alpha^2} \cdot r_\alpha$$

where $V_\alpha = V_n - V_\infty$, $r_\alpha = r$.

- $V_a^* \sim T_a^*$ is a trajectory with a specified period of revolution around the sun. In particular, if $T_a^* = K \cdot T_n$, where T_n is the period of revolution of the flyby planet, $K = n/m$, n and m are integer numbers, the second encounter with the planet will take place after $n \cdot T_n$ periods;
- $V_{ar}^* = 0$ are heliocentric orbits, for which the flyby point is the apsidal point, i.e. $r = r_\pi$ or $r = r_\alpha$ (depending on the sign of inequality $V_a^* \gtrless V_{cr}$).

Note that for such orbits, the argument of pericenter ω, measured from the orbital plane of the planet, amounts to $\omega = 0$ ($V_{\infty b} > 0$) or $\omega = \pi$($V_{\infty b} < 0$), i.e. the apsides lie within the orbital plane of the planet (or almost in the ecliptic plane). This condition is also the boundary separating the trajectories which, after flyby, move away from or toward the sun.

- $V_{ar}^* = V_{nr} \pm V_\infty$: two trajectories corresponding to maximum local velocities of moving away from and toward the sun (for $V_{nr} < V_\infty$);
- $V_{am}^* = 0$ is possible if $V_\infty \geqslant V_{nm}$, then we have the spacecraft trajectories with an inclination $i_a = 90$ to the orbital plane of the flyby planet;
- $V_{ab}^* = 0$: heliocentric trajectories lying within the orbital plane of the flyby planet ($i_a = 0$);
- $V_{am1}^* = 0$ corresponds to two conditions $V_{ab}^* = 0, V_{am}^* = 0$; it is possible when $V_\infty \geqslant V_{nm}$, which yields rectilinear orbits. In the general case, specifying the value

of V_{am1}^* is equivalent to specifying the parameter P_a of the post-flyby heliocentric orbit.

7.2.3 Choice of inclination

Suppose that we are to provide a given inclination of the post-flyby heliocentric trajectory with respect to a certain reference plane specified by a unit vector $\bar{C}^0\{C_r^0, C_m^0, C_b^0\}$ in the planetocentric orbital coordinate system. This condition corresponds to the system of vector equations

$$\frac{\bar{r}\times(\bar{V}_n+\bar{V}_\infty)\cdot}{|\bar{r}\times(\bar{V}_n+\bar{V}_\infty)|,}\cdot\bar{C}^0=\cos i_a^*,$$
$$\bar{V}_\infty\cdot\bar{V}_\infty=V_\infty^2. \tag{7.54}$$

Using the relationships (7.47) we obtain the components of the kinetic momentum vector of the orbit after the flyby:

$$\bar{r}\times(\bar{V}_n+\bar{V}_\infty)=\{0,\ -r\cdot V_{\infty b},\ r\cdot(V_{nm}+V_{\infty m})\}.$$

Substituting this into (7.54) yields the equation of the second-order surface

$$(C_b^{02}-\cos^2 i_a^*)(V_{nm}+V_{\infty m})^2+2C_m^0C_b^0(V_{nm}+V_{\infty m})V_{\infty b}+(C_m^{02}-\cos^2 i_a^*)V_{\infty b}^2=0$$

which split into two planes passing through the $\bar{r}^0$ axis:

$$V_{\infty b}(C_m^{02}-\cos^2 i_a^*)+(C_m^0C_b^0\pm\sqrt{\sin^2 i_a^*-C_r^{03}}\cdot\cos i_a^*)V_{\infty m}$$
$$=(-C_m^0\cdot C_b\mp\sqrt{\sin i_a^*-C_r^{02}}\cdot\cos i_a^*)V_{nm} \tag{7.55}$$

$\sin^2 i_a^*>C_r^{02}$ is the condition of existence of solutions.

Thus, the system (7.54) is equivalent to two systems of vector equations of the type

$$\bar{r}_{1,2}^0\cdot V_\infty^0=K_{1,2},$$
$$\bar{V}_\infty^0\cdot\bar{V}_\infty^0=1 \tag{7.56}$$

where $\bar{r}_{1,2}^0$ are the unit vectors of the vectors

$$\bar{r}_{1,2}=\{0, C_m^0\cdot C_b^0\pm\sqrt{\sin^2 i_a^*-C_r^{02}}\cdot\cos i_a^*, C_m^{02}-\cos^2 i_a^*\}$$

and

$$K_{1,2}=\frac{(-C_m^0C_b^0\mp\sqrt{\sin^2 i_a^*-C_r^{02}}\cdot\cos i_a^*)V_{nm}}{|\bar{r}_{1,2}|V_\infty}.$$

System (7.56) determines a circle on the unit sphere of the departure velocities. Variations in i_a^* yield two families of misaligned circles formed by a bundle of planes passing through the $\bar{r}^0$ axis.

Since the system (7.56) is equivalent to the problem (7.37), the isolines of equal inclinations in the B-plane are also either circles or straight lines (7.39)–(7.42).

When selecting interplanetary probe trajectories, the value of i_a^* is specified as a rule with respect to the ecliptic plane. Taking into account the fact that the planetary orbits have a small inclination to the ecliptic, we will consider the problem with specified inclination to the orbital plane of the flyby planet, which allows a significant simplification of the expressions (7.55)–(7.56). For this case, $C_r^0=C_m^0=0$, $C_b^0=1$, and equations of the planes will take the form $-V_{\infty m}\sin i_a^* \pm V_{\infty b}\cos i_a^* = V_{nm}\sin i_a^*$, and for the system (7.56)

$$\bar{r}^0_{1,2}=\{0, -\sin i_a^*, \pm\cos i_a^*\},$$
$$K_{1,2}=\frac{V_{nm}}{V_\infty}\sin i_a^*, \tag{7.57}$$

$\bar{r}^0_{1,2}$ specifies the center of a circle of radius β, where $\cos\beta=V_{nm}\sin i_a^*/V_\infty$. The condition for implementation of the solution is

$$V_\infty > V_{nm}\sin i_a^*.$$

Consider the geometry of a family of circles on the sphere of departure velocities and its mapping onto the B-plane depending on the relation between V_∞ and V_{nm}:

(i) $V_\infty < V_{nm}$. We have two types of orbits, for which the flyby point is an ascending node ($\bar{r}_1^0$) and a descending node ($\bar{r}_2^0$) of the orbit. Here, $0 \leqslant i_a^* \leqslant i_{max}$, where $\sin i_{max}=V_\infty/V_{nm}$. Figure 7.16a demonstrates this case and shows the section of the sphere by the plane m^0, b^0. Varying i_a^* we obtain in the B-plane two families of nonintersecting circles with points corresponding i_{max}. The circle $i=0$ belongs to both the families.

(ii) $V_\infty = V_{nm}$. ,In this case $0 \leqslant i_a^* \leqslant \pi/2$. We obtain two families of circles with a common point of tangency corresponding to the vertical orbit (Figure 7.16b).

(iii) $V_\infty \geqslant V_{nm}$. Any inclination can be implemented. We have two families of circles, which intersect in two points corresponding to vertical orbits. These points divide the circle into two arcs corresponding to the sign of kinetic momentum, i.e. to the preservation of the angle i_a^* and its changing to $\pi - i_a^*$. In the case where i_a^* changes to $\pi - i_a^*$, the ascending node of the orbit changes to the opposite one (Figure 7.16c). In the implementation of vertical orbits, the initial heliocentric velocity of the spacecraft after the flyby will be

$$V_{ar1,2}=V_{nr}\pm\sqrt{V_\infty^2-V_{nm}^2},\quad V_{am}=0,\quad V_{ab}=0.$$

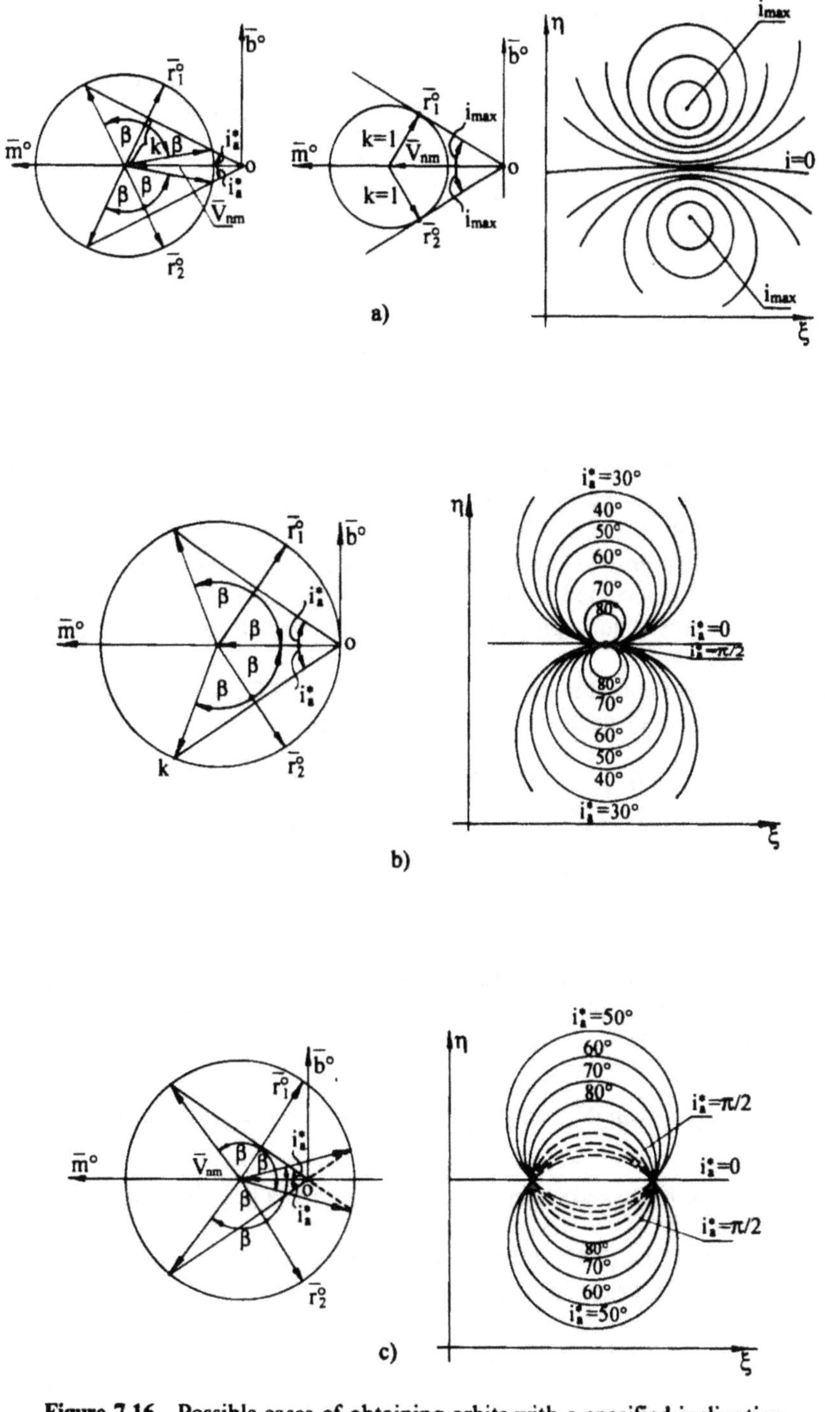

Figure 7.16 Possible cases of obtaining orbits with a specified inclination.

7.2.4 Determination of the pericenter altitude

Heliocentric trajectories with specified values of the pericenter or apocenter altitude are associated with the rotational surfaces in the space of velocities.

$$r_\pi^*: -\frac{V_r^2}{p^2}+\frac{V_m^2}{q^2}+\frac{V_b^2}{q^2}=1, \tag{7.58}$$

$$r_\alpha^*: \frac{V_r^2}{p^2}+\frac{V_m^2}{q^2}+\frac{V_b^2}{q^2}=1, \tag{7.59}$$

$$p^2=\left|\frac{2\mu}{r}\cdot\frac{\rho-r}{r}\right|;\quad q^2=\frac{2\mu}{r}\cdot\frac{\rho}{r+\rho}. \tag{7.60}$$

Here, the value of ρ is equal to r_π^* or r_α^*, respectively.

These equations stem immediately from the integrals of areas and energies [9]. Note that the value of q is equal to the velocity in the apsidal point r, provided that the second apsidal point is ρ.

The case r_π^* corresponds to the trajectory options for probes aimed at studying near-solar space. The specified value of r_α^* determines the orbits probing external space beyond the flyby planet, provided that the orbit remains elliptical, i.e. returns into the internal zone.

Consider in more detail the case $r_\pi=r_\pi^*$. Combining Equation (7.58) with the Equation of the sphere of departure velocities we get

$$\begin{gathered}-\frac{(V_{nr}+V_{\infty r})^2}{p^2}+\frac{(V_{nm}+V_{\infty m})^2}{q^2}+\frac{V_{\infty b}^2}{q^2}=1,\\ V_{\infty r}^2+V_{\infty m}^2+V_{\infty b}^2=V_\infty^2.\end{gathered} \tag{7.61}$$

Relationships (7.61) determine the spatial curve $r_\pi=r_\pi^*$ of intersection of a sphere centered in the origin with a rotational hyperboloid of one sheet (rotational axis is $\bar{r}^0$) shifted by r,m in the plane $-V_{nr}, -V_{nm}$. The projection of this curve onto the plane r,m is the parabola:

$$V_{\infty m}=\frac{1}{2p^2V_{nm}}[(p^2+q^2)V_{\infty r}^2+2q^2V_{nr}V_{\infty r}-p^2(V_{nm}^2+V_\infty^2)+q^2(V_{nr}^2+p^2)]. \tag{7.62}$$

If $r_\pi^* \ll r$, and since V_{nr} usually is also small, the parabola is almost symmetric with respect to the axis $\bar{m}^0$.

Thus, the relationships (7.62) allow us to represent the curve in the parametric form, which is suitable for calculations:

$$\begin{gathered}V_{\infty m}=f(V_{\infty r}),\\ V_{\infty b}=\pm\sqrt{V_\infty^2-V_{\infty r}^2-f^2(V_{\infty r})}.\end{gathered} \tag{7.63}$$

Since the curve is symmetric with respect to the plane r, m, one $V_{\infty m}$ and two $V_{\infty b}$, differing only by their signs, will correspond to each value of $V_{\infty r}$.

The range of $V_{\infty r}$ variations will be:

$$V_{\infty r}^2 + f^2(V_{\infty r}) \leqslant V_{\infty}^2.$$

Applying transformation (7.36) in each point of the curve (7.63) we will obtain the isoline $r_{\pi} = r_{\pi}^*$ in the B-plane.

Note that if $V_{nr} \approx 0$ and $V_{\infty} < V_n$, then at $V_{\infty m} = -V_{\infty}$ we will obtain the case $V_a^* = V_{a\min}$ corresponding to $r_{\pi}^* = r_{\pi\min}$. This technique can be used to assess the minimum possible pericentral altitude of the heliocentric orbit after the planetary flyby.

7.2.5 Analysis of the maximum possible distance from the orbital plane

The distance Z between any point of the trajectory and some reference plane will be

$$Z = p\frac{\sin(\omega + v)}{1 + e\cos v} \cdot \sin i \tag{7.64}$$

where p, e are the parameter and eccentricity of the orbit; ω is the argument of pericenter; i is inclination; v is the true anomaly of the point. Values of ω and i are determined with respect to a certain reference plane.

Differentiating (7.64) with respect to v and equating the derivative to zero, we obtain the following equations for determining the point v_m most distant from the plane:

$$\begin{aligned} &\cos(\omega + v_m) + e\cos\omega = 0, \\ &v_m = \arccos(-e\cos\omega) - \omega. \end{aligned} \tag{7.65}$$

This equation is presented in [171].

From (7.65) we have:

$$\sin(\omega + v_m) = K\sqrt{1 - e^2\cos^2\omega},$$

$$\cos v_m = -e\cos^2\omega + K\sqrt{1 - e^2\cos^2\omega} \cdot \sin\omega.$$

Here $K = \pm 1$, which reflects the duality of solution (7.65). Substituting these expressions into (7.64), we get the maximum distance

$$Z_m = \frac{p\sin i}{K\sqrt{1 - e^2\cos^2\omega} + e\sin\omega}. \tag{7.66}$$

Now note that if we take the orbital plane of the flyby planet as the reference plane, the flyby point r will be either the ascending node ($r = r_{\Omega}$), or the descending node ($r=r_{\mho}$) of the orbit, and the value of ω is equal to the true anomaly of the flyby point on the heliocentric trajectory.

Express Z_m in terms of the velocity components. Taking into account the well-known relationships between ω and v we can write:

$$\left.\begin{aligned} e\sin\omega &= K_1\frac{V_r V_{m1}}{\mu}\cdot r \\ e\cos\omega &= K_1\left(\frac{rV_{m1}^2}{\mu}-1\right) \end{aligned}\right\}\quad K_1=\begin{cases}-1 & r=r_{\Omega} \\ +1 & r=r_{\Omega}\end{cases}.$$

$V_{m1}=V_m^2+V_b^2$ is the transversal component of velocity in the spacecraft orbital plane after the flyby. Substituting into (7.66) yields

$$Z_m=\frac{rV_{m1}}{K\sqrt{2\mu/r-V_{m1}^2}+K_1V_r}\cdot\sin i.$$

However, in particular at $V_r=0$, we obtain an evident relationship $Z_m=\pm b\sin i$, where b is the minor semi-axis of the post-flyby heliocentric orbit.

From (7.57) it follows that

$$\sin i=\frac{V_b}{\sqrt{V_m^2+V_b^2}}=\frac{V_b}{V_{m1}}$$

and finally we have

$$Z_m=\frac{rV_b}{K\sqrt{2\mu/r-V_{m1}^2}+K_1V_r}. \tag{7.67}$$

Note that Z_m is meaningful only for elliptic heliocentric orbits.

Now let the value of Z_m be specified. Then from (7.67) we obtain the surface $Z_m=Z_m^*$:

$$V_r^2+V_m^2+(1+\mathscr{H}^2)V_b^2-2\mathscr{H}V_bV_r=V_{par}^2 \tag{7.68}$$

where

$$\mathscr{H}=K_1\frac{r}{Z_m^*};\quad V_{par}=\sqrt{\frac{2\mu}{r}}.$$

Invariants and eigenvalues of the form (7.68) will be as follows:

$$I=3+\mathscr{H}^2;\quad J=3+\mathscr{H}^2;\quad D=1;\quad A=-V_{par}^2$$

$$\lambda_2=2;\quad \lambda_{1,3}=\frac{2+\mathscr{H}^2\pm\mathscr{H}\sqrt{4+\mathscr{H}^2}}{2}$$

and the eigendirections

$$\lambda=\lambda_2:\ r^0=0,\quad m^0=1,\quad b^0=0$$

$$\lambda=\lambda_{1,3}:(1-\lambda_{1,3})\cdot r^0_{1,3}-\mathscr{H}b^0_{1,3}=0,\quad m_{1,3}=0.$$

Thus, the surface $Z_m=Z^*_m$ is a real ellipsoid; one of its principal axes coincides with $\bar{m}^0$ and two others lie in the plane $\bar{r}^0,\bar{b}^0$ and are separated by an angle determined by:

$$\frac{b^0_{1,3}}{r^0_{1,3}}=\frac{1-\lambda_{1,3}}{\mathscr{H}}.$$

The semiaxes of the ellipsoid can be found from eigenvalues $\lambda_{1,2,3}$, where $\lambda_1>\lambda_2>\lambda_3$.

Thus, the problem of constructing the flyby trajectories with $Z_m=Z^*_m$ is reduced to determining the line of intersection of the ellipsoid (7.68) with the sphere:

$$\begin{gathered}(V_{nr}+V_{\infty r})^2+(V_{nm}+V_{\infty m})^2+(1+\mathscr{H}^2)V^2_{\infty b}-2\mathscr{H}(V_{nr}+V_{\infty r})V_{\infty b}=V^2_{\text{par}},\\ V^2_{\infty r}+V^2_{\infty m}+V^2_{\infty b}=V^2_{\infty}.\end{gathered}\tag{7.69}$$

From these equations it follows:

$$2V_{nm}V_{\infty m}+2(V_{nr}-\mathscr{H}V_{\infty b})V_{\infty r}-2\mathscr{H}V_{nr}V_{\infty b}+\mathscr{H}^2V_{\infty b}+V^2_n-V^2_{\text{par}}+V^2_{\infty}=0.$$

Considering $V_{\infty b}$ as a parameter for the curve (7.69) we get:

$$V_{\infty m}=\alpha V_{\infty r}+\beta,$$

$$\alpha=\frac{\mathscr{H}V_{\infty b}-V_{nr}}{V_{nm}},$$

$$\beta=\frac{1}{2V_{nm}}[2\mathscr{H}V_{nr}V_{\infty b}-\mathscr{H}^2V_{\infty b}+V^2_{\text{par}}-(V^2_n+V^2_{\infty})].\tag{7.70}$$

Substituting (7.70) into the equation of the sphere we obtain the second-order equation:

$$(1+\alpha^2)V^2_{\infty r}+2\alpha\beta V_{\infty r}+\beta^2+V^2_{\infty b}-V^2_{\infty}=0$$

and the curve (7.69) can be written in the parametric form:

$$V_{\infty r1,2}=\frac{-\alpha\beta\pm\sqrt{\alpha^2\beta^2-(1+\alpha^2(\beta^2+V_{\infty b}^2-V_{\infty}^2))}}{(1+\alpha^2)};$$

$$V_{\infty m}=\alpha V_{\infty r}+\beta; \qquad (7.71)$$

$$\alpha=\alpha(V_{\infty b}); \quad \beta=\beta(V_{\infty b}).$$

The range of variations of the obtained curve will be

$$V_{\infty b}^2+V_{\infty r}^2(V_{\infty b})\leqslant V_{\infty}^2.$$

Normalizing the components of the vector $\bar{V}_{\infty}$ obtained from (7.71) by V_{∞} and applying the transformation (7.36) we get the mapping of condition $Z_m=Z_m^*$ into the B-plane.

7.2.6 Determining the probe orbit based on two parameters

As mentioned above, the target points for two specified flyby conditions can be easily found provided that these conditions are represented by two circles on the sphere of departure velocities (for example, two constant-energy conditions or one constant-energy and the specified inclination). In the case of more complicated conditions, the solution can be obtained by one of two methods. One of these is to search for a general equation, which proves to have the degree not lower than four; the roots of this equation are to be found and analyzed as to their physical feasibility. This method is used in [171] to find a trajectory with $r_\pi=r_\pi^*$ and $i^*=90°$ by solving a four-degree equation. The other method consists in a one-dimensional search for zeros of the specified function using relationships suggested in this chapter. Consider the most typical problems.

1. $r_\pi=r_\pi^*$, $i=i^*$
   (i) By motion along the circles (7.56), their intersections with the hyperboloid (7.58) are sought;
   (ii) By motion along the curves (7.63), their intersections with the planes (7.55) are sought.
2. $r_\pi=r_\pi^*, T=T^*$
   (i) By motion along the circles (7.48), their intersections with the surfaces (7.58) are sought;
   (ii) By motion along the curve (7.63), their intersection with the sphere

   $$(V_{nr}+V_{\infty r})^2+(V_{nm}+V_{\infty m})^2+V_{\infty b}^2=V_a^{*2}$$

   is sought.
3. $r_\pi=r_\pi^*$, $Z=Z^*$. The points of intersection of the curves (7.71) and (7.63) are sought on the sphere of departure velocities.

Applying the methods for constructing intersection points has a number of advantages: the search is made within a limited interval, each point of intersection is physically feasible, thus simplifying the analysis, and finally, the common character of the algorithms of analysis and search for the solutions is used.

7.3 Examples of application of the methods of planetocentric orbit analysis

The methods presented in Sections 7.1 and 7.2 have been widely used both to analyze the missions of automatic interplanetary stations "Mars", "Venera", "Vega", "Phobos", and "Luna", and to control the flights of spacecrafts.

We will demonstrate the methods using as an example planetocentric orbits with respect to Jupiter. This case is even more interesting considering that Jupiter's gravity field is strong enough to allow one to obtain qualitatively different trajectories of interplanetary probes. In addition, the planet has a system of natural satellites.

Let us consider a typical interplanetary trajectory of a flight to Jupiter with a launch date which is almost optimal in terms of energy (the departure velocity from Earth at infinity is 9.22 km/s) and a flight time of 815 days. In this case the Sun–Jupiter–Earth angle attains its maximum value of 10.6°.

The trajectory parameters at the moment of arrival at the planet are as follows:

- velocity at infinity $V_\infty = 6.62$ km/s;
- distance from Earth to Jupiter $r_{SJ} = 790$ million km;
- angle between the vectors $\bar{V}_\infty$ and $\bar{r}_{SJ}$ is 56°;
- inclination of vector $\bar{V}_\infty$ to the ecliptic is 1°;
- longitude of vector $\bar{V}_\infty$ in the ecliptic measured from the vernal equinox of the Earth is 183°.

Calculations were performed in the B-planes constructed based on vectors $\bar{V}_\infty$, $\bar{r}_{SJ}$.

To illustrate the analysis of hitting trajectories, Figure 7.17 presents the B-plane on which parallels of Jupiter, the Equator, North and South poles are shown. Concentric circles show the entry angles into Jupiter's atmosphere at the height of 1000 km ($R_J = 69880$ km).

Also shown here are the subsolar S and sub-Earth T points (points where vectors Jupiter–Sun and Jupiter–Earth intersect the planetary surface), the terminator, and the boundary of the zones of direct observability of the Earth. The aiming points lying to the right of the axis $p_S - p_N$ correspond to the trajectories entering in the direction of Jupiter's rotation, and the points lying to the left correspond to the trajectories with the opposite direction. The velocity of entry into the atmosphere with respect to a non-rotating planet is 60 km/s. The Sun–Jupiter–Earth angle is small, the terminator and the boundary of the zone of direct observability are close to one another, so the landing site, when on the illuminated side of the planet, will always occur on the visible part of Jupiter.

Figure 7.18 shows the isolines of angular elements of the orbits of flyby trajectories, where i is the inclination of the planetocentric orbit to Jupiters orbital plane; Ω is the longitude of the ascending node on the orbital plane of Jupiter measured

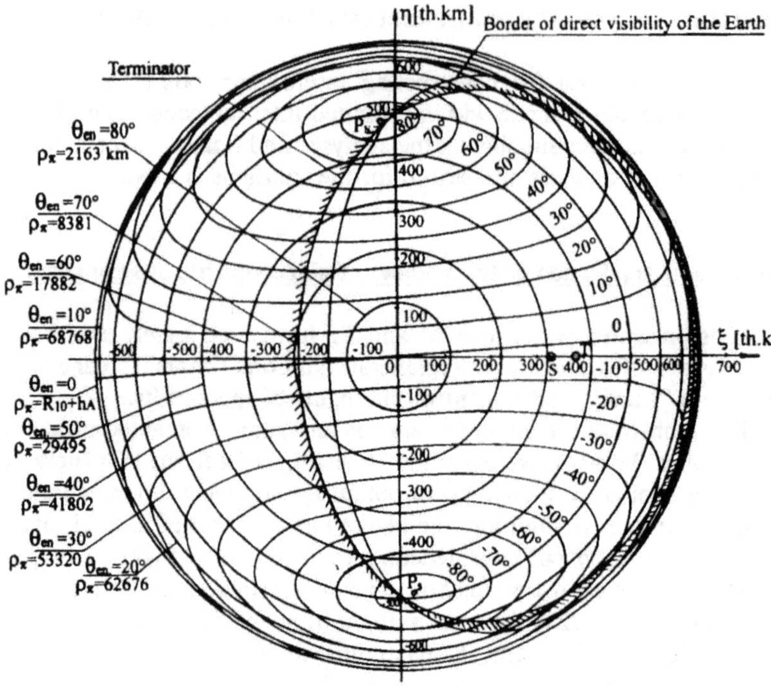

Figure 7.17 Isolines of the pericentral radius, the entry angle of the hitting trajectories ar isolines of joviographic latitude in B-plane.

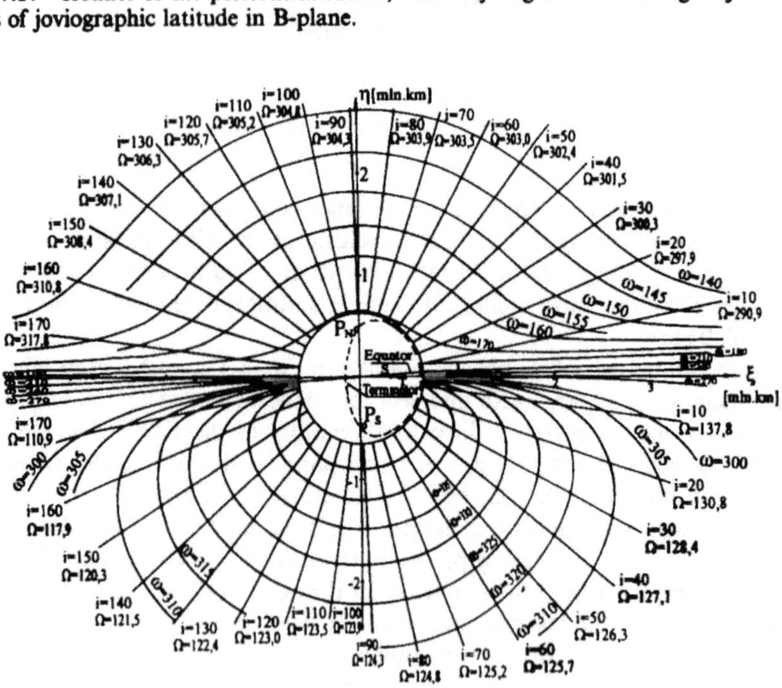

Figure 7.18 Isolines of inclination and pericentral argument of the trajectory approachi Jupiter (in B-plane).

from the radius Jupiter–Sun in the counterclockwise direction viewed from the North pole of the ecliptics; ω is the pericentral argument measured from Ω in the direction of motion. These relationships allow an estimate of the orbital elements of artificial satellites of Jupiter for different aiming points for coplanar and apsidal transfer.

The mapping of the orbits of the Galilean satellites onto the B-plane is presented in Figure 7.19. Arrows show the direction of satellite movement.

The orbit of each satellite is shown twice (by solid and broken lines), because each point of the satellite orbit can be reached by flying around the planet in two different directions (Section 7.1). In this case, the aiming distances differ in magnitude, but the flyby planes coincide. Note that most of the orbital points lie along the same straight line. This can be explained by the vector $\bar{V}_\infty$ lying virtually in Jupiter's equatorial plane (its inclination $\bar{V}_\infty$ to the equator is $\sim 0.2°$). Each point of the curves is associated with a certain true anomaly of the satellite motion and the time the satellite passes through this point. This allows one to easily determine the encounter point and the velocity vector $\bar{V}_{\infty sat}$ with respect to the gravity center of the satellite.

Figures 7.20 and 7.21 show the mapping of shadowed zones with respect to the directions to the sun and Canopus Star (shown by hatching). The zones near "ξ" and "η" axis correspond to the first solutions, i.e. entry into the zone is accomplished mainly on the incoming legs of the hyperbolas. For the opposite zones, intersection with the cylinder ($\alpha^* = 0$) or cone ($\alpha^* = 20°$) of the shadow takes placé after the pericenter has been passed. Mappings of the circles of entry or leaving the shadow, corresponding to the given distance from the center of Jupiter, are shown within the zones.

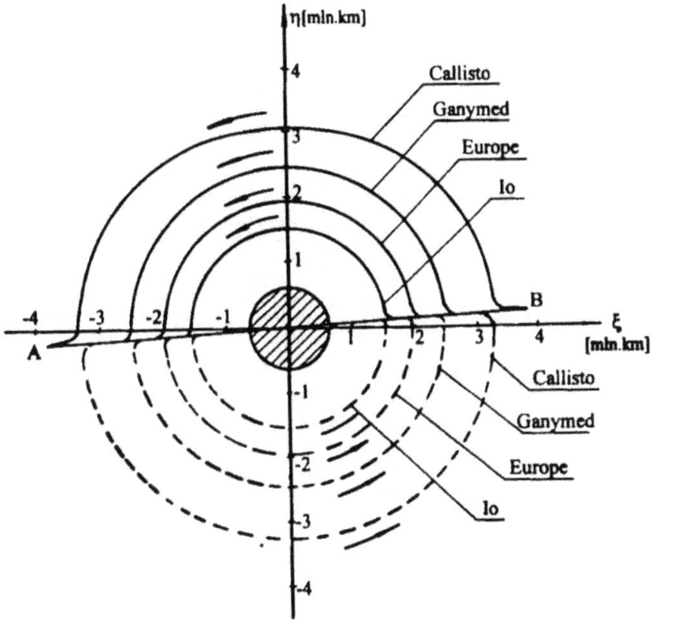

Figure 7.19.

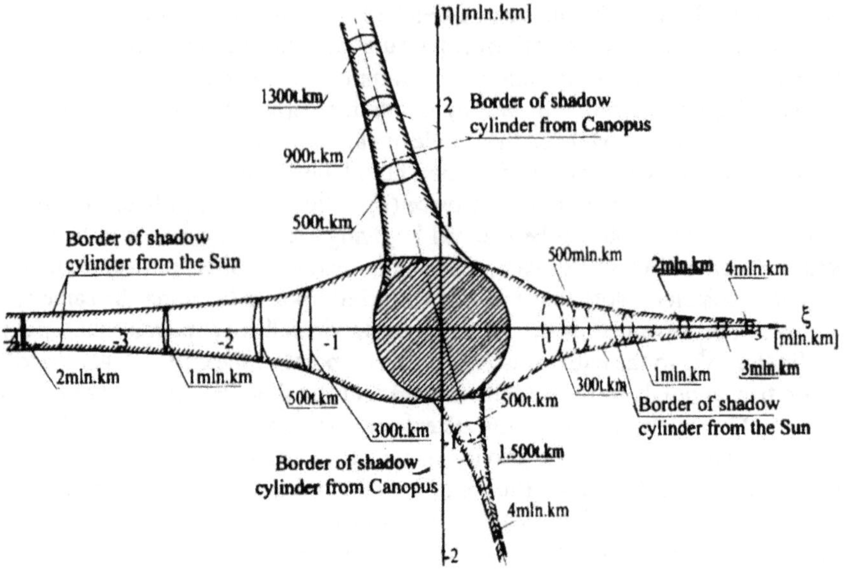

Figure 7.20 Zones of shadow from the Sun and Canopus, $\alpha^* = 0$.

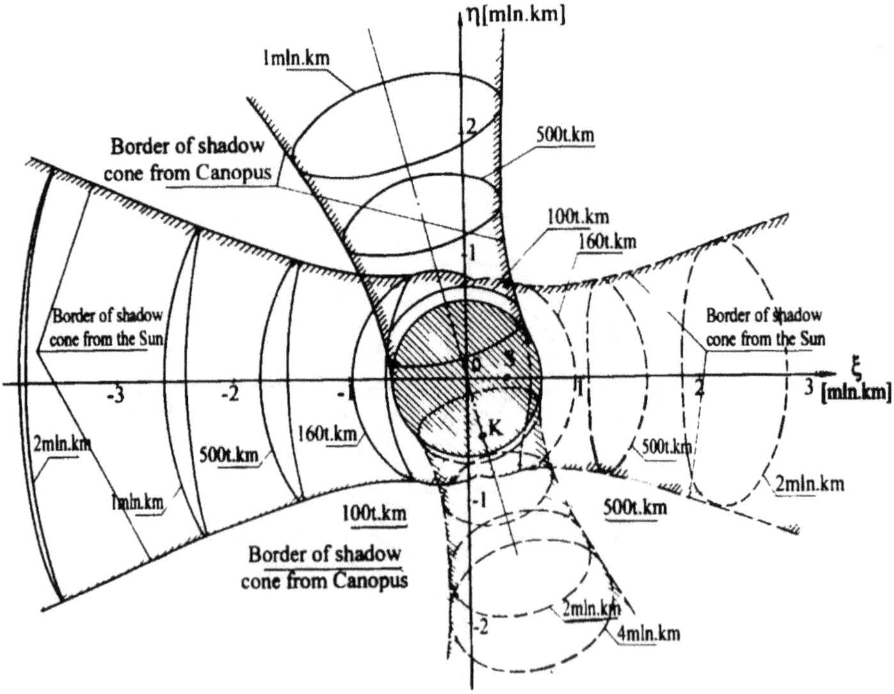

Figure 7.21 Zones of shadow from the Sun and Canopus, $\alpha^* = 20°$.

Now let us consider possible orbits of the interplanetary probe.

Figure 7.22 presents a mapping onto the B-plane of the parameters of heliocentric trajectories after Jupiter flyby; the following designations are used: T is the period of rotation; ρ_π is the pericentral altitude; $i_\Omega, i_\mho$ are the inclinations with respect to the Jupiter orbital plane (i_Ω and $i_\mho$ implies that the flyby point is an ascending and descending nodes, respectively); $V_r = 0$ is the geometric locus of the points for which the heliocentric velocity vector is perpendicular to the Sun–Jupiter vector, i.e. the flyby point is an apsidal one; V_{cir}, V_{par} are the trajectories for which the post-flyby velocity is circular or parabolic, respectively; $V'_{\infty max}$ is the orbit with maximum asymptotic velocity of departure from the solar system; $V_{r max}$ is the orbit with maximum radial post-flyby velocity directed toward the Sun.

For the Earth–Jupiter trajectories that are optimal in terms of energy, as can be seen from Figure 7.22, the following characteristics of the interplanetary probe are possible:

- minimum flyby altitude with respect to the Sun amounts to ~100 million km, the probe rotation period being ~5.5 years, that is, the probe remains virtually within the Jupiter orbital plane, and the flyby of Jupiter is accomplished from the dark side with the aiming distance of ~10 million km;
- for the probe to return into the vicinity of Earth (r_π ~150 million km) the flyby will also be accomplished from the dark side, but the aiming distance is ~4 million km, and the inclination increases to ~20°;

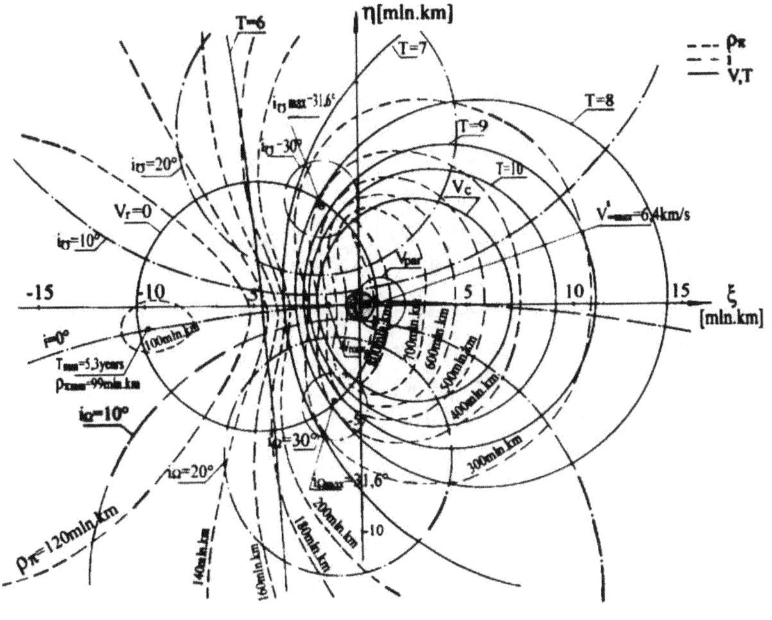

Figure 7.22 Isolines of parameters of heliocentric trajectories after flyby of Jupiter (in B-plane), $T = 815$ days.

- maximum inclination to the Jupiter orbital plane amounts to ~31.5°, the orbital apocenter after flyby remains close to the distance between the Sun and Jupiter, and the pericenter amounts to ~400 million km;
- for the spacecraft to become a galactic probe, the flyby of Jupiter can be performed from the illuminated side as near to the planetary surface as is wished.

Similar constructions were made for two types of accelerated trajectories to Jupiter with the same starting dates but different flight times:

$$2.\ T_{fl}=650 \text{ days};\quad V_{\infty E}=9.65\text{ km/s};\quad V_{\infty J}=9.21\text{ km/s};$$

$$3.\ T_{fl}=500 \text{ days};\quad V_{\infty E}=10.97\text{ km/s};\quad V_{\infty J}=13.89\text{ km/s}.$$

The approach velocity at Jupiter along the latter trajectory is somewhat greater than the velocity of the planetary motion along its orbit, which is equal to V_J 13.05 km/s.

As can be seen from Figure 7.23 the minimum flyby distance with respect to the Sun for the trajectory with $T_{fl}=650$ days decreased to ~28.5 million km, and the maximum inclination increased to ~47.5°. However, if the flyby of the Sun is accomplished within ~50 million km, the orbital inclination of ~35° can be provided; the aiming distance in this case will fall to 2–3 million km. Flybys in the immediate vicinity of Jupiter from the illuminated side always results in hyperbolic trajectories.

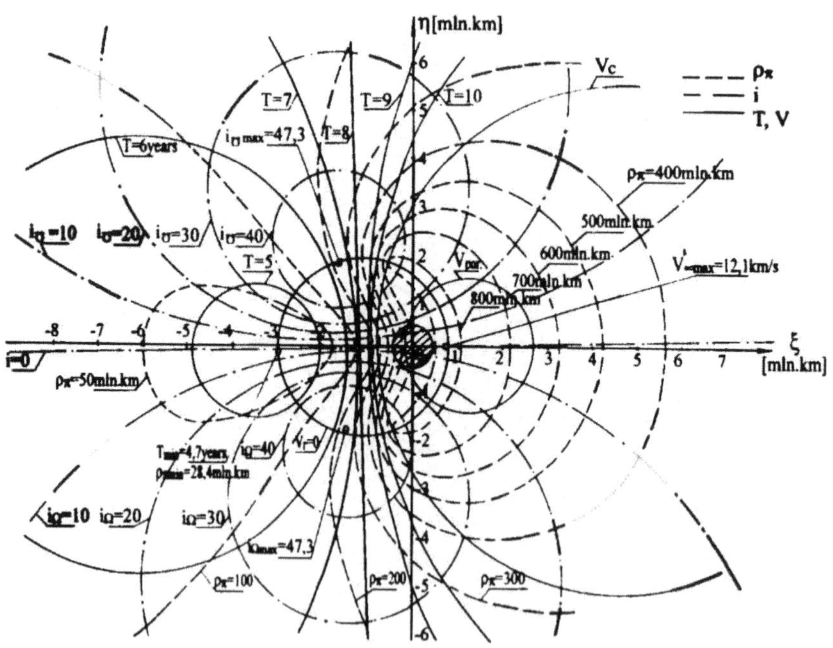

Figure 7.23 Isolines of parameters of heliocentric trajectories after flyby of Jupiter (in B-plane), $T=650$ days.

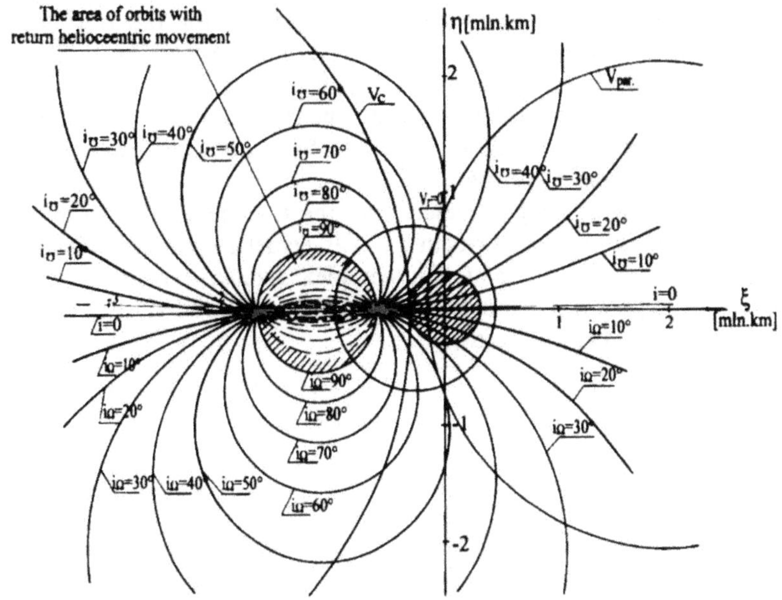

Figure 7.24 Isolines of parameters of heliocentric trajectories after flyby of Jupiter (in B-plane), $T = 500$ days.

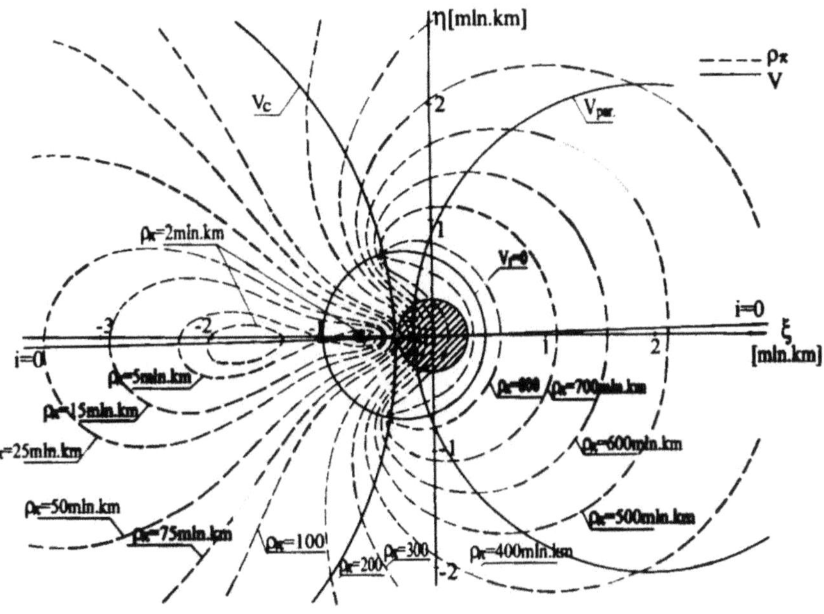

Figure 7.25 Isoline of parameters of heliocentric trajectories after flyby of Jupiter (in B-plane), $T = 500$ days.

A radical change in the isoline pattern is observed for the trajectories with $T_{fl}=500$ days, because $V_{\infty J}>V_J$. As can be seen from Figures 7.24–7.25, rectilinear orbits (points of isolines intersection $i_{\Omega}, i_{\mho}$) and orbits with reverse orbital motion (the domain within $i=90°$) can be obtained. The flyby distance with respect to the sun can be made as small as is wished, and in the field of isolines, two minima appear which lie near the rectilinear orbits.

The above mapping of the parameters onto the B-plane allows one to perform a comprehensive analysis of trajectories in the near-planetary part of the flight. Among others, such analysis can be made for complex combined schemes, taking into account various requirements and limitations on design and ballistic characteristics of the spacecraft and its components.

REFERENCES

1. Abalakin, V.K., Aksenov, E.P., Grebennikov, E.A., Demin, V.G. and Ryabov, Yu.A., 1976. *Handbook on Celestial Mechanics and Astrodynamics* [in Russian], Moscow: Nauka, 864 pp.
2. Battin, R.H., 1964. *Astronautical Guidance*, New York, McGraw-Hill.
3. Battin, R.H., 1987. *An Introduction to the Mathematics and Methods of Astrodynamics*, AIAA Education Series, AIAA, New York.
4. Brown, Ch.D., 1991. *Spacecraft Mission Design*, AIAA Education Series, AIAA, New York.
5. Davletshin, G.Z., 1980. *Powered and Gravity Assist Maneuvers of Spacecrafts* [in Russian], Moscow: Mashinostroenie, 256 pp.
6. Ephemeris of Small Planets for 1986, Leningrad: Nauka.
7. Egorov, V.A. and Gusev, L.I., 1980. *Dynamics of Flights from Earth to Moon* [in Russian], Moscow: Nauka, 543 pp.
8. Ehricke, K.A., 1960. *Space Flight*, v.1, D.Van Nostrand Concentration, Princeton.
9. El'yasberg, P.E., 1965. *Introduction to the Theory of Flight of Earth's Artificial Satellites* [in Russian], Moscow: Nauka, 540 pp.
10. Escobal, P.R., 1964. *Methods of Astrodynamics*, John Wiley and Sons, New York, 1964.
11. Fiacco, A.V. and McCormic, G.P., 1968. *Nonlinear Programming: Sequential Unconstained Minimization Techniques*, Wiley, N.Y.
12. "Principles of the Theory of Spacecraft Flights", G.S. Narimanov and M.K. Tikhonravov, Eds. [in Russian], Moscow: Mashinostroenie, 1972, 608 pp.
13. Griffin, M.D. and French, J.R., 1991. *Space Vechicle Design*, AIAA Education Series, AIAA, New York, 465 pp.
14. Grodzovskii, G.L., Ivanov, Yu.N. and Tokarev, V.V., 1975. *Mechanics of Space Flight (Problems of Optimization)* [in Russian], Moscow: Nauka, 702 pp.
15. Herrick, S., 1971. *Astrodynamics*, v.1, D. Van Nostrand Reinhold Co., London.
16. Hohmann, W., 1925. "Die Erreichbarkeit der Himmelskorper", München-Berlin.
17. Ivanov, N.M., Dmitrievskii, A.A. and Lysenko, L.N., 1986. *Ballistics and Navigation of Space Vehicles* [in Russian], Moscow: Mashinostroenie, 295 pp.
18. Ivashkin, V.V., 1975. *Optimization of Spacecraft Maneuvers with Constraints on the Distances from Planets* [in Russian], Moscow: Nauka, 392 pp.
19. Konstantinov, M.S., Kamenkov, E.F., Perelygin, B.P. and Bezverbyi, V.K., 1989. *Space Flight Mechanics* [in Russian], Moscow, Mashinostroenie, 407 pp.
20. Kubasov, V.I. Dashkov, A.A., 1979. *Interplanetary Flights* [in Russian], Moscow: Mashinostroenie, 272 pp.
21. Lawden, D.F., 1963. *Optimal Trajectories for Space Navigation*, Butterworths, London.
22. Levantovskii, V.I., 1980. *Space Flight Mechanics in Elementary Presentation* [in Russian], Moscow, Nauka, 511 pp.
23. Markus, M. and Mink, Kh., 1972. *Review of the Theory of Matrices and Matrix Inequalities* [in Russian], Moscow: Nauka.
24. Marsden, B.G., 1982. *Catalogue of cometary orbits IAU Central Bureau for Astronautical Telegrams*, fourth edition.

25. Mishin, V.P., Bezverbyi, V.K., Pankratov, B.M. and Shcheverov, D.N., 1985. ***Principles of Designing Flying Vehicles (Transportation Systems)*** [in Russian], Moscow, Mashinostroenie, 360 pp.
26. Okhotsimsky, D.E. and Sikharulidze, Yu.G., 1990. ***Foundations of Spaceflight Mechanics*** [in Russian], Moscow: Nauka.
27. Rozenfel'd, B.A., 1966. ***Multidimensional Spaces*** [in Russian], Moscow, Nauka.
28. Ruppe, H.O., 1966. ***Introduction to Astronautics***, Vol. 1, Academic Press, New York and London, 612 pp.
29. Ruskol, E.L., 1986. ***Natural Planetary Satellites*** [in Russian], ser. Astronomiya Vol. 28, (Itogi nauki i tekhniki. VINITI AN SSSR). Moscow.
30. Solov'ev, Ts.V. and Tarasov, E.V., 1973. ***Forecasting of Interplanetary Flights*** [in Russian], Moscow, Mashinostroenie, 400 pp.
31. Tarasov, E.V., 1970. ***Algorithm of Optimal Spacecraft Designing*** [in Russian], Moscow, Mashinostroenie, 364 pp.
32. Tarasov, E.V., 1977. ***Kosmonautics*** [in Russian], Moscow, Mashinostroenie, 215 pp.
33. Tarasov, E.V., Konstantinov, M.S., Labunsky, A.V. *et al.*, 1972. ***Variational Methods in Spacecraft Designing*** [in Russian], MAI-Publisher, 220 pp.
34. Tsander, F.A., 1964. ***Flights to Other Planets (Theory of Interplanetary Flights) 1924–25.*** In the book ***Pioneers of Space Technology***, Selected Works [in Russian], Moscow, Nauka, 671 pp.
35. Tsiolkovskii, K.E., 1911–12. ***Space Exploration by Jet Devices*** [in Russian], Sobr. soch. vol. II, Moscow, 455 pp.
36. Crocco, C.A., 1956. ***One-year Exploration Trip Earth–Mars–Venus–Earth***, Proc VII Intern. Astron. Congress, Rome.
37. Battin, R.H., 1959. The Determination of Round-Trip Planetary Reconnaisance Trajectories, *Journal of the Aero/Space Sciences*, **26**(9), 545–567.
38. Sohn, R.L., 1964. Venus swingby mode for manned Mars mission, *Journal of Spacecraft and Rockets*, **1**(5), p. 565.
39. Hollister, W.M. and Prussing, J.E., 1965. "Optimum Transfer to Mars via Venus", *AIAA Preprint No. 65-700.*
40. Minovich, M.A. Jr., 1963. The determination and characteristics of ballistic interplanetary trajectories under the influence of multiple planetary attractions, *JPL Tech. Rep.* N **32**, 464.
41. Gedeon, G.S., 1962. "Round Trip Trajectories to Mars and Venus" *Nortrop Space Laboratories, NSL 62-83.*
42. Young, A.C., 1966. *Multiple Planet Flyby Missions to Venus and Mars in 1975 to 1980 Time Period*, TM X-53511, NASA.
43. Mityaev, Yu.I., 1973. *Trajectories of Close Passage of Mars and Venus with Return to Earth*, Report at the Conf. on Celestial Mechanics and Astrodynamics [in Russian], Moscow: 1967. Collection of Papers "Present-Day Problems of Celestial Mechanics and Astrodynamics", Moscow: Nauka, 256 pp.
44. Gillespie, R.W. and Ross, S., 1967. "Venus-swingby mission mode and its role in the manned exploration of Mars", *Journal of Spacecraft and Rockets*, **4**(2), p. 170.
45. Van der Veen, A.A., 1969. "Triple planet ballistic flyby of Mars and Venus", *Journal of Spacecraft and Rockets*, **6**(4), p. 383.
46. Ross, S., 1963. *A Systematic Approach to the Study of Non-stop Interplanetary Round Trips*, AAS Paper 63-007.
47. Ehricke, K.A., 1965. "Interplanetary Maneuvers in Manned Helionautical Missions", *AIAA Preprint No. 65-695.*

48. Hollister, W.M., 1969. Periodic Orbits for Planetary Flight *J. Spacecraft and Rockets*, **6**(4), pp. 366–369.
49. Hollister, W.M. and Menning, M.D., 1970. "Periodic Swing-by Orbits between Earth and Venus" *J. Spacecraft and Rockets*, **7**(10), pp. 1193–1199.
50. Hollister, W.M., 1969. "Castles in Space", *Astronautica Acta*, **14**(2), pp. 311–315
51. Ragsac, R.V. and Titus, R.R., 1963. Optimization of interplanetary stopover missions, *AIAA J.* N **1**, pp. 1861–1864.
52. Breakwell, J.V., Gillespie, R.W. and Ross, S., 1961. Researches in Interplanetary Transfer, *ARS Journal*, **31**(2), pp. 201–208.
53. Titus, R.R., 1965. Powered Flyby of Mars, *Astronautica Acta*, **11**(5).
54. Sturms, F.M. and Cutting, E., 1965. Trajectory Analysis of a 1970 Mission to Mercury via a Close Encounter with Venus, *AIAA Preprint No. 65-90*.
55. Casal, F.G. and Ross, S., 1965. The Use of Close Venusian Passages During Solar Probe Missions, *AAS Preprint 65-31*.
56. Niehoff, J.C., 1966. Gravity-assist trajectories to solar system targets, *J. Spacecraft and Rockets*, **3**(9).
57. Clarke, V.C., Jr., Bolman, W.E., Fetis, P.H. and Roth, R.Y., 1966. "Design Parameters for Ballistic Interplanetary Trajectories, Part II: One-Way Transfers to Mercury and Jupiter" *Jet Propulsion Laboratory, Pasadena, CA, Technical Report 32-77*.
58. Roberts, D.L., 1967. "The requirements of unmanned space missions to Jupiter". *Raumfahrtforschung*, Bd. **11**, No. 1.
59. Bauze, V.R., Dashkov, A.A. and Kubasov, V.N., 1968. Trajectories with Planetary Flybys with Return to Earth [in Russian], *Kosmich. Issl.*, **4**(6).
60. Ross, St., 1965. *Synthesis of Trajectories for Studying Interplanetary Operations*, in *Recent Developments in Space Flight Mechanics*, AAS Science and Technology Series, Proceedings of the AAAS/AAS Special Aeronautics Symposium, Bercley, California.
61. Kislik, M.D., 1964. Spheres of Activity of Large Planets and Moon [in Russian], *Kosmich. issled*. **2**(6).
62. Lawden, D.F., 1954. "Perturbation maneuveres" *J. of the British Interplanetary Society*, **13**, pp. 329–334.
63. Gobets, F.B., 1963. Optimal transfer between hyperbolic asymptotes, *AIAA J.* **1**(9).
64. Walton, J.M., Marchal, C. and Clup, R.D., 1975. Synthesis of the Optimal Transfers between Hyperbolic Asymptotes, *AIAA Journal*, **13**(8), pp. 980–988.
65. Bschorr Oscar, 1966. "Bahntransitionen in bewegen Schwerefeldern" *Jahrb. Wiss. Ges. Luft-und Raumfahrt*, Braunschweig, pp. 253–257.
66. Labunsky, A.V. and Sukhanov, K.G., 1979. Some Problems of Forming Spacecraft Trajectories with Gravity Assist Maneuver, in Problems of Automatization of Designing Flying Vehicles [in Russian], *Trudy MAI*, iss. **503**, pp. 72–77.
67. Broucke, R.A., 1988. *The celestial mechanics of gravity assist*, AIAA/AAS Astrodynamics conference, Minneapolis, Minnesota, 69–77.
68. Ronald, A., Randolph, J., 1990. Hypersonic Maneuvering to Provide Planetary Gravity Assist, *AIAA Pap.* N **0539**, pp. 1–11.
69. Lohar Fayyaz, A., Matescu Dan, 1992. *Aero-gravity assist for planetary missions*. 43rd Congress of the IAF.
70. Sukhanov, A.A., 1993. Double Planetary Swingby in Space Mission [in Russian], Preprint IKI RAN, N 1858, pp. 1–38.
71. Sohn, R.L., 1959. A proposed Kepler Diagram, *ARS Jour.* **29**: 51–54.
72. Altman, S.P. and Pistiner, J.S., 1961. Hodograph Analysis of the Orbital Transfer Problem for Coplanar, Nonaligned Elliptical Orbits, *ARS Jour.* **31**: 1217–1225.

73. Jensen, J., Townsend, G., Kork, J. and Kraft, D., 1962. *Design Guide to Orbital Flight*, McGraw Hill Book Company, N.Y.
74. Heacock, F.A., 1964. *Graphics in Space Flight*, McGraw Hill Book Company, N.Y.
75. Belyakov, A.I., 1973. Graphical–Analytical Method for Studying Spacecraft Motion [in Russian], Mashinostroenie, Moscow.
76. Egorov, V.A., 1958. Certain Problems of Moon Flight Dynamics Russian Literature on Satellites. Part 1, International Physical Index, Inc., N.Y.
77. Breakwell, J.V. and Perko, L.M., 1966. *Matched Asymptotic Expansions, Patched Conics and the Computation of Interplanetary Trajectories* AIAA Progress in Astronautics and Aeronautics. Methods in Astrodynamics and Celestial Mechanics. Vol. 17, ed. by R.L. Duncombe and V.G. Szebenhely, Academic Press, N.Y., pp. 159–192.
78. Stumpff, K. and Weiss, E.H., 1968. Applications of an N-Body Reference Orbit *J. Astronaut. Sci.* **15**(5), 257–261.
79. Byrnes, D.V. and Hooper, H.L. 1970. Multi-conic: a fast and accurate method of computing space flight trajectories, *AIAA Paper*, N 1062.
80. Hazelrigg Jr., George, A., 1970. "Optimal interplanetary trajectories for chemially propelled spacecraft" AIAA Paper, N 1039, 1–8.
81. Wilson, S.W., 1970. A Pseudostate Theory for the Approximation of Three-Body Trajectories AIAA aper N 70-1071, AIAA Astrodynamics Conference, Santa Barbara, California.
82. Bayliss Stephen, 1971. Precision targeting for multiple swingby interplanetary trajectories *AIAA Paper, 1971*, N 191, pp. 1–9.
83. Tarasov, E.V. and Labunsky, A.V., 1973. *Optimization of Ballistic-Design Characteristics of an Interplanetary Spacecraft to Be Orbited as an Artificial Satellite of the Target Planet* [in Russian], in *Proc. of VI Conference Dedicated to the Memory of K.E. Tsiolkovsky*, Kaluga, Section: Space-Flight Mechanics, Moscow: IIET AN SSSR.
84. Lancaster, J.E. and Alleman, R.A., 1972. Numerical analysis of the asymptotic two-point boundary value solution for N-body trajectories, *AIAA Paper*, N **49**, pp. 1–10.
85. D'Amario and Edelbaum, T.N., 1974. Minimum Impulse Three-Body Trajectories *AIAA J.* **12**(4), upper 455–462.
86. Friedlander, A.L., 1974. "(MULIMP) Multiple Impulse Trajectory and Mass Optimization Programm" Science Application, Inc. Report N SAI I-120-383-74.
87. Sukhanov, K.G., 1975. *A Method for Constructing Isoline Fields For Trajectories with Gravity Assist Flybys of Planets*" [in Russian], in *Determination of Spacecraft Motion*, Nauka, Moscow.
88. Labunsky, A.V. and Leshchenko, A.V., 1976. "Algorithm for the Calculation and Optimization of Recurrent Trajectories Containing Flybys" [in Russian], in "Proc. of X Conference Dedicated to the Memory of K.E. Tsiolkovsky, Kaluga, 1975, Section: Space-Flight Mechanics, Moscow: IIET AN SSSR.
89. D'Amario, L.A., Byrnes, D.V., Sacket, L.L. and Stanford, R.H., 1979. Optimization of Multiple Flyby Trajectories, *AAS Paper* **79**, 162.
90. D'Amario, L.A., Byrnes, D.V. and Stanford, R.H., 1981. A new Method for Optimizing Multiple Flyby Trajectories, *J. of Guidance and Control*, **4**(6), pp. 591–596.
91. Kryczynski, L.R. and Boden, D.G., 1982. Multiple intercept trajectories "Astrodyn. 1981. Proc. AAS/AIAA Astrodyn. Conf., North Lake Tahoe, Nev., Aug. 3–5. 1981 Pt.2 San Diego Calif. pp. 547–561.
92. Solander, J.K. and Fiengold, H., 1982. "A systematic method generating Galilean satellite transfer for orbiter/lander missions" "Astrodyn. 1981. Proc. AAS/AIAA Astrodyn. Conf., North Lake Tahoe, Nev., Aug. 3–5. 1981 Pt.1" San Diego Calif. pp. 223–236.

93. D'Amario, L.A., Byrnes, D.V. and Stanford, R.H., 1982. Interplanetary Trajectory Optimization with Application to Galileo, *J. of Guidance and Control*, **5**(5), pp. 465–471.
94. Labunsky, A.V., Kotin, V.A., Papkov, O.V. and Sukhanov, K.G., 1982. *Graphical–Analytical Method of Studying Spacecraft Trajectories with Gravity Assist Maneuvering* [in Russian], in *Designing of Spacevehicles and Systems*, Trudy MAI, Moscow, pp. 15–20.
95. Sergeyevsky, A.B., Byrnes, D.V. and D'Amario, L.A. "*Application of the rectilinear impact pseudostate method to modeling of third-body effects on interplanetary trajectories*" N.Y. Rep./AIAA; N83-0015.
96. Konstantinov, M.S., Fedotov, G.G., Labunsky, A.V., Cherepanov, N.V. *et al.*, 1985. *Analog System for Analysis of Interplanetary Flight Trajectories* [in Russian], Proc. of IV Conference Dedicated to the Memory of Yu.A. Gagarin, Moscow, Section "Flight Mechanics", Moscow: Nauka, pp. 56–57.
97. Labunsky, A.V., 1984. "*An Algorithm for Automated Selection of Spacecraft Path with Gravity Assist Maneuvers*" Tr. MAI, *Designing of Spacecrafts and Their Systems* [in Russian], Moscow: MAI, pp. 54–57.
98. Page, J., Rodriguez, J., Verges, M. and Villa, R., 1986. "Generation of asteroids flyby and rendezvous trajectories" *IAF Pap.*, N **221**, 1–8.
99. Wetzel, D., 1986. Optimization de missions interplanetaries. Application aux missions delta V-EGA *Aeron. et Astronaut.*, N **6**, 41–47.
100. Bonneau, F. and Fontaine, L., 1987. Optimization of interplanetary trajectories *IAF Prepr.* N 331 1–8.
101. Sauer C.G., "Optimization of Interplanetary Trajectories with Unprowered Planetary Swingbys" AAS Paper 87-424, AAS/AIAA Astrodynamics Specialists Conference, Kalispell, Montana, August 10–13, 1987.
102. Kiederon, K. and Sweester, T.H., "A comparision between one step and other multiconic trajectory propagation methods" AIAA/AAS Astrodynamics conference, Minneapolis, Minnesota, Aug. 15–17, 1988, 391–400.
103. Labunsky, A.V., 1988. "A Universal Algorithm for Desining Complicated Spacecraft Path in the Solar System" [in Russian], Izd. MAI, Moscow.
104. Byrnes, D.V., 1989. "Application of the Pseudostate Theory to the Three-Body Lambert Problem", *J.Astronaut. Sci.* **37**(3), pp. 221–232.
105. D'Amario Lois, A., 1989. "Trajectory Optimization Software for Planetary Mission Design", *J. Astronaut. Sci.* **37**(3), pp. 213–220.
106. Sauer Carl., Jr., 1989. MIDAS: Mission Design and Analysis Software for the Optimization of Ballistic Interplanetary Trajectories, *J. Astronaut. Sci.* **37**(3), pp. 251–259.
107. Sweetser, T.H., 1989. "Some Notes on Applying the One-step Multiconic Method of Trajectory Propagation, *J. Astronaut. Sci.* **37**(3), pp. 233–250.
108. Skinner David. L., Bass Laura. E., Byrnes Dennis, V., Cheng Jeannie, T., Fordyce Jes, E., Knocke Philip, C., Lyons Daniel, T., Pojman Joan L., Stetson Dauglas, S. and Wolf Aron. A., 1990. *Mission Design applicationsof QUICK* AIAA/AHA Astrodyn. Conf., Portland, Ore. 950–954.
109. Longuski, J.M. and Williams, S.N., 1991. Automated design of gravity-assist trajectories to Mars and outer planets, *Celest. Mech. and Dyn. Astron.* **52**(3), pp. 207–220.
110. Sokolov, L.L. and Titov, V.B., 1991. Spacecraft Trajectories with Gravity Assist Maneuvers [in Russian], *Vestn. LGU. Ser. 1.* N **3**. pp. 111–114.
111. Abramovich, S.K., Ageeva, T.D., Akim, E.L., Zaslavskii, G.S., Ivanov, N.M., Kazanskii, M.A., Lyaskovskaya, V.I., Morskoi, I.M., Papkov, O.V., Polyakov, V.S., Stepan'yants, V.A., Sukhanov, K.G., Tikhonov, V.F. and Kheifets, V.N., 1979. Ballistics and

Navigation of Interplanetary Probes "Venera-11" and "Venera-12" [in Russian], *Kosmich. issledovaniya*, vol. **XVII**, iss. 5, pp. 670–677.

112. Ageeva, T.D., Akim, E.L., Zaslavskii, G.S., Ivanov, N.M., Kazanskii, M.A., Kolyuka, Yu.F., Luk'yanov, S.S., Lyaskovskaya, V.I., Morskoi, I.M., Mottsulev, B.P., Papkov, O.V., Polyakov, V.S., Stepan'yants, V.A., Sukhanov, K.G., Tikhonov, V.F., Kheifets, V.N. and Khristoforov, Zh.I., 1983. Ballistics and Navigation of Interplanetary Probes "Venera-13" and "Venera-14" [in Russian], *Kosmich. issledovaniya*, vol **XXI**, ss. 2, pp. 154–162.
113. Kremnev, R.S., Rogovskii, G.N. and Sukhanov, K.G., 1988. Fobos: Perfect Readiness [in Russian], *Nauka, v SSSR*, no. **1**, pp. 2–8.
114. Williams, S.N. and Longuski, J.M., 1991. Low Energy Trajectories to Mars via Gravity Assist from Venus to Earth, *J. Spacecraft and Rockets*. **28**(4), pp. 486–488.
115. Rall, C.S. and Hollister, W.M., 1991. Periodic Swing-by Orbits Connecting Earth and Mars *J. Spacecraft and Rockets*, **8**(10), pp. 1017–1020.
116. Friedlander, A.L., Niehoff, J.C., Byrnes, D.V. and Longuski, J.M., 1986. "Circulating Transportation Orbits Between Earth and Mars", *AIAA Paper 86-2009*.
117. Labunsky, A.V., 1988. *Use of Earth–Mars–Earth and Earth–Venus–Earth Trajectories for the Formation of Recurrent–Periodical Paths in the Solar System* [in Russian], MAI, in *Applied Celestial Ballistics and Spacecraft Designing*, pp. 23–27.
118. Labunsky, A.V., 1991. *An Investigation of orbits with multiple flybys of the Earth and Mars* [in Russian], *Cosmich. Issled*. **29**(3), pp. 390–396.
119. Byrnes, D.V., Longuski, J.M. and Aldrin, B., 1993. Cycler orbits between Earth and Mars, *J. Spacecraft and Rockets* **30**(3), pp. 334–336.
120. Uphoff, C.W. and Crouch, M.A., 1993. Lunar cycler orbits with alternating semi-monthly transfer windows, *J. Astronaut. Sci.* **41**(2), pp. 189–195.
121. Ivashkin, V.V. and Tupitsyn, N.N., 1971. On the Use of Lunar Gravity Field for Orbiting a Spacecraft as an Artificial Satellite of Earth [in Russian], *Kosmich. issled. 1971*, **9**(2), p. 163.
122. Ross, D.J., 1980. Material capture by double lunar gravity assist, *AIAA Pap.* N **1673**, 7 pp.
123. Kogan, A., 1986. "Orbits with Periodic Flights Around the Moon and their Use in Very Long Baseline Interferometry", *Cosmic Research*, **24**(1), pp. 43–48 (English).
124. Marsh, S.M. and Howell, K.C., 1988. *Double lunar swingby trajectory design*, AIAA/AAS Astrodynamics conference, Minneapolis, Minnesota, pp. 69–77.
125. Hesugi, K., Hayashi, T. and Matsuo, H., 1988. Muses-A double lunar swingby mission, *Acta Astronaut*., **17**(5), pp. 495–501.
126. Farquhar, R.W., 1990. Halo-Orbit and Lunar-Swing by Missions of the 1990's 41st Congress of the IAF., Germany.
127. Lidov, M.L., Lyakhova, V.A. and Teslenko, N.M., 1992. Transfer trajectories Earth Moon-halo-orbit near the L2 point in the Earth–Sun System [in Russian], *Cosmich. Issled*. **30**(4), pp. 435–454.
128. Smirnov, V.V., Egorov, V.A. and Sazonov, V.V., 1995. Papameter Optimization of a Spacecraft Perturbation Maneuver in the Sphere of Lunar Activity [in Russian], *Cosmich. Issled*. **33**(3), pp. 298–306.
129. Uesugi, K., Kawaguchi, J., Shuto, M., Ishii, N., Kamimura, M., Ishii, S., Kimura, M. and Tanaka, K., 1990. *Trajectory design and control of a multiple lunar swingby orbit for MUSES-A*, International Simposium on Space Technology and Science, Tokyo.
130. Kawaguchi, J., Yamakawa, H., Uesugi, T. and Matsuo, H., 1995. "On making use of lunar and solar gravity assists in LUNAR-A, PLANET-B missions", *Acta Astronaut*. **35**. Suppl. pp. 633–646.

131. Sukhanov, K.G., 1975. Determination of Coordinates and Orientation of a Space Vehicle on Planetary Surface [in Russian], *Kosm. Issledov.*, **13**(5).
132. Bender, D.F., 1976. "Ballistic trajectories for Mercury orbiter missions using optimal Venus flybys, a systematic search", *AIAA Pap.*, N **796**, pp. 1–12.
133. Yen, C.L., 1989. Ballistic Mercury Orbiter Mission via Venus and Mercury Gravity Assists, *J. Astronaut. Sci.* **37**(3), p. 416.
134. Solov'ev, Ts.V., Shmakova, N.F., 1974. Optimization of the Trajectory to Fly to Mercury in the Gravity Field of Venus [in Russian], *Kosmich. issled.* **12**(6).
135. Nelson, R.M., Horn, L.J., Weiss, J.R. and Smythe, W.D., 1995. Hermes global orbiter: a discovery mission in gestation, *Acta Astronaut.* **35**. Suppl. pp. 387–395.
136. Bond, V.R. and Anson, K.W., 1972. Trajectories that fly by Jupiter and Saturn and return to Earth, *J. Spacecraft and Rockets*, **9**(6).
137. Isakovich, A.A. and Kirpichnikov, S.I., 1974. Some Cases of Interplanetary Flights Using Gravity Assist [in Russian], *Kosmich. issled.* **12**(5).
138. Hollenbeck, G.R., 1975. *New Flight Techniques for Outer Planet Missions*, AAS/AIAA Astrodynamics Conference, Nassau, Bahamas, Paper N 25-087.
139. Kotin, V.A., Labunsky, A.V. and Leshchenko, A.V., 1978. *Some Solutions to the Problems of Interplanetary Flight along Earth – Mars – Jupiter Trajectory* [in Russian], Proc. XI Conference Dedicated to the Memory of K.E.Tsiolkovsky, Section "Space Flight Mechanics", IIET AN SSSR Moscow, p. 6.
140. Georgiev, K.G. and Papkov, O.V., 1978. Trajectories of Flight to Jupiter with the Use of the Gravity Field of Mars [in Russian], *Kosmich. issled.* **16**(1).
141. Georgiev, K.G. and Papkov, O.V., 1980. Examination of a Trajectory of Flight to Jupiter with a Flyby of Earth [in Russian], *Kosmich. issled.* **18**(1).
142. Nock, K.T., 1980. "Interplanetary trajectory options for project Gallileo", *AIAA Pap.*, N **1697**.
143. Sergeyevsky, A.B. and Snyder, G.C., 1982. *Interplanetary Mission Design Handbook, Earth to Jupiter Ballistic Mission Opportunities, 1985–2000*, Jet Propulsion Laboratory, Pasadena, CA., Publication 82–43, Vol. 1, Part 3.
144. Yokota Hiroki, Tanabe Toru, 1983. *Interplanetary mission designs* by *multiple Earth–Venus–Jupiter swingbys*. Inst. Space and Astronaut. Sci. Rept. N S.P.I., pp. 73–85.
145. Labunsky, A.V., 1983. *On the Interplanetary Flight Trajectories with a Gravity Assist Maneuver at the Launch Planet* [in Russian], in *Dynamics, Control, Navigation*, MAI, Moscow, pp. 10–14.
146. Tolyarenko, N.V. and Chumakov, V.A., 1983. *On the Problem of the Use of Multipurpose Spacecrafts for Flights to Outer Planets* [in Russian], in *Ideas of F.A.Tsander in the Development of Rocket and Space Science and Engineering*, Moscow: Nauka, pp. 166–170.
147. D'Amario, Louis, A., Byrnes, Dennis, V., Johannesen, Jennie, R. and Nolan Brian, G., 1989. Galileo 1989 VEGA Trajectory Design, *J. Astronaut. Sci.*, **37**(3), pp. 281–306.
148. Longmann, R.W. and Schneider, A.M., 1970. "Use of Jupiter's moons for gravity assist" *J. Spacecraft and Rockets*, **7**(5).
149. Uphoff, C., Roberts, Ph. and Friedman, L., 1974. "Orbit Design Concept for Jupiter Orbiter Missions", *AIAA Pap. 74-781*.
150. Beckman, J.C., Hyde, J.R. and Rasool, S.I., 1974. Exploring Jupiter and its satellites with an orbiter, *Astronaut. and Aeronaut.*, **12**(9), 24–35.
151. Beckman, J.C. and Miner, E.D., 1975. Jovian system science issues and implications for Mariner Jupiter orbiter mission, *AIAA pap. n 1141*.
152. Friedman Louis, D. and Nunamaker Robert, R., 1975. *Misson design of a Pioner Jupiter orbiter AIAA Pap.*, N **1135**, 1–10.

153. Labunsky, A.V., 1978. Problems of Ballistic Designing of Spacecraft for Flights to Natural Satellite of a Planet [in Russian], in *Designing and Construction of Flying Vehicles*, iss. 445, 4 pp.
154. Diehl, R.E. and Nock, K.T., 1979. "Galileo Jupiter Ecounter and Satellite Tour Trajectory Design", *AIAA Paper N 79-141*, AAS/AIAA Astrodynamics Specialist Conference, Provincetown, Massachusetts.
155. O'Neil, W., 1984. Galileo mission, *AIAA Pap.* N 2, 159 pp.
156. Wolf, A.A., 1984. A generalized strategy for tour design for the Galileo mission, *AIAA Pap.* N 2015 9 pp.
157. Kubarev, V.V. and Labunsky, A.V., 1984. *On a Scheme for Implementation of Flights to a Natural Satellite of Jupiter* [in Russian], Proc. XIII Conference Dedicated to the Memory of K.E. Tsiolkovsky, Section "Space Flight Mechanics", IIET AN SSSR Moscow, pp. 32–37.
158. Roberts Phillip, H.Jr., 1975. Trajectory design for Saturn orbiter missions in the mid-1980s, *AIAA Pap.*, N 1136, pp. 1–8.
159. Georgiev, K.G. and Papkov, O.V., 1981. Analysis of Some Schemes for Flight to Saturn [in Russian], *Kosmich. Issled.* 19(2).
160. Sergeyvsky, A.B., 1982. *Voyager 2: a grand tour of the giant planets* Astrodyn. 1981. Proc. AAS/AIAA Astrodyn. Conf., North Lake Tahoe, Nev., Pt. 2 San Diego, Calif. pp. 769–790.
161. Labunsky, A.V. and Saprykin, O.A., 1987. *On a Scheme for Flight to Saturn with the Use of Gravity Assist Maneuvers* [in Russian], in Proc. XX Conference Dedicated to the Memory of K.E.Tsiolkovsky, Kaluga, Section "Space Flight Mechanics", Moscow: pp. 28–31.
162. D'Amario, L.A., Byrnes, D.V., Diehl, R.E., Bright, L.E. and Wolf, A.A., 1989. Preliminary Design for a Proposed Saturn Mission with a Second Galileo Spacecraft, *J. Astronaut. Sci.*, 37(3), pp. 307–332.
163. Hendrics, T.C., Satin, A.L. and Tindle, E., 1976. Mission to Titan (1983–2000): an analysis of orbitters and entry vehicles, *AIAA Pap.*, No. 799.
164. Diehl, R.E., 1982. *Touring the satellites of Saturn*, Astrodyn. 1981. Proc. AAS/AIAA Astrodyn. Conf., North Lake Tahoe, Nev., Pt. 2 San Diego, Calif. 791–810.
165. Sergeyvsky, A.B, Keridge, S.J. and Stetson, D.S., 1987. *Cassini-A Missions to the Saturnian System*, AAS Paper 87-423 presented at AAS/AIAA Astrodynamics Specialists Conference, Kalispele, Montana.
166. Schilling, K. and Scoon, G., 1989. Mit Cassini zum Saturn. Kapsel landet auf Titan. *Luft-und Raumfahrt* 10(1), p. 6.
167. Huchins, E.K. and Spehalski, R.J., 1995. *The Cassini/Hugens Mission to Saturn*, 46th Congress IAF, Oslo, Norway.
168. Cesarone, R.J., Gray, D.E., Francis, K. and Potts, C.L., 1985. Voyager 2 Uranus targeting strategy, *Astrodynamics*, Proc. AAS/AIAA Astrodyn. Conf., Vail, (Colorado) Pt 2. San Diego (Calif), pp. 1331–1354.
169. Callies, R., 1991. Optimal design of a mission to Neptune, *Tagungsber.J Math.Forschunginst.*, Oberwolfach. N 23, p. 6.
170. Sims, J.A., Longuski, J.M. and Patel M.R., 1995. Aerogravity-assist trajectories to the outer planets, *Acta Astronaut.* 35, Suppl. 297–306.
171. Khavenson, I.G. and El'yasberg, P.E., 1972. On the Possibility to Use the Gravity Field of Jupiter to Fly at the Given Distance from the Sun and Leaving the Ecliptic Plane, *Kosmich. issled.*, 10(2), pp. 159–166.
172. Labunsky, A.V., 1974. *Some Solutions to the Problems of Interplanetary Flights with Gravity Assist Maneuver with Allowance Made for the Sizes of the Planetary Gravisphere*

[in Russian], Proc. II Conference Dedicated to the Memory of F.A. Tsander. Section "Astrodynamics" IIET AN SSSR.
173. Bender, D.F., 1976, Out-of-ecliptic missions using Venus or Earth assists, *AIAA Pap.*, N **189**, pp. 1–9.
174. "Development and Coordination of basic data for the preliminary design of the Mission to the Sun: Project TSIOLKOVSKY" [in Russian], Izd. MAI 1990.
175. Rijov, Yu.A., Kovtunenko, V.M., Malyshev, V.V., Morozov, N.A. and Usachov, V.E., 1994, TSIOLKOWSKI Automatic Spacecraft for the Sun and Outer Planets Exploration, *Aerospace MAI Journal*, **1**, pp. 4–10.
176. Mc. Nutt, R.L. Jr., Krimigis, S.M., Cheng, A.F., Gold, R.E., Farquhar, R.W., Roelof, R.C., Coughlin, T.B., Santo, A., Bokulic, R.S., Reynolds, E.L., Williams, B.D. and Willey C.E., 1995. Mission to the Sun: The Solar Pioneer, *Acta Astronaut*. **35**, Suppl. pp. 247–255.
177. Labunsky, A., Kotin, V., Papkov, O. and Sukhanov, K., 1995. *Mission design to the Sun: The comparative analysis of alternative routes*. 46th Congress IAF, Oslo, Norway.
178. Kovtunenko, V.M., Kremnev, R.S., Pichkhadze, K.M., Sukchanov, K.G., Papkov, O.V., Yakovlev, B.D., Kotin, V.A., Tuljakov, V.A., Galeev, A.A., Vaisberg, O.L., Oraevski, V.N. and Randolf, I., 1975. *Russian–American Project FIRE: Exploration of the Sun Corona* 46th Congr. IAF, Oslo, Norway.
179. Keridge, S., Evans, M. and Tsurutani, B., 1995. Cost-effective mission design for a small solar probe, *Acta Astronaut*. **35**. Suppl. pp. 257–266.
180. Maccone, C., 1996. *Solar Foci Missions*, IAA 2nd International Conference on low-cost Planetary Missions, Laurel, Maryland, USA.
181. Ericke, K.A., 1972. Saturn–Jupiter Rebound—a Method of High-Speed Spacecraft Ejection from Solar system, *J. of Brit. Interplan. Society*, **25**, pp. 561–571.
182. Surdin, V.G., 1985. Launching a Galactic Probe wih the Use of Multiple Gravity Assist Maneuvers [in Russian], *Astron. vestnik*. **19**(4), pp. 354–358.
183. Nock, K., 1987. *TAU—A Mission to a Thousand Astronomical Units* pap. AIAA- -87-1049.
184. Matloff, G. and Parks, K., 1988. Interstellar Gravity Assist Propulsion: a Correction and a New Application, *J. of Brit. Interplan. Society*, **41**, pp. 519–526.
185. Farquhar, R.W. and Ness Norman, F., 1972. Two early missions to the comets, *Astronaut. and Aeronaut*., **10**, pp. 32–37.
186. Stancati, M.L. and Soldner, J.K., 1981. Near Earth Asteroids a Survey of Ballistic Rendezvous and Sample Return Missions, *AAS/AIAA Rap*. 81–185.
187. Kazakova, R.K., Kotin, V.A., Papkov, O.V., Platonov, A.K. and Sukhanov, K.G., 1981. Studies of Direct Flight Trajectories to Some Periodic Comets [in Russian], *Kosmich. issled*. **19**(2), pp. 377–391.
188. Labunsky, A.V., Pankratov, A.V. and Margorin, O.K., 1981. *Prediction of Some Schemes for Flights to Comets* [in Russian], Proc. XV Conference Dedicated to the Memory of K.E.Tsiolkovsky, Section "Space Flight Mechanics", IIET AN SSSR Moscow, pp. 42–46.
189. Farquhar, R.W. and Dunham, D.W., 1982. Earth—Return Trajectory Options for the 1985–86 Halley Opportunity, *J. of Astronautical Sciences*, **30**, pp. 307–328.
190. Kotin, V.A., Labunsky, A.V., Papkov, O.V. and Sukhanov, K.G., 1982. *On the Possible Schemes of Flight to the Comets of Jupiterian Group* [in Russian], Proc. VII Confenrence Dedicated to the Memory of F.A. Tsander, Section "Astrodyanmics", Kuibyshev, 1981 IIET AN SSSR, Moscow, pp. 96–103.
191. Kotin, V.A., Labunsky, A.V., Papkov, O.V., Sukhanov, K.G. and Pigurnov, V.A., 1982. *Studies of Trajectories for Flights to Comets with Flybys of Planets* [in Russian], "Proc. XII Conference Dedicated to the Memory of Yu.A. Gagarin", Moscow, Section "Space Flight Mechanics", Nauka, Moscow, 1984, pp. 138–139.

192. Morozova, E.I., Kostin, S.N., Grishin, S.D., Uspenskii, G.R., Tselin, A.V. and Tsimbalyuk, N.M., 1982. Trajectories of Spacecraft Flights for Studying Comets [in Russian], *Bull. Inst. Theor. Astronom. AN SSSR*, **15**(4) pp. 222–224.
193. Tien Khang Uu, 1983. La navigation interplanetarie on le "flipper" cosmique, *Navigation. (Fr)*, **31**(122), pp. 227–235.
194. Uphoff, C.W., Glukman, R.E. and Stuart, J.R., 1983. Characteristics of a dual mission concept for intensive use of Moon and Mars or Moon and asteroids, *AIAA Paper* 83-0349.
195. Labunsky, A.V., Papkov, O.V. and Sukhanov, K.G., 1983. *On the Possibility of Reaching Halley's Comet* [in Russian], in *Methods of Selecting Design Parameters of Flying Vehicles*, Moscow, pp. 28–32.
196. Livanov, L.B., 1983. Flights to Asteroids with Spacecraft Maneuvers at Venus, Earth, Mars, Jupiter [in Russian], *Kosmich. issled.* **21**(6), 1983, p. 861.
197. Byrnes, D.V. and D'Amario, L.A., 1984. "Asteroid/comet mission possibilities using a Galileo spacecraft" *Advances in Astronomical Sciences*, V. **54**. Part 1 p. 71 (AAS Paper 83-309).
198. Muhonen, D., Davis, S. and Dunham, D., 1984. Alternative gravity-assist sequences for the ISEE 3 escape trajectory, *AIAA Pap.* N **1977** 9 pp.
199. Yen, C.L., 1984. Ballistic comet exploration missions options, *AIAA Pap.* N **2028** 19 pp.
200. Penzo, P.A. and Mayer, H.L., 1984. Tethers and asteroids for artifical gravity assist in the Solar system, *AIAA Pap.* N **2056** 8 pp.
201. Akim, E.L., Belyaev, N.A., Kazakova, R.K., Papkov, O.V., Platonov, A.K., Savchenko, V.V. and Sukhanov, K.G., 1984. Ballistic Aspects of Exploration of Comets [in Russian], *Doklady AN SSSR*, **275**(2), pp. 323–327.
202. Labunsky, A.V., 1984. *Analysis of Gravity Assist Potentialities of Venus in the Schemes of Flights to Asteroids*, in *Designing and Construction of Space Vehicles* [in Russian], Moscow: MAI, 4 pp.
203. Venus–Halley mission, MNTK, Paris, 1985.
204. Wallace, R.A., Blume, W.H., Hulcower, N.D. and Yen C.L., 1985. Interplanetary missions for the late twentieth century, *Journal of Spacecraft and Rockets* **22**(3), p. 316.
205. Kazakova, R.K., Kotin, V.A., Papkov, O.V., Platonov, A.K. and Sukhanov, K.G., 1985. Some Problems of Planning Flights to Comets [in Russian], *Kosmich. issled.* **23**(2), pp. 296–307.
206. Labunsky, A.V., 1985. Exploration of Space Vehicle Paths to Asteroids with Planetary Flybys [in Russian], Preprint of MAI, *Space Flight Mechanics* Moscow: izd. MAI, pp. 51–57.
207. Labunsky, A.V., 1986. *A Scheme of Spacecraft Flight to Small Celestial Bodies with the Use of the Gravitational Fields of Venus or Mars and Some Results of Its Examination*, [in Russian], Preprint of MAI, in "Space Flight Mechanics", Moscow, pp. 29–33.
208. Livanov, L.B., 1986. Flights to Asteroids with the use of Paths Optimal in Terms of Energy with Gravity Assist, Powered, and Aerodynamical Maneuvers at Venus, Earth, Mars, and Jupiter, *Kosmich. issled.* **24**(4).
209. Farquhar, R.W., Dunham, D.W. and Hsu S.-C., 1987. A Voyager—style tour of comets and asteroids 1994–2005, *J. Astronaut. Sci.*, **35**(4), 399–417.
210. Dolgopolov, V.P., Karyagin, V.P., Kovtunenko, V.M., Kremnev, R.S., Papkov, O.V., Pichkhadze, K.M., Rogovskii, G.K., Sagdeev, R.Z. and Sukhanov, K.G., 1987. Automated Space Probes Vega-1 and Vega-2. Operating of Probes in the Encounter with Halley's Comet [in Russian], *Kosmich. issled.* **19**(3), pp. 377–391.
211. Labunsky, A.V., 1987. *On the Formation of Rational Flight Schemes to Small Celestial Boll Celestial Bodies with Gravity Assist Maneuvers* [in Russian], Preprint of MAI, in *Control, Stabilization, and Orientation of Flying Vehicles*, Moscow, pp. 24–27.

212. Grard, R., 1988. The Vesta mission, *ESA Bulletin*, No 8, p. 36.
213. Myers, M.R., Stetson, D.S. and Byrnes, D.V., 1988. *Trajectory design for the comet rendezvouz asteroid fly-by 1995–96 opportunties*, AIAA/AAS Astrodynamics conference, Minneapolis, Minnesota, pp. 379–390.
214. Stetson, D., 1988. *Near-nucleus trajectory design for comet penetrator-lander data relay*, AIAA/AAS Astrodynamics conference, Minneapolis, Minnesota, pp. 391–400.
215. Vaning, W., 1988. *The Fast, Asteroid Reconnaissance Mission*, AIAA/AAS Astrodynamics conference, Minneapolis, Minnesota, pp. 329–337.
216. Sagdeev, R.Z.(Ed.), 1989. TV Surveyng of Comet Halley. Nauka, Moscow, 295 pp.
217. Yen, C.L., 1989. Main-belt asteroid exploration: mission options for the 1990s, *J. Astronaut. Sci.* 37(3), p. 333.
218. Yen, C.L., 1989. Mission Opportnity Maps for Rendezvous With Earth-Crossing Asteroids, *J. Astronaut. Sci.* 37(3), p. 399.
219. Zhirnov, V.A. and Lidov, M.A., 1989. On the Problem of Flybys of Several Asteroid [in Russian], *Kosmich. issled.*, 27(1), pp. 3–8.
220. Anselmo, L., 1990. *PIAZZI: a Probe to the Apollo–Amor Asteroids*, International Symposium on Space Technology and Science, Tokyo.
221. Labunsky, A.V. and Kireev, E.A., 1989. *Multipurpose Missions of Spacecraft Flights to Small Bodies of the Solar System* [in Russian], in Proc. XXIII Conference Dedicated to the Memory of K.E. Tsiolkovsky, Kaluga, Sep. 1989, Section "Space Flight Mechanics", Moscow, 1990, pp. 27–31.
222. Williams, S.N., Knocke, P.C. and Bright, L.E., 1991. *Interplanetary mission Design for the Comet Rendezvous Asteroid Flyby Mission* AAS/AIAA Astrodynamics Specialists Conference, AAS 91–396, Dunrango,Colorado.
223. Wertz, J.R. and Larson, W.J.(Ed.), 1995. *Space Mission Analysis and Design*, Space Technology Library, New York.
224. Farquhar, R.W., Dunham, D.W. and Mac Adams, J.V., 1995. NEAR Missions Overview and Trajectory Design, *J. of Astronautical Sciences*, v. 43(4).
225. Sheers, D.J., Williams, B.G., Bollman, W.E., Davis, R.P., Helfrich, C.E., Synoff, S.P. and Yeomans, D.K., 1995. Navigation for low-cost mission to small Solar-system bodies, *Acta Astronaut*. 35. Suppl. pp. 211–220.
226. Veverka, J., Adams, G.L., Binzel, R.P., Brown, R.H., Evans, L., Gafey, M.J.,Gavin, T., Klaasen, K., Miller, S.L., Squyers, S., Thomas, P.C., Trombka, J. and Yeomans, D.K., 1995. MASTER: a discovery-class mission for the detailed study of large Mainbelt-asteroids, *Acta Astronaut*. 35. pp. 201–210.
227. Veverka, J., Farquhar, R.W., Reynolds, E., Belton, M.J.S., Cheng, A., Clark, B., Kissel, J., Malin, M., Niemann, H., Thomas, P.C. and Yeomans, D.K., 1995. Comet nuclens tour, *Acta Astronaut*. 35. Suppl. pp. 181–191.
228. Villefranche, P., Evans, J., Faye, F., Witteveen, P. and Mecke, G., 1995. *Rosetta: The ESA comet rendezvous mission*, The 46th IAF Congress. Oslo, Norway.
229. Kisza, M.E. and Casani, E.K., 1995. *The New Millennium Program*, The 46th IAF Congress, Oslo, Norway.
230. Norberg, O., Lundin, R., Barabash, S., Yamauchi, M., Sukhanov, A., Zakharov, A., Marklund, G., Lagerqvist, C.-I., Magnusson, P., Woch, J., Rathsman, P. and Grahn S., 1996. *Hannes: a mission to Asteroid belt*, IAA 2nd International Conference on low-cost Planetary Missions, Laurel, Maryland, USA.
231. Sims, J.A., Longuski, J.M. and Stangler, A.J., 1996. *Trajectory options for low-cost missions to asteroids*, IAA 2nd International Congress on low-cost Planetary Missions, Laurel, Maryland, USA.

AUTHOR INDEX

SUBJECT INDEX

Milton Keynes UK
Ingram Content Group UK Ltd.
UKHW040446071024
449327UK00020B/1027

9 780367 579234